TypeScript
图形渲染实战
2D架构设计与实现

步磊峰 ◎ 编著

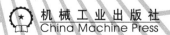

图书在版编目（CIP）数据

TypeScript图形渲染实战：2D架构设计与实现/步磊峰编著. —北京：机械工业出版社，2019.2

ISBN 978-7-111-61924-6

Ⅰ. T… Ⅱ. 步… Ⅲ. JAVA语言–程序设计 Ⅳ. TP312.8

中国版本图书馆CIP数据核字（2019）第025527号

本书使用微软最新的TypeScript语言，以面向接口及泛型的编程方式，采用HTML 5中的Canvas2D绘图API，实现了一个2D动画精灵系统，并在该精灵系统上演示了精心设计的与图形数学变换相关的Demo。通过阅读本书，读者可以系统地掌握TypeScript语言、面向接口和泛型的编程方式、Canvas2D API绘图、图形数学（向量、点与基本形体的碰撞检测、矩阵及贝塞尔曲线）、数据结构（队列、栈、树）及重要的设计模式。

本书共10章，分为4篇。第1篇TypeScript篇，主要介绍了如何构建TypeScript开发、编译和调试环境，以及如何使用TypeScript实现Doom 3词法解析器；第2篇Canvas2D篇，主要介绍了动画与Application类，以及如何使用Canvas2D绘图；第3篇图形数学篇，主要介绍了坐标系变换、向量数学及基本形体的点选、矩阵数学及贝塞尔曲线；第4篇架构与实现篇，主要介绍了精灵系统、优美典雅的树结构及场景图系统。

本书内容丰富，讲解由浅入深，特别适合对图形、游戏和UI开发感兴趣的读者阅读，也适合需要系统学习图形开发技术的人员阅读。另外，本书还适合JavaScript程序员及想从C/C++、Java、C#等语言转HTML 5开发的程序员阅读。编程爱好者、高校学生及培训机构的学员也可以将本书作为兴趣读物。

TypeScript图形渲染实战：2D架构设计与实现

出版发行：机械工业出版社（北京市西城区百万庄大街22号 邮政编码：100037）

责任编辑：欧振旭 李华君　　　　　　　　责任校对：姚志娟

印　　刷：中国电影出版社印刷厂　　　　版　　次：2019年3月第1版第1次印刷

开　　本：186mm×240mm　1/16　　　　印　　张：24.5

书　　号：ISBN 978-7-111-61924-6　　　　定　　价：99.00元

凡购本书，如有缺页、倒页、脱页，由本社发行部调换

客服热线：（010）88379426 88361066　　　投稿热线：（010）88379604

购书热线：（010）68326294 88379649 68995259　　读者信箱：hzit@hzbook.com

版权所有·侵权必究

封底无防伪标均为盗版

本书法律顾问：北京大成律师事务所　韩光/邹晓东

前言

TypeScript 是由微软公司开发的一种开源编程语言，主要为大型应用而设计。它是 JavaScript 的一个超集，扩展了 JavaScript 的语法，任何现有的 JavaScript 程序都可以不加改变地在 TypeScript 下工作。

本书使用最新的 TypeScript 语言，以面向接口及泛型的编程方式，采用 HTML 5 中的 Canvas2D 绘图 API，实现了一个 2D 动画精灵系统，并在该精灵系统上演示了精心设计的与图形数学变换相关的 Demo。该精灵系统是一个具备最小运行环境（更新、重绘、裁剪，以及事件分发和响应），并支持精确点选（点与点、线段、矩形、圆形、椭圆、三角形及凸多边形），采取了享元设计模式，基于场景图管理且兼容非场景图类型，而且易于扩展的系统。

本书有何特色

1. 全程使用TypeScript面向接口的编程语言

本书通过 TypeScript 语言提供的面向接口编程功能实现了：
- Doom3 引擎中文本格式的词法解析器接口；
- 接口一致，且使用了树结构管理（场景图）和线性结构管理（非场景图）的精灵系统，并且能够让场景图类型的精灵系统兼容运行非场景图类型的精灵系统。

2. 剖析Canvas2D的底层原理

本书中的 Canvas2D 相关章节实现了如下几个重要例子：
- 模拟了 Canvas2D 中的渲染状态堆栈和矩阵变换堆栈，并且在实例中使用自己的矩阵变换堆栈来运行相关例子；
- 实现了类似于 Canvas2D 文本绘制的各种对齐算法；
- 实现了加强版的 drawImage 方法，使其支持 repeat/repeat-x/repeat-y 填充模式的算法（类似于 Canvsas2D 中的图案填充），并且使用加强版的 drawImage 实现了九宫格缩放算法（Scale9Grid）。

3. 详解图形数学及点与基本几何形体的碰撞检测算法

图形编程的最大"拦路虎"是涉及图形数学变换。书中以图形数学变换为核心实现了以下例子：
- 将平移、缩放和旋转的不同顺序组合绘制出来；
- 通过绘制各种轨迹来观察和理解物体原点变换的奥秘所在；
- 贝塞尔曲线原理动画；
- 坦克方向正确地朝着鼠标指针位置移动（三角函数版）；
- 坦克方向正确地朝着鼠标指针位置移动（使用向量缩放来避免使用 sin / cos 三角函数）；
- 坦克方向正确地朝着鼠标指针位置移动（使用从两个方向以向量构建的旋转矩阵来避免使用三角函数 atan2，并且用自己实现的矩阵堆栈来替换 Canvas2D 中的矩阵堆栈）；
- 坦克沿着多条贝塞尔曲线围成的封闭路径朝向正确地运动；
- 点投影到向量上的动画效果（涉及向量的所有操作）；
- 使用鼠标精确点选旋转中的精灵（测试点与线段、矩形、圆形、椭圆、三角形，以及凸多边形碰撞检测算法及凸多边形判断算法）；
- 太阳系模拟动画；
- 骨骼层次精灵动画。

4. 提供了几何原理图的生成源代码

本书图形数学相关章节中的几何原理图都是使用 Canvas2D 直接绘制出来的，并提供了绘制源代码。主要有如下图示：
- 向量方向与大小概念图示；
- 向量的加减法图示；
- 负向量图示；
- 向量与标量乘法图示；
- 向量点乘图示；
- 点与三角形关系图示；
- 凹多边形、凸多边形及三角形扇形化图示；
- 旋转矩阵推导图示。

5. 精准地使用设计模式

本书中用到的设计模式如下：
- 使用工厂模式生产各种接口（精灵系统的精灵 ISprite 和所有基本形体 IShape 都是由 SpriteFactory 制造生产的）；

- 使用迭代器模式抽象所有具有迭代功能的类或接口（IDoom3Tokenizer、贝塞尔曲线迭代器、树结构深度优先和广度优先的 8 个线性遍历迭代器）；
- 使用模板方法设计模式（Application 类通过模板方法模式规定了整个入口类的运行流程）；
- 使用适配器模式（在树结构中为了统一线性遍历算法，将队列和栈数据结构适配成一致的操作接口）；
- 使用享元设计模式（实现的精灵系统使用享元模式，多个精灵可以共享同一个形体，但是可以有不同的渲染状态、大小、位置和方向等）。

6．详尽地描述了树的数据结构要点

本书专门用一章来介绍树的数据结构，并且详尽地描述了树结构的各种重要算法。
- 基于队列的广度优先方式的 4 种非递归遍历算法（先根、后根、从左到右、从右到左）；
- 基于栈的深度优先方式的 4 种非递归遍历算法；
- 基于"父亲+儿子"数组方式的深度优先的 4 种递归遍历算法；
- 基于"父亲+儿子兄弟"方式的深度优先的 4 种递归遍历算法；
- 基于"父亲+儿子兄弟"方式的深度优先的非递归遍历算法；
- 基于 JSON 的序列化和反序列化算法。

7．提供完善的技术支持和售后服务

本书提供了专门的售后服务邮箱 hzbook2017@163.com。读者在阅读本书的过程中若有疑问，可以通过该邮箱获得帮助。

本书内容及知识体系

第1篇　TypeScript篇（第1、2章）

第 1 章构建 TypeScript 开发、编译和调试环境，主要介绍了 TypeScript 语言的开发、编译和调试环境的搭建，最终形成一个支持源码自动编译、模块自动载入、服务器端热部署及具有强大断点调试功能的 TypeScript 快捷开发环境。

第 2 章使用 TypeScript 实现词法解析器，用 TypeScript 语言，以面向接口的方式实现了一个 Doom3（原 id Software 公司毁灭战士 3 游戏引擎）词法解析器；并在此基础上实现了工厂和迭代器两种设计模式，使其支持接口的生成及使用迭代方式进行 Token 解析输出；同时封装了 XMLHttpRequest 类，用来支持从服务器端下载要解析的文件。本章还系统地介绍了后续章节中要用到的一些 TypeScript 常用的语法知识。

第2篇　Canvas2D篇（第3、4章）

第 3 章动画与 Application 类，详细介绍了 requestAnimationFrame 方法与屏幕刷新频率之间的关系；并在此基础上封装了一个支持基于时间的刷新、重绘，以及事件的分发和处理的 Application 类；还实现了能正确处理 CSS 盒模型的坐标变换功能；最后添加了支持不同帧率运行的计时器，模拟了 setTimeout 和 setInterval 方法的实现。

第 4 章使用 Canvas2D 绘图，主要介绍了 Canvas2D 中矢量图形、文本、图像及阴影绘制的相关内容，实现了一个本章及后续章节都要使用的基于 Canvas2D 的演示和测试环境。本章需要读者重点关注渲染状态堆栈的实现原理，以及各种文本对齐方式的算法和支持 Repeat 模式的 drawImage 实现等内容。

第3篇　图形数学篇（第5～7章）

第 5 章 Canvas2D 坐标系变换是本书的精华，通过多个例子演示了 Canvas2D 中局部坐标系变换的相关知识点，让读者知道变换顺序的重要性，理解变换及掌握原点变换的几种方式；并且通过太阳自转和月亮自公转的例子，介绍了 Canvas2D 中矩阵堆栈层次变换的用法；最后通过一个坦克跟随鼠标指针朝向正确地运动 Demo，深入讲解了一些常用三角函数的应用。

第 6 章向量数学及基本形体的点选，首先讲述了向量的一些基本操作，然后通过向量的加法和缩放操作替换第 5 章坦克 Demo 中使用的 sin/cos 函数。为了演示向量的一些基本操作，特意实现了点投影到向量的动画效果 Demo；给出了点与线段、圆、矩形、椭圆、三角形及凸多边形等基本形体之间的碰撞检测算法；最后给出了本章所有几何图示的生成源代码，便于读者更加深入地理解向量各个操作背后的几何含义。

第 7 章矩阵数学及贝塞尔曲线，首先讲述了矩阵的相关知识，重点推导了旋转矩阵；然后将第 5 章中的坦克 Demo 用矩阵方式重写，演示如何通过两个单位向量构建旋转矩阵，从而消除对 atan2 函数的使用；接着模拟了 Canvas2D 中的矩阵堆栈，并用自己实现的矩阵堆栈重写坦克 Demo；最后介绍了贝塞尔曲线多项式的推导过程，并实现了一个曲线动画的 Demo。

第4篇　架构与实现篇（第8～10章）

第 8 章精灵系统，以面向接口编程的方式实现了一个精灵系统，并在该系统上实现了一个 Demo，用来测试系统的点与各个基本形体之间的精确碰撞检测。该系统具有必要的功能（更新、绘制、鼠标和键盘事件的分发与响应），使用了非场景图类型，支持精确点选，基于保留模式，并采用了享元设计模式。

第 9 章优美典雅的树结构，主要介绍了树结构的增、删、改、查，以及各种遍历算法，最后实现了树结构的 JSON 序列化和反序列化算法。本章重在灵活应用 TypeScript 泛型编程，涉及不少泛型编程的细节。

第 10 章场景图系统，融合前面章节所讲知识，以面向接口的编程方式实现了一个精灵系统。该系统具有必要的功能（更新、重绘、裁剪及事件分发和响应），使用了场景图类型（建立在第 9 章的树结构上），支持精确点选，基于非立即渲染模式（保留模式），采用了享元设计模式，兼容运行第 8 章的非场景图类型。在此基础上，通过骨骼层次精灵动画，演示了场景图的层次变换功能及享元模式的优点；最后实现了坦克沿着贝塞尔路径朝向正确运行的 Demo，以演示该精灵系统的综合特点。

本书配套资源获取方式

本书涉及的源代码文件和 Demo 需要读者自行下载。请登录华章公司网站 www.hzbook.com，在该网站上搜索到本书，然后单击"资料下载"按钮即可在页面上找到"配书资源"下载链接。

运行书中的源代码需要进行以下操作：
（1）按照本书第 1 章中的介绍下载并安装 Node.js 和 VS Code。
（2）在 VS Code 的终端对话框中输入 npm install 命令，自动下载运行依赖包。
（3）下载好依赖包后继续输入 npm run dev 即可自动运行 Demo。

本书读者对象

- 对图形、游戏和 UI 开发感兴趣的技术人员；
- 想转行到图形开发领域的技术人员；
- 需要全面学习图形开发技术的人员；
- 想从 C/C++、Java、C#、Objective-C 等语言转 HTML 5 开发的技术人员；
- JavaScript 程序员；
- 想学习 TypeScript 的程序员；
- 想提高编程水平的人员；
- 在校大学生及喜欢计算机编程的自学者；
- 专业培训机构的学员。

本书阅读建议

- 没有图形框架开发基础的读者，建议从第 1 章顺次阅读并演练每一个实例。
- 有一定图形开发基础的读者，可以根据实际情况有重点地选择阅读各个模块和项目案例。对于每一个模块和项目案例,先思考一下实现思路,然后阅读,效果更佳。
- 可以先阅读书中的模块和 Demo,再结合配套源代码理解并调试,这样更加容易理解,而且也会理解得更加深刻。

本书作者

本书由步磊峰编写。感谢在本书编写和出版过程中给予了笔者大量帮助的各位编辑!

由于笔者水平所限,加之写作时间较为仓促,书中可能还存在一些疏漏和不足之处,敬请各位读者批评指正。

步磊峰

目录

前言

第1篇　TypeScript 篇

第1章　构建 TypeScript 开发、编译和调试环境 2
- 1.1 TypeScript 简介 2
- 1.2 安装 TypeScript 开发环境 3
  - 1.2.1 安装 Node.js 3
  - 1.2.2 安装 VS Code 4
  - 1.2.3 NPM 全局安装 TypeScript 6
  - 1.2.4 第一个 TypeScript 程序 7
- 1.3 使用 TypeScript 编译（转译）器 13
  - 1.3.1 生成 tsconfig.json 文件 13
  - 1.3.2 解决生成 tsconfig.json 文件后带来的常见问题 13
  - 1.3.3 自动编译 TypeScript 文件 15
- 1.4 模块化开发 TypeScript 15
  - 1.4.1 tsconfig.json 文件中的 target 和 module 命令选项 16
  - 1.4.2 编写 Canvas2D 类导出给 main.ts 调用 16
  - 1.4.3 使用 lite-server 搭建本地服务器 17
- 1.5 使用 SystemJS 自动编译加载 TypeScript 18
  - 1.5.1 NPM 本地安装 TypeScript 库和 SystemJS 库 18
  - 1.5.2 SystemJS 直接编译 TypeScript 源码 19
- 1.6 使用 VS Code 调试 TypeScript 源码 20
  - 1.6.1 安装及配置 Debugger for Chrome 扩展 20
  - 1.6.2 VS Code 中单步调试 TypeScript 20
- 1.7 本章总结 22

第2章　使用 TypeScript 实现 Doom 3 词法解析器 24
- 2.1 Token 与 Tokenizer 24
  - 2.1.1 Doom3 文本文件格式 26

	2.1.2	使用 IDoom3Token 与 IDoom3Tokenizer 接口 ·············	26
	2.1.3	ES 6 中的模板字符串 ··································	28
	2.1.4	IDoom3Token 与 IDoom3Tokenizer 接口的定义 ··········	29
2.2	IDoom3Token 与 IDoom3Tokenizer 接口的实现 ·················		30
	2.2.1	Doom3Token 类成员变量的声明 ·························	30
	2.2.2	Doom3Token 类变量初始化的问题 ·······················	31
	2.2.3	IDoom3Token 接口方法的实现 ··························	32
	2.2.4	Doom3Token 类的非接口方法实现 ·······················	33
	2.2.5	Doom3Tokenzier 处理数字和空白符 ·····················	34
	2.2.6	IDoom3Tokenizer 接口方法实现 ························	34
	2.2.7	Doom3Tokenizer 字符处理私有方法 ·····················	35
	2.2.8	核心的 getNextToken 方法 ·····························	36
	2.2.9	跳过不需处理的空白符和注释 ···························	37
	2.2.10	实现_getNumber 方法解析数字类型 ····················	38
	2.2.11	实现_getSubstring 方法解析子字符串 ··················	40
	2.2.12	实现_getString 方法解析字符串 ······················	41
	2.2.13	IDoom3Tokenizer 词法解析器状态总结 ··················	42
2.3	使用工厂模式和迭代器模式 ······································		43
	2.3.1	微软 COM 中创建接口的方式 ····························	43
	2.3.2	Doom3Factory 工厂类 ··································	43
	2.3.3	迭代器模式 ···	44
	2.3.4	模拟微软.NetFramework 中的泛型迭代器 ·················	44
	2.3.5	IDoom3Tokenizer 扩展 IEnumerator 接口 ·················	45
	2.3.6	修改 Doom3Tokenizer 源码 ······························	45
	2.3.7	使用 VS Code 中的重命名重构方法 ·····················	46
	2.3.8	使用迭代器解析 Token ·································	46
	2.3.9	面向接口与面向对象编程的个人感悟 ·····················	47
2.4	从服务器获取资源 ···		47
	2.4.1	HTML 加载本地资源遇到的问题 ·························	48
	2.4.2	从服务器加载资源 ·····································	48
	2.4.3	使用 XHR 向服务器请求资源文件 ·······················	49
	2.4.4	TypeScript 中的类型别名 ······························	50
	2.4.5	使用 doGet 请求文本文件并解析 ·······················	51
	2.4.6	解决仍有空白字符输出问题 ·····························	52
	2.4.7	实现 doGetAsync 异步请求方法 ·························	52
	2.4.8	声明 TypeScript 中的回调函数 ·························	54
	2.4.9	调用回调函数 ···	55
2.5	本章总结 ···		57

第 2 篇　Canvas2D 篇

第 3 章　动画与 Application 类 60
3.1　requestAnimationFrame 方法与动画 60
3.1.1　HTML 中不间断的循环 60
3.1.2　requestAnimationFrame 与监视器刷新频率 62
3.1.3　基于时间的更新与重绘 65
3.2　Application 类及其子类 67
3.2.1　Application 类体系结构 67
3.2.2　启动动画循环和停止动画循环 68
3.2.3　Application 类中的更新和重绘 69
3.2.4　回调函数的 this 指向问题 70
3.2.5　函数调用时 this 指向的 Demo 演示 71
3.2.6　CanvasInputEvent 及其子类 73
3.2.7　使用 getBoundingRect 方法变换坐标系 75
3.2.8　将 DOM Event 事件转换为 CanvasInputEvent 事件 77
3.2.9　EventListenerObject 与事件分发 77
3.2.10　让事件起作用 79
3.2.11　Canvas2DApplication 子类和 WebGLApplication 子类 79
3.3　测试及修正 Application 类 80
3.3.1　继承并覆写 Application 基类的虚方法 80
3.3.2　测试 ApplicationTest 类 81
3.3.3　多态（虚函数动态绑定） 82
3.3.4　鼠标单击事件测试 83
3.3.5　CSS 盒模型对_viewportToCanvasCoordinate 的影响 84
3.3.6　正确的_viewportToCanvasCoordinate 方法实现 86
3.4　为 Application 类增加计时器功能 90
3.4.1　Timer 类与 TimeCallback 回调函数 90
3.4.2　添加和删除 Timer（计时器） 91
3.4.3　触发多个定时任务的操作 93
3.4.4　测试 Timer 功能 95
3.5　本章总结 96

第 4 章　使用 Canvas2D 绘图 98
4.1　绘制基本几何体 98
4.1.1　Canvas2DApplication 的绘制流程 98
4.1.2　绘制矩形 Demo 99
4.1.3　模拟 Canvas2D 中渲染状态堆栈 100

		4.1.4	线段属性与描边操作（stroke）	103
		4.1.5	虚线绘制（交替绘制线段）	105
		4.1.6	使用颜色描边和填充	108
		4.1.7	使用渐变对象描边和填充	110
		4.1.8	使用图案对象描边和填充	113
		4.1.9	后续要用到的一些常用绘制方法	115
	4.2	绘制文本		117
		4.2.1	封装 fillText 方法	117
		4.2.2	文本的对齐方式	119
		4.2.3	自行实现文本对齐效果	121
		4.2.4	计算文本高度算法	122
		4.2.5	嵌套矩形定位算法	122
		4.2.6	fillRectWithTitle 方法的实现	125
		4.2.7	自行文本对齐实现 Demo	126
		4.2.8	font 属性	128
		4.2.9	实现 makeFontString 辅助方法	129
	4.3	绘制图像		130
		4.3.1	drawImage 方法	131
		4.3.2	Repeat 图像填充模式	133
		4.3.3	加强版 drawImage 方法的实现	134
		4.3.4	加强版 drawImage 方法效果演示	136
		4.3.5	离屏 Canvas 的使用	137
		4.3.6	操作 Canvas 中的图像数据	138
	4.4	绘制阴影		141
	4.5	本章总结		142

第 3 篇　图形数学篇

第 5 章　Canvas2D 坐标系变换 146

	5.1	局部坐标系变换		146
		5.1.1	准备工作	146
		5.1.2	平移操作演示	149
		5.1.3	平移和旋转组合操作演示	150
		5.1.4	绘制旋转的轨迹	152
		5.1.5	变换局部坐标系的原点	154
		5.1.6	测试 fillLocalRectWithTitle 方法	156
		5.1.7	彻底掌控局部坐标系变换	158
		5.1.8	通用的原点变换方法	166

5.1.9　公转（Revolution）与自转（Rotation） 169
　　　5.1.10　原点变换的另一种方法 171
　5.2　坦克 Demo 173
　　　5.2.1　象限（Quadrant）文字绘制 174
　　　5.2.2　坦克形体的绘制 175
　　　5.2.3　坦克及炮塔的旋转 178
　　　5.2.4　计算坦克的朝向 179
　　　5.2.5　坦克朝着目标移动 182
　　　5.2.6　使用键盘控制炮塔的旋转 183
　　　5.2.7　初始朝向的重要性 184
　　　5.2.8　朝向正确的运行 187
　　　5.2.9　坦克朝着目标移动效果的生成代码 189
　5.3　本章总结 190

第 6 章　向量数学及基本形体的点选 192

　6.1　向量数学 192
　　　6.1.1　向量的概念 192
　　　6.1.2　向量的大小与方向 194
　　　6.1.3　向量的加减法及几何含义 196
　　　6.1.4　负向量及几何含义 198
　　　6.1.5　向量与标量乘法及几何含义 198
　　　6.1.6　向量标量相乘取代三角函数 sin 和 cos 的应用 200
　　　6.1.7　向量的点乘及几何含义 201
　　　6.1.8　向量的夹角及朝向计算 203
　6.2　向量投影 Demo 203
　　　6.2.1　Demo 的需求描述 205
　　　6.2.2　绘制向量 205
　　　6.2.3　向量投影算法 207
　　　6.2.4　投影效果演示代码 208
　　　6.2.5　向量 getAngle 和 getOrientation 方法的区别 210
　6.3　点与基本几何形体的碰撞检测算法 211
　　　6.3.1　点与线段及圆的碰撞检测 211
　　　6.3.2　点与矩形及椭圆的碰撞检测 213
　　　6.3.3　点与三角形的碰撞检测 213
　　　6.3.4　点与任意凸多边形的碰撞检测 215
　6.4　附录：图示代码 217
　　　6.4.1　图 6.1 向量概念图示源码 217
　　　6.4.2　图 6.2 和图 6.3 向量加减法图示源码 218
　　　6.4.3　图 6.4 负向量图示源码 218

	6.4.4	图 6.5 向量与标量相乘图示源码	219
	6.4.5	图 6.6 向量的点乘图示源码	219
	6.4.6	图 6.11 点与三角形的关系图示源码	220
	6.4.7	图 6.12 和图 6.13 凹凸多边形图示源码	221
6.5	本章总结	222

第 7 章 矩阵数学及贝塞尔曲线 223

7.1 矩阵数学 223

- 7.1.1 矩阵乘法 223
- 7.1.2 单位矩阵 225
- 7.1.3 矩阵求逆 225
- 7.1.4 用矩阵变换向量 226
- 7.1.5 平移矩阵及其逆矩阵 227
- 7.1.6 缩放矩阵及其逆矩阵 228
- 7.1.7 旋转矩阵及其逆矩阵 230
- 7.1.8 从两个单位向量构建旋转矩阵 233
- 7.1.9 使用 makeRotationFromVectors 方法取代 atan2 的应用 234
- 7.1.10 仿射变换 237
- 7.1.11 矩阵堆栈 237
- 7.1.12 在坦克 Demo 中应用矩阵堆栈 239
- 7.1.13 图 7.1 旋转矩阵推导图示绘制源码 243

7.2 贝塞尔曲线 245

- 7.2.1 Demo 效果 245
- 7.2.2 使用 Canvas2D 内置曲线绘制方法 246
- 7.2.3 伯恩斯坦多项式推导贝塞尔多项式 249
- 7.2.4 贝塞尔曲线自绘版 251
- 7.2.5 鼠标碰撞检测和交互功能 253
- 7.2.6 实现贝塞尔曲线枚举器 255

7.3 本章总结 257

第 4 篇 架构与实现篇

第 8 章 精灵系统 260

8.1 精灵系统的架构与接口 260

- 8.1.1 应用程序的入口与命令分发 262
- 8.1.2 IRenderState、ITransformable 和 ISprite 接口 265
- 8.1.3 IDrawable、IHittable 和 IShape 接口 267

8.2 实现非场景图类型精灵系统 268

- 8.2.1 Transform2D 辅助类 269

        8.2.2　ISprite 接口的实现 ··················· 269
        8.2.3　Sprite2DManager 管理类 ··········· 272
    8.3　IShape 形体系统 ···································· 275
        8.3.1　线段 Line 类 ···························· 276
        8.3.2　BaseShape2D 抽象基类 ············· 278
        8.3.3　Rect 类和 Grid 类 ····················· 280
        8.3.4　Circle 类和 Ellipse 类 ··············· 281
        8.3.5　ConvexPolygon 类 ···················· 282
        8.3.6　Scale9Grid 类 ··························· 283
        8.3.7　SpriteFactory 生产 IShape 产品 ··· 289
    8.4　精灵系统测试 Demo ······························ 290
        8.4.1　Demo 的运行流程 ···················· 291
        8.4.2　创建各种 IShape 对象 ··············· 292
        8.4.3　创建网格精灵和事件处理函数 ··· 293
        8.4.4　非网格精灵的事件处理函数 ······ 295
        8.4.5　Demo 的入口代码 ···················· 296
    8.5　本章总结 ··· 296
第 9 章　优美典雅的树结构 ································ 298
    9.1　树的数据结构 ··· 298
        9.1.1　树结构简介 ································ 298
        9.1.2　树节点添加时的要点 ················· 300
        9.1.3　树节点 isDescendantOf 和 remove 方法的实现 ··· 300
        9.1.4　实现添加树节点方法 ················· 303
        9.1.5　树结构的层次关系查询操作 ······ 304
    9.2　树数据结构的遍历 ································· 308
        9.2.1　树结构遍历顺序 ························ 308
        9.2.2　树结构线性遍历算法 ················· 309
        9.2.3　树结构遍历枚举器 ···················· 310
        9.2.4　树结构枚举器的实现 ················· 311
        9.2.5　测试树结构枚举器 ···················· 316
        9.2.6　深度优先的递归遍历 ················· 320
        9.2.7　使用儿子兄弟方式递归遍历算法 ··· 321
        9.2.8　儿子兄弟方式非递归遍历算法 ··· 323
    9.3　树数据结构的序列化与反序列化 ············ 327
        9.3.1　树节点自引用特性导致序列化错误 ··· 328
        9.3.2　树节点的序列化和反序列化操作 ··· 328
    9.4　队列与栈的实现 ····································· 331
    9.5　本章总结 ··· 332

第 10 章 场景图系统 ... 334

10.1 实现场景图精灵系统 ... 334
10.1.1 非场景图精灵系统的不足之处 ... 334
10.1.2 树结构场景图系统 ... 336
10.1.3 矩阵堆栈和场景图 ... 338
10.1.4 实现场景图精灵系统概述 ... 338
10.1.5 核心的 SpriteNode 类 ... 338
10.1.6 实现 SpriteNode 类的接口方法 ... 340
10.1.7 SpriteNode 的 findSprite 方法实现 ... 342
10.1.8 递归的更新与绘制操作 ... 343
10.1.9 SpriteNodeManager 类 ... 344
10.1.10 修改 Sprite2D 类的 getWorldMatrix 方法 ... 346
10.1.11 让 Sprite2DApplication 类支持场景图精灵系统 ... 347

10.2 骨骼层次精灵 Demo ... 348
10.2.1 实现骨骼形体 ... 348
10.2.2 SkeletonPersonTest 类 ... 349
10.2.3 事件处理程序 ... 351
10.2.4 使用 renderEvent 事件 ... 354
10.2.5 本节总结 ... 356

10.3 坦克沿贝塞尔路径运动 Demo ... 357
10.3.1 实现 BezierPath 形体类 ... 358
10.3.2 需求描述 ... 359
10.3.3 Demo 的场景图 ... 361
10.3.4 TankFollowBezierPathDemo 类初始化 ... 361
10.3.5 创建锚点、控制点及连线精灵 ... 363
10.3.6 创建二次贝塞尔路径及坦克精灵 ... 364
10.3.7 键盘事件处理方法 ... 365
10.3.8 鼠标事件处理方法 ... 366
10.3.9 坦克沿路径运动的核心算法 ... 367
10.3.10 让坦克动起来 ... 368

10.4 让精灵系统支持裁剪操作 ... 370
10.5 本章总结 ... 372

第 1 篇
TypeScript 篇

▶▶ 第 1 章 构建 TypeScript 开发、编译和调试环境

▶▶ 第 2 章 使用 TypeScript 实现 Doom 3 词法解析器

第 1 章 构建 TypeScript 开发、编译和调试环境

毋庸置疑，探索和实践是学习的最佳方法。本书首章就为广大读者讲解功能完备的 TypeScript 开发环境，使读者可以在以后各章的学习过程中使用实例进行探索和实验。

本书使用的 TypeScript 开发、编译和调试环境需要安装 Node.js、Visual Studio Code（简称 VS Code）代码编辑器、TypeScript 语言编译器、lite-server 服务器、SystemJS，以及 Debugger for Chrome 调试插件。在本章中，将循序渐进地讲解这些工具的安装方式和用途，最后读者将得到一个具有自动编译、自动部署和强大的 Debug 能力的 TypeScript 开发环境。

1.1 TypeScript 简介

如图 1.1 所示，TypeScript 是一门由微软开发的开放源代码的编程语言，它扩展了 JavaScript 的语法，是 JavaScript 语言的超集。根据官网（http://www.typescriptlang.org/）的描述，可以了解到：

- Typed：是一门强类型的语言。
- superset of JavaScript：是 JavaScript 的超集。
- that compiles to plain JavaScript：TypeScript 最终是被编译（转译）成 JavaScript 代码。
- Any browser，Any host：可以在任何浏览器、任何宿主环境中运行，即跨平台。
- Open source：开源的，广大读者可以自由地阅读源码。

图 1.1 TypeScript 官方说明

1.2 安装 TypeScript 开发环境

可以通过两种方式来安装 TypeScript 的开发环境，一种是通过 Node.js 包管理器（NPM）。如果已经安装微软的 Visual Studio 2012 及以上版本，则可使用另一种安装方式，即通过插件方式安装 TypeScript 编译器。

本书中，仅使用 Node.js 包管理器（NPM）来安装 TypeScript 开发环境。

1.2.1 安装 Node.js

Node.js 官方（https://nodejs.org/）说明如图 1.2 所示。

图 1.2 Node.js 官方说明

安装 Node.js 时，一般会提供两个版本：LTS 版和 Current 版。其中，LTS 是 Long Term Support（官方长期支持版）的缩写，在生产环境中，请使用 LTS 版。

本书以 Mac OS（x64）系统为演示环境。Windows 和 Linux 安装和使用与 Mac OS 具有相似性。

如图 1.2 所示，选择 8.11.2 LTS 选项开始下载 Node.js 安装包，本书编写时，最新稳定版是 8.11.2 LTS。

下载完 Mac OS 系统安装包后，双击安装文件进入如图 1.3 所示安装界面，一直单击"继续"按钮，直到安装完成。Node.js 安装包会将 Node 及 NPM 程序安装到/usr/local/bin/系统目录下。

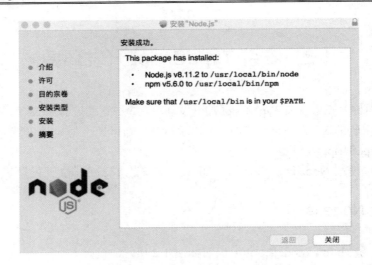

图 1.3　Node.js 安装界面及安装目录

打开 Mac OS 中的终端程序，输入相应命令：node -v 及 npm -v。若成功安装，会显示如图 1.4 所示的内容。

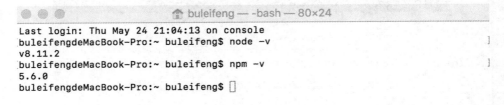

图 1.4　测试 Node.js 和 NPM 版本

至此，已经成功安装 Node.js，下面来安装 VS Code。

1.2.2　安装 VS Code

从官方资料可知，VS Code 是一个运行在桌面系统上、并且可用于 Windows、Mac OS X 和 Linux 平台的轻量级且功能强大的源代码编辑器。它直接内置了对 JavaScript、TypeScript 和 Node.js 的支持，并具有其他语言（C/C++、C♯、Python 和 PHP 等）的扩展，以及提供了一个丰富的生态系统。

对于 VS Code 更加详细的描述，读者可以访问官网（https://code.visualstudio.com/）自行了解。

在此想强调的是 VS Code 的 3 个令笔者愉悦的体验：

- VS Code 是 VS（Visual Studio，VS）家族中第一个支持 Mac OS X、Linux 和 Windows 的跨平台开发工具。

- 强大的智能代码补全功能。
- 通过插件方式提供编辑—编译—调试各种语言的能力。开发者不只是写代码，他们还要不断调试程序，而 VS Code 直接内置了 Node.js 的调试器，并且通过扩展来支持 Python、PHP、C/C++、C#和 TypeScript 等编程语言的调试。

在 Mac OS X 中安装 VS Code 是极其容易的一件事，你所要做的仅仅是：单击"下载"按钮进行下载。下载完成后，会获得一个名为 VSCode-darwin-stable.zip 的压缩文件，双击该压缩文件，解压后出现 VS Code 应用。然后拖曳 VS Code 到程序坞（Dock），单击右键，在弹出的快捷菜单中选择"程序坞中保留"命令。

到目前为止，双击程序坞中的 VS Code 图标，就能启动 VS Code。但是，想通过终端命令行输入 code 运行 VS Code，则需要将 code 命令添加到$PATH 环境变量中，具体步骤如下所述。

（1）选择"查看"|"命令面板"命令（或使用 Shift+Ctrl+P 快捷键）打开命令面板。

注：Mac 键盘上的 Ctrl 键是 Command 键。

（2）在命令面板中输入 shell command 命令，如图 1.5 所示。

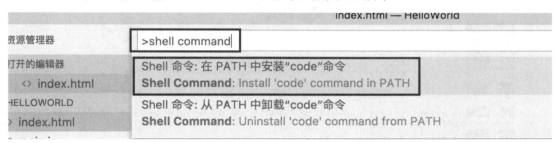

图 1.5　在命令面板中通过 shell command 安装 code 命令

（3）然后选择"在 PATH 中安装'code'命令"选项，得到如图 1.6 所示的内容，只要按照指示一直单击"确定"按钮就可以了。

图 1.6　确认获得管理员权限

（4）在成功安装后，就可以在终端中直接使用 code 命令启动 VS Code，如图 1.7 所示。

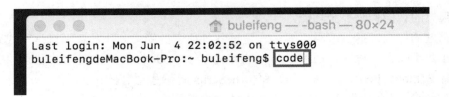

图 1.7　使用命令行启动 VS Code

1.2.3　NPM 全局安装 TypeScript

到目前为止，安装 TypeScript 前的所有必要环境已经搭建好了。接下来就可以安装 TypeScript 编译器了，在 Mac OS 的终端中输入：npm install -g typescript，然后按回车键，得到如图 1.8 所示的结果，可以看出 NPM 全局安装 TypeScript 出现了权限问题。

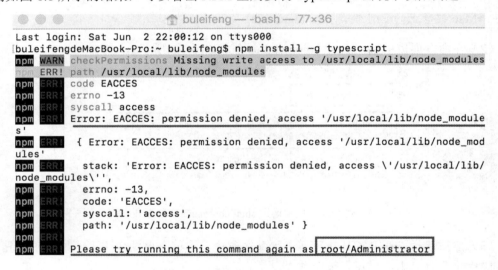

图 1.8　NPM 全局安装 TypeScript 出现权限问题

出现 permission denied 的原因是：在终端执行命令行命令时没有获得管理员权限。

解决办法是，在命令前面加上 sudo，然后输入计算机的管理员密码，操作即可完成。

注意：在 Linux 中和 Mac OS 一样，需要使用 sudo 命令进行 NPM 安装。

在 Mac OS 的终端中继续输入：sudo npm install -g typescript，然后按回车键，并且输入管理员密码，顺利完成 TypeScript 编译器的安装，如图 1.9 所示。

如果正确完成 TypeScript 安装，则在 Mac OS 的终端中输入 tsc-v 命令后，会显示当前安装的 TypeScript 的版本号，如图 1.10 所示。

第 1 章　构建 TypeScript 开发、编译和调试环境

图 1.9　完成 NPM 全局安装 TypeScript

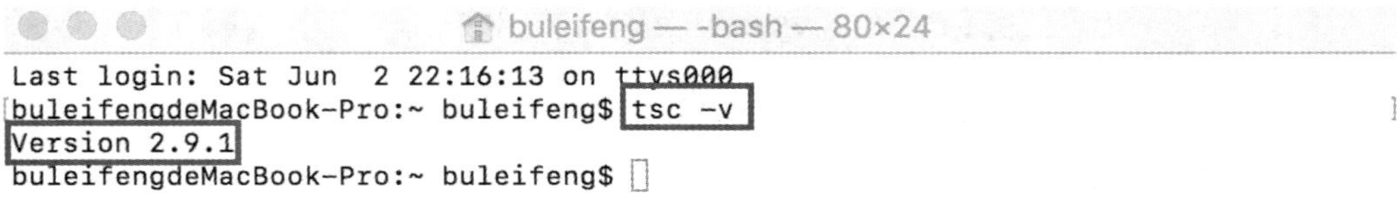

图 1.10　显示 TypeScript 的版本号

本书编写时最新版本的 TypeScript 是 2.9.1。其中，tsc 命令是 TypeScript Compiler 的缩写，用于将 TypeScript 源码编译（转译）成 JavaScript 源码。

NPM 安装命令中-g 表示全局安装，安装在系统目录/usr/local/lib/node_modules/下。

1.2.4　第一个 TypeScript 程序

作为本书的第一个 Demo，我们使用 TypeScript 构建一个简单的 Web 应用程序：使用 Canvas2D 居中绘制 Hello World。

首先在桌面上单击右键，在弹出的菜单中选择新建文件夹选项，命名为 HelloWorld。然后将新建的文件夹拖曳到程序坞中的 VS Code 图标上，启动 VS Code 应用程序。单击新建文件按钮，创建一个名为 index.html 的文件，如图 1.11 所示。

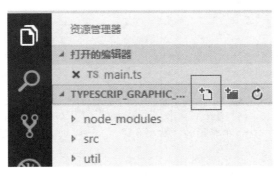

图 1.11　新建 index.html 文件

· 7 ·

接着打开 index.html 文件，使用 Shift + Ctrl + P 快捷方式打开命令面板，在输入框中输入 insert snippet 命令后，会出现"插入代码片段"的命令，如图 1.12 所示。

图 1.12　使用 VS Code 的插入代码片段功能

选择"插入代码片段"命令后再选择 html，就会自动生成如下 HTML 代码：

```html
<!DOCTYPE html>
<html>
  <head>
    <meta charset = "utf-8" />
    <meta http-equiv = "X-UA-Compatible" content = "IE=edge" >
    <title> Page Title </title>
    <meta name = "viewport" content = "width=device-width,initial-scale=1" >
    <link rel = "stylesheet" type = "text/css" media = "screen" href = "main.css" />
    <script src = "main.js"> </script>
  </head>
  <body>
  </body>
</html>
```

将<title>标签中的文本内容替换为 Hello World，<link>和<script>标签内容保持不变，后续会使用 TypeScrip Compiler 生成 main.js 源文件。

在 VS Code 中创建一个名为 main.css 的文件，输入如下代码：

```css
/*
CSS 选择器：# 表示 ID 选择器
*/
#canvas {
    background : #ffffff ;
    margin-left : 10px ;
    margin-top : 10px ;
    -webkit-box-shadow : 4px 4px 8px rgba ( 0 , 0 , 0 , 0.5 ) ;
    -moz-box-shadow : 4px 4px 8px rgba ( 0 , 0 , 0 , 0.5 ) ;
    box-shadow :  4px 4px 8px rgba ( 0 , 0 , 0 , 0.5 ) ;
}
```

注意：CSS 中仅可以使用/* */进行注释。

在 VS Code 中创建一个名为 main.ts 的文件，输入如下代码：

```typescript
class Canvas2DUtil {
    //声明 public 访问级别的成员变量
    public context : CanvasRenderingContext2D ;
    // public 访问级别的构造函数
    public constructor ( canvas : HTMLCanvasElement ) {
        this . context = canvas . getContext ( "2d" ) ;
    }
    // public 访问级别的成员函数
    public drawText ( text : string ) : void {
        // Canvas2D 和 webGL 这种底层绘图 API 都是状态机模式
        //每次绘制前调用 save 将即将要修改的状态记录下来
        //每次绘制后调用 restore 将已修改的状态丢弃，恢复到初始化时的状态
        //这样的好处是状态不会混乱
        //假设当前绘制文本使用红色，如果你没有使用 save/restore 配对函数的话
        //则下次调用其他绘图函数时，如果你没更改颜色，则会继续使用上次设置的红色进行绘制
        //随着程序越来越复杂，如不使用 save/restore 来管理，最后整个渲染状态会极其混乱
        //请时刻保持使用 save / restore 配对函数来管理渲染状态
        this . context . save ( ) ;
        //让要绘制的文本居中对齐
        this . context . textBaseline = "middle" ;
        this . context . textAlign = "center" ;
        //计算 canvas 的中心坐标
        let centerX : number = this . context . canvas . width * 0.5 ;
        let centerY : number = this . context . canvas . height * 0.5 ;
        //红色填充
        // this . context . fillStyle = " red " ;
        //调用文字填充命令
        this . context . fillText ( text , centerX , centerY ) ;
        //绿色描边
        this . context . strokeStyle = "green";
        //调用文字描边命令
        this . context . strokeText ( text , centerX , centerY ) ;
        //将上面 context 中的 textAlign, extBaseLine, fillStyle, strokeStyle
        状态恢复到初始化状态
        this . context . restore ( ) ;
    }
}
```

接下来调用 Canvas2DUtil 类，绘制居中对齐的文字。具体代码如下：

```typescript
let canvas : HTMLCanvasElement | null = document . getElementById ('canvas')
as HTMLCanvasElement ;
if ( canvas === null ) {
    alert ( "无法获取 HTMLCanvasElement !!! " ) ;
    throw new Error ( "无法获取 HTMLCanvasElement !!! " ) ;
}
let canvas2d : Canvas2DUtil = new Canvas2DUtil ( canvas ) ;
canvas2d . drawText ( "Hello World" ) ;
```

选择"查看|集成终端"菜单项（或使用 Ctrl+`快捷键）打开集成终端面板，使用 tsc

命令将 TS 代码编译（转译）成 JS 代码：

```
tsc main.ts
```

当按回车键后会发现 VS Code 左侧的资源管理中多了一个名为 main.js 的文件，如图 1.13 所示，该 main.js 文件就是使用 main.ts 编译（转译）后的结果。

图 1.13 通过 tsc 命令将 TS 文件转译成 JS 文件

接下来，在 index.html 中的 body 标签对内声明一个 canvas 标签，代码如下：

```
<body>
    <canvas id = "canvas"  width = "800" height = "600" > </canvas>
</body>
```

该 canvas 标签的 ID 名称必须和 main.css 中的 CSS ID 选择器名称一致，这样才能使 #canvas 中设置的 CSS 属性生效。

接着使用 canvas 元素的 width / height 属性进行尺寸设置，当前设置的尺寸是 800×600 像素。如果没有设置 canvas 元素的 width / height 属性，默认情况下，浏览器所创建的 canvas 元素是 300 像素宽，150 像素高。

为了能够方便地运行 HTML 文件，引入一个名为 open in browser 的 VS Code 扩展插件。请单击 VS Code 最左侧的活动栏中的扩展图标（或使用 Shift + Ctrl + X 快捷键）打开扩展面板，输入 Open in Browser 后下载安装，如图 1.14 所示。

安装完 Open in Browser 插件后，在 VS Code 中定位到 HTML 文件处，单击右键显示上下文菜单，即可使用该扩展，如图 1.15 所示。

其中，Open In Default Browser（Alt + B 快捷键）可直接在默认的浏览器中运行 HTML 文件。而 Open in Other Browsers（Shift + Alt + B 快捷键）则显示如图 1.16 所示面板，可以选择想要运行的浏览器。

第 1 章 构建 TypeScript 开发、编译和调试环境

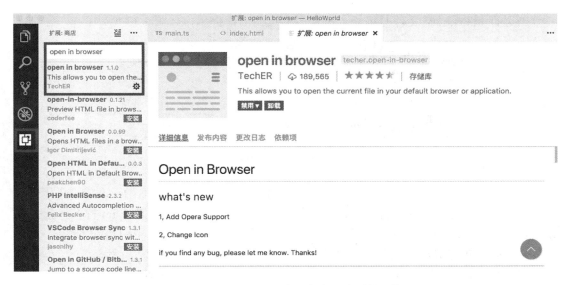

图 1.14 从 VS Code 应用商店下载安装插件

图 1.15 使用 Open in Browser 插件

图 1.16 Mac OS X 中可运行的浏览器

当运行 index.html 文件时可能会出现运行页面出错提示，如图 1.17 所示。

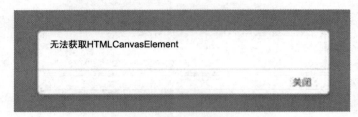

图 1.17　运行页面出错

之所以出现这个问题，是浏览器解析标签顺序导致的。参考上面的 index.html 代码，会发现 script 标签在 canvas 标签前面，浏览器解析 HTML 时，先遇到 script 标签，就去下载 script 标签中的 main.js 脚本，然后开始运行这段脚本，此时脚本中引用的 canvas 元素实际并没有解析完成，因此会弹出"无法获取 HTMLCanvasElement"提示。

有两种修改代码方式来解决这个问题，第一种修改方式是将 script 标签放到 canvas 标签后面。具体代码如下：

```
<body>
    <canvas id = "canvas" width = "800" height = "600" > </canvas>
    <script src = "main.js" > </script>
</body>
```

第二种方式不用调整代码顺序，仍旧让 script 放在<head> </head>中，并使用 defer 属性。具体代码如下：

```
<head>
    <script src = "main.js" defer > </script>
</head>
```

defer 属性会让脚本的执行延迟到整个文档加载后再运行，这样就能避免上面遇到的问题。

修正了上述问题后，继续运行 Open In Default Browser 命令，程序正常运行，文字居中绘制，具体效果如图 1.18 所示。

图 1.18　Hello World 居中绘制

1.3 使用 TypeScript 编译（转译）器

在上一节中，详细地了解了整个 TypeScript 开发环境的构建，并且通过一个简单的 Demo，演示了如何以命令行方式将 TS 代码编译（转译）成 JS 代码。

手动编译方式虽然很灵活，但是存在两个问题：
- 命令行编译方式比较麻烦，tsc 命令参数众多，手动输入浪费时间。
- 如果修改了 TS 源码，不得不再次进行手动编译。

本节就是来解决上面提到的问题，实现 TypeScript 的自动编译（转译）功能。

1.3.1 生成 tsconfig.json 文件

本节以 1.2 节的 HelloWorld Demo 为例，生成自动编译（转译）所需的 tsconfig.json 文件。使用 Control（Ctrl）+ ` 的快捷方式打开 VS Code 的集成终端面板，并输入 tsc --init 命令，然后按回车键，自动在 Hello World 项目的根目录中生成一个名为 tsconfig.json 的文件，如图 1.19 所示。

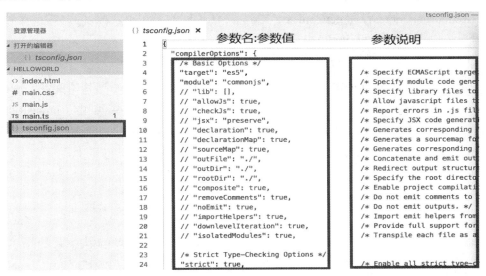

图 1.19 tsconfig.json 文件

1.3.2 解决生成 tsconfig.json 文件后带来的常见问题

如图 1.20 所示，当生成 tsconfig.json 文件后，原本的 main.ts 文件会被标记为红色，并显示有一个错误。之所以这样,是因为在默认状态下,tsconfig.json 中的 strict 命令被设置为 true,此时 VS Code 会对所有 TS 源码进行严格的类型检测(strict type-checking)。

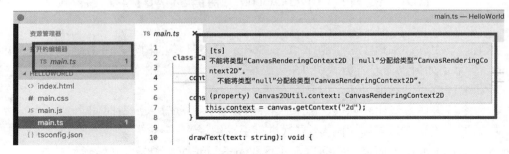

图 1.20　TS 源码错误提示

解决这个问题有两种方式:
- 简单地将 strict 命令设置为 false,停止 VS Code 的严格类型检测。
- 根据 VS Code 的错误提示修改源码,使之符合 VS Code 的检测要求。

来看一下 VS Code 的错误提示:不能将类型" CanvasRenderingContext2D | null "分配给类型"CanvasRenderingContext2D",这是因为 canvas . getContext ("2d")的返回类型是 CavasRenderingContext | null,而在类 Canvas2DUtil 中声明的成员变量 context 的类型是 CanvasRenderingContext,和 getContext ("2d")方法返回的类型不匹配。因此修正代码如下:

```
class Canvas2DUtil {
    public context : CanvasRenderingContext2D | null ;
}
```

在 TypeScript 中允许使用"|"操作符来表示联合类型,联合类型是用来表示一个值具有多种不同的类型。上述代码中使用联合类型意味着可以将类型 CanvasRenderingContext2D 或 null 赋值给 context 变量。当完成上述修改后,会产生更多的错误,如图 1.21 所示。

图 1.21　更多错误提示

VS Code 提示的错误都是"对象可能为 null",因此我们要做的是在调用 this.context 之前进行 null 值检查,就能消除 VS Code 的错误提示。

```
If ( this . context !== null ) {
    this . context . save ( ) ;
    this . context . textBaseline = "middle" ;
    this . context . textAlign = "center" ;
    . . . . . . .
    . . . . . . .
    this . context. restore ( ) ;
}
```

在笔者使用 TypeScript 的过程中,"类型不匹配"和"需要 null 值检测"是被提示最多的信息,因此值得花点时间和篇幅来了解解决方案。在实际开发过程中,使用严格类型检测的确会增加一些代码,但是其所带来的优势也是非常明显的,即让程序更加健壮,这也是 TypeScript 的魅力所在。

1.3.3 自动编译 TypeScript 文件

如下面的代码所示,在使用 tsc --init 命令生成的 tsconfig.json 文件中,默认情况下,在顶级属性 compilerOptions 中设置了 4 个(target、module、strict 和 esModuleInterop)编译选项命令,其中 strict 命令在上一节中已有所了解。

```
{
    "compilerOptions" : {
        "target" : "es5" ,
        "module" : "commonjs" ,
        "strict" : true ,
        "esModuleInterop" : true,
        "watch" : true
    }
}
```

至于 target、module 及 esModuleInterop 这 3 个命令,与模块的导入/导出及模块代码生成相关,将在后面章节中介绍。

tsconfig.json 文件还有很多编译命令选项,请读者参考 TypeScript 官网的相关说明。在此仅重点关注一下 watch 这个命令,该命令默认情况下是关闭的。将其设置为 true,然后直接在使用 VS Code 内置的集成终端中输入 tsc 命令后,则会在监视模式下运行编译(转译)器。此时 TypeScript 编译器会监视所有后缀名为 ts 的文件,当这些 ts 文件有任何变化时,TypeScript 编译器会自动重新编译(转译)。

1.4 模块化开发 TypeScript

像笔者这种从 C++转到 TypeScript/JavaScript 开发的程序员,除了强类型外,最喜欢的就是模块化开发，本节就来介绍 TypeScript 模块化开发相关的内容。

1.4.1　tsconfig.json 文件中的 target 和 module 命令选项

target 命令选项规定了将 TypeScript 代码编译（转译）成哪个 ECMAScript（简称 ES）标准，标准有 ES 3（default）、ES 5、ES 2015、ES 2016、ES 2017、ES 2018 和 ESNEXT，一般将其设置为 ES 5 标准。

module 命令选项指定生成哪个模块系统代码，模块系统有 commonjs、amd、system，以及 umd 和 es2015，为了在 HTML 中使用模块化功能，必须选择使用 ES 2015 选项。

tsconfig.json 模块化编译选项代码如下：

```
{
    "compilerOptions" : {
    "target" : "es5" ,
    "module" : "es2015" ,
    "strict" : true ,
    "esModuleInterop" : true ,
    "watch" : true
    }
}
```

设置好 tsconfig.json 后，在 VS Code 的集成终端中输入 tsc 命令以监控方式启动 TypeScript 编译（转译）器。

1.4.2　编写 Canvas2D 类导出给 main.ts 调用

在根目录下创建一个名为 canvas2d 的文件夹，在该文件夹中建立 Canvas2D.ts 文件，使用 export 关键词导出 Canvas2D 类。具体代码如下：

```
export class Canvas2D {
   // 源码同 Canvas2DUtil 类
}
```

在 main.ts 中使用 import 关键词导入 Canvas2D 类：

```
import { Canvas2D } from "./canvas/Canvas2D"

let canvas2d : Canvas2D = new Canvas2D ( canvas ) ;
canvas2d . drawText ( " Hello World From Module ! " ) ;
```

当使用 Open In Browser 插件运行 index.html 后，会发现文字并没有绘制出来，通过浏览器控制台得到的错误提示是"Uncaught SyntaxError：Unexpected token {"。这是因为 export / import 属于 ES 2015 标准的内容，浏览器加载 ES 2015 模块化代码也是使用<script>标签，但是要加入 type = "module" 属性。

在<script>标签中加入 type = "module"属性运行后，继续得到错误提示"Access to Script at 'file:///XXX/HelloWorld/main.js' from origin null' has been blocked by CORS policy : Invalid response. Origin 'null' is therefore not allowed access. ",这个很明显是跨域问题。

从 HTTP 服务器加载文件就能解决这个问题,可以使用 Tomcat 和 Apache 等开源 HTTP 服务器,但是既然已经安装了 Node.js,那么使用 NPM 包管理器安装本地服务器才是最佳解决方案。

1.4.3 使用 lite-server 搭建本地服务器

引用官方的描述,lite-server 是一个轻量级的、基于 Node.js 运行环境的服务器,目前仅支持 Web App。当运行 lite-server 后,它能够打开浏览器,当 HTML、CSS 或 JavaScript 文件变化时,它会识别到并自动刷新浏览器页面,当路由没有被找到时,它将自动后退页面。lite-server 自动刷新的功能对于开发调试来说是非常方便的。

lite-server 官方推荐 NPM 本地安装,其流程如下所述。

(1)打开 VS Code 集成终端,输入 npm init -f 命令,在项目的根目录下生成 package.json 文件。

(2)继续在集成终端中输入 npm install lite-server --save-dev 命令进行本地安装。

当完成 lite-server 安装后,打开项目根目录下的 package.json 文件,找到 scripts 项,输入如下命令:

```
"scripts" : {
    "dev" : "lite-server"
}
```

打开 VS Code 中的集成终端,输入 npm run dev 命令就能运行 lite-server 了。

当从服务器(http:// localhost:3000)载入 index.html 页面时,浏览器控制台会显示: GET http://localhost:3000/canvas2d/Canvas2D404(Not Found)。

很明显,服务器没找到 Canvas2D 文件,这是因为 TypeScript 导入 Canvas2D 时不允许使用.ts 扩展名,如图 1.22 所示。

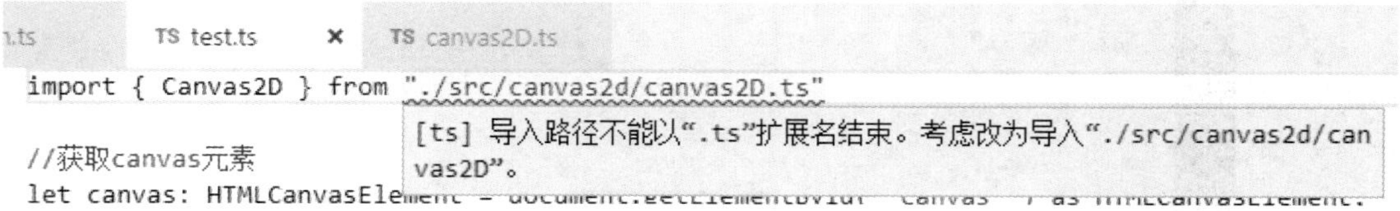

图 1.22 导入路径不能使用.ts 扩展名

通过 tsc 命令编译(转译)为 JS 代码后,也没有扩展名,而 JavaScript 模块化导入需要以.js 为扩展名,因此需要手动更改一下 main.js 文件:

```
// import { Canvas2D } from "/ canvas/Canvas2D" 增加.js 扩展名
import { Canvas2D } from "/canvas/Canvas2D.js " ;
```

当保存好修改的内容后,lite-server 会自动刷新页面,文字将会正确地显示出来。

1.5 使用 SystemJS 自动编译加载 TypeScript

到目前为止，虽然能自动编译（转译）TS 并通过 lite-server 进行热部署资源，但仍有两个不足之处：
- 需要手动修改 import 代码来支持模块加载。
- Index.html 的 Script 引入的是编译（转译）后的 main.js 文件。

比如，我们的需求是直接在浏览器中进行 TS 源码编译（转译），那么 SystemJS 就能实现。SystemJS 是一个通用的模块加载器，它能在浏览器或者 Node.js 上动态加载模块，并且支持 CommonJS、AMD、全局模块对象和 ES 6 模块。

SystemJS 的另一个优点是，它建立在 ES 6 模块加载器之上，所以其语法和 API 在将来很可能是语言的一部分，这会让代码更不会过时。

注意：关于 CommonJS、AMD 和全局模块等相关内容，请读者自行查阅资料。

1.5.1 NPM 本地安装 TypeScript 库和 SystemJS 库

为了使用 SystemJS 加载 TypeScript，最好本地安装 TypeScript 库。在 VS Code 的集成终端中输入 npm install typescript@2.8.3 --save 命令，将 TypeScript 库安装到根目录下的 node_modules 子目录（若有需要，NPM 包安装器会自动创建该目录）中。

安装好 TypeScript 库后，继续在集成终端中输入 npm install systemjs@0.19.47 --save 命令安装 SystemJS 库。

最终的项目结构和 package.json 文件如图 1.23 所示。

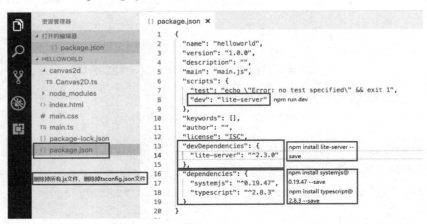

图 1.23　Hello World 项目最终结构与 package.json 文件

为了更好地了解 SystemJS，将资源管理器中的所有.js 文件及 tsconfig.json 文件都删除。

1.5.2　SystemJS 直接编译 TypeScript 源码

在 index.html 中配置 SystemJS 编译（转译）TS 源码，代码如下：

```
<head>
    <meta charset = " utf-8 " />
    <meta http-equiv = " X-UA-Compatible " conten t= " IE = edge ">
    <title> Hello Wrold </title>
    <meta name = " viewport " content = " width = device-width , initial-scale = 1 ">
    <link rel = " stylesheet " type = " text / css " media = " screen " href = " main.css " />
    <script src = "node_modules / systemjs / dist / system.js" > </script>
    <script src= "node_modules / typescript / lib / typescript.js" >
    </script>
    <script>
        System.config ( {
            transpiler : ' typescript ' ,    //使用 TypeScript 进行编译（转译）
            packages : {
                ' / ' : {
                    defaultExtension : 'ts'    //编译（转译）当前目录（./）下的所有
                                                 TS 文件,如果不设定 defaultExtension,
                                                 则程序无法正确运行
                }
            }
        } ) ;
        System
            .import ( './main.ts' )            //入口文件
            .then ( null , console.error.bind ( console ) );
                        //如果出现错误,转发到浏览器的 console 中显示错误信息
    </script>
</head>

<body>
    <canvas id = "canvas" > </canvas>
</body>

</html>
```

设置好 SystemJS 后，在集成终端中运行 npm run dev 后，会发现程序顺利地运行起来了。关于 SystemJS 更多的用法，请参考 https：//www.npmjs.com/package/systemjs。

在使用 lite-server 服务器热部署及 SystemJS 后：

- 不再需要使用基于客户端的 open in browser 插件了。
- 不再需要在<script>标签中使用 defer 关键词。
- 不再需要设置 type = "module"了，一切由 SystemJS 来掌控。
- 不再需要 tsconfig.json 文件了。

- 不再需要在集成终端中运行 tsc 命令来自动编译（转译）JS 了，仅需要一次性地调用 npm run dev 命令启动 lite-server 服务器。

但是还是建议读者使用 tsc --init 命令生成 tsconfig.json，用于在编码时让 VS Code 进行严格类型检查，增强代码的健壮性。此时 tsconfig.json 仅作为类型检测使用，实际并不参与 TS 的编译（转译）工作。

1.6 使用 VS Code 调试 TypeScript 源码

VS Code 提供了强大的调试功能，可以非常方便地调试相关代码。本节来介绍 VS Code 中如何断点调试 TypeScript 源代码。

1.6.1 安装及配置 Debugger for Chrome 扩展

要调试 HTML 页面中的 TypeScript（或 JavaScript）代码，必须要安装 Debugger for Chrome 扩展。可以单击 VS Code 最左侧的活动栏中的扩展图标（或使用 Shift + Ctrl + X 快捷键）打开扩展面板，输入 Debugger for Chrome 后下载安装。

当安装好 Debugger for Chrome 后，按 F5 键进入调试状态，此时 VS Code 会显示如图 1.24 所示界面。

选择 Chrome 选项后，VS Code 会自动在当前项目的根目录下生成一个名为 .vscode 的文件夹，并且同时在该文件夹内生成 launch.json 文件，具体内容如下：

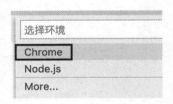

图 1.24 使用 Debugger for Chrome

```
{
    "version" : "0.2.0" ,
    "configurations" : [
    {
      "type" : "chrome" ,
      "request" : "launch" ,
      "name" : "启动 Chrome 并打开 localhost" ,
"url" : "http: //localhost:8080" ,
"webRoot" : "${workspaceFolder}"
    } ]
}
```

在默认情况下，launch.json 中的 url 属性使用 8080 端口，而 lite-server 服务器使用的端口是 3000，因此更改 url 为 htt://localhost:3000。由此可见，调试 TypeScript / JavaScript 源码需要服务器提供服务。

1.6.2 VS Code 中单步调试 TypeScript

当再次按 F5 键启动 Debugger for Chrome 后（当前 lite-server 服务器已开启），会出

现调试界面，运行 index.html 页面，如图 1.25 所示。

图 1.25　VS Code 调试按钮界面

关于调试按钮说明，如表 1.1 所示。

表 1.1　调试按钮说明

调　试　按　钮	快　捷　键	说　　明
暂停/继续（pause / continue）	F5	遇到断点，程序中断。若存在下一个断点，按F5键运行到下一个断点
单步跳过（step over）	F10	单步执行，不会进入子函数，而是将子函数执行完再停止
单步调试（step in）	F11	单步执行，遇到子函数就会进入子函数，然后继续单步执行
单步跳出（step out）	Shift + F11	当单步执行到子函数内后，使用step out可以执行完子函数的剩余部分，并返回到上一层函数
重启（restart）	Ctrl + Shift + F5	重新调试运行整个程序
停止 （stop）	Shift + F5	退出调试程序

以前面的 Hello World Demo 为例，来看一下如何使用表 1.1 中的调试命令。

如图 1.26 所示，在 maint.ts 中添加一个断点（第 51 行）。

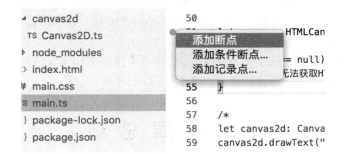

图 1.26　添加调试断点

按 F5 键启动 Debugger for Chrome 调试器，此时显示如图 1.25 所示的调试界面，并且程序在断点处（第 51 行）中断，如图 1.27 所示。

按 F10 键（单步跳过），会发现程序单步执行到下一行代码（此时程序在第 52 行代码处中断）。

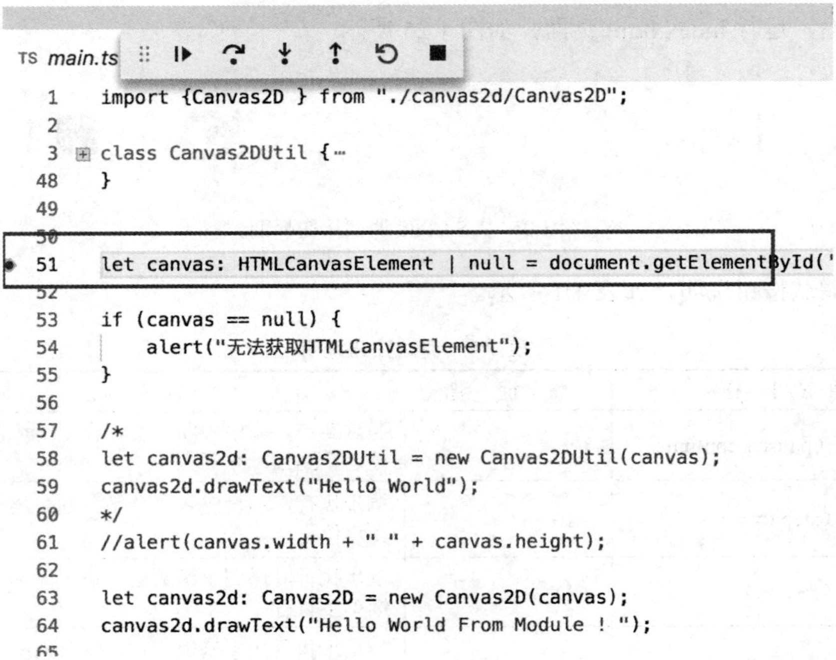

```
TS main.ts
 1  import {Canvas2D } from "./canvas2d/Canvas2D";
 2
 3  class Canvas2DUtil {…
48  }
49
50
51  let canvas: HTMLCanvasElement | null = document.getElementById('
52
53  if (canvas == null) {
54      alert("无法获取HTMLCanvasElement");
55  }
56
57  /*
58  let canvas2d: Canvas2DUtil = new Canvas2DUtil(canvas);
59  canvas2d.drawText("Hello World");
60  */
61  //alert(canvas.width + " " + canvas.height);
62
63  let canvas2d: Canvas2D = new Canvas2D(canvas);
64  canvas2d.drawText("Hello World From Module ! ");
65
```

图 1.27 按 F5 键启动调试后程序在断点处中断

如果现在不想单步执行了，那么可以按 F5 键（继续）运行程序，如果后续代码没有断点，则程序会正常执行到全部结束为止。如果后续代码有另外一个断点（例如在第 64 行代码处再增加一个断点），那么按 F5 键后会发现，程序再一次在第 64 行处中断。

如果现在想在第 64 行代码处进入 drawText 方法，那很简单，可以按 F11 键（单步进入）进入 drawText 方法内。如果不使用 F11 键而是用 F10 键（单步跳过），则会跳过 drawText 方法，执行到第 65 行代码处中断。

现在假设已经在 drawText 方法体内，可以继续使用 F10 键进行单步执行，也可以使用 Shift + F11 键跳出 drawText 方法，并返回到第 65 行代码处。

1.7 本章总结

本章主要介绍了如何构建 TypeScript 的开发、编译和调试环境，使读者了解到 TypeScript 语言本身和开发工具是两个不同的概念。

首先 TypeScript 语言本身需要使用 Node.js 的包管理器 NPM 来获取，因此先从 Node.js 官网获取其安装包，然后通过 NPM 来全局安装 TypeScript 语言。

然后为了更好地体验 TypeScript 的开发环境，安装了 TypeScript 最佳搭档 VS Code 代码编辑器。

有了 TypeScript 语言和 VS Code 代码编辑器后，随后又构建了一个简单的开发和运行环境，实现了本书的第一个 TypeScript 程序，用来了解如何使用 TypeScript 编译（转译）器将单个 TS 文件编译（转译）成相应的 JS 文件。

有了 JS 文件后，就能在 HTML 中运行 JS 脚本，这时面临着如何快速启动 HTML 文件的问题。但是不想每次都要定位到 HTML 所在目录，然后通过右键菜单选择要运行的浏览器，因此此时引入了一个 VS Code 的插件：Open In Default Browser。通过该插件就能加速打开并运行 HTML 文件。

在 TS 源文件比较少的情况下，可以使用 tsc 命令然后输入多个 TS 源文件生成 JS 文件；如果 TS 源文件非常多的话，手动输入每个 TS 源文件，并且每次源文件修改后都要重新编译，这种情况非常不便。

我们会发现，通过 tsc --init 生成 tsconfig.json 文件，配置相关属性，使用 tsc 命令直接运行编译（转译）器后，任何源文件发生改变时会自动编译（转译）成 JS 文件。

至此，获得了一个方便、好用的 TypeScript 开发和运行环境。但是对于程序来说，模块化、解决资源自动加载跨域问题，以及 TypeScript / JavaScript 断点调试也是不可或缺的需求。要实现这些需求，需要一个 HTTP 服务器。原本笔者使用 Tomcat，但是涉及复杂的 Java 运行和配置环境，后来发现了 lite-server 这个基于 Node.js 实现、使用 NPM 安装的轻量级服务器，它无缝地并入当前的开发环境，由此获得自动加载、热部署，以及潜在的调试功能。

配置好上述内容后，依旧存在一些问题（请参考 1.5 节介绍的两个不足之处），为了解决这两个问题，引入了 SystemJS 来自动编译和加载 TypeScript 源码，并且解决了存在的问题。

最后，我们构建了 TypeScript / JavsScript 基于 VS Code 的断点调试环境，先明确的一点是本书使用 Chrome 浏览器作为开发和测试环境，因此先通过 NPM 安装 Debugger for Chrome 调试插件，然后仅需配置一次，就能在 VS Code 中直接通过 F5 键启动 HTML 页面，并支持断点调试功能。

至此，搭建了一个完整的 TypeScript 开发、编译和调试的环境，能够支持如下几个功能：

- 直接启动 HTML 页面。
- 自动重新编译（转译）TypeScript。
- 解决资源跨域问题，支持服务器热部署。
- 模块化开发和加载 TypeScript。
- TypeScript 严格类型检测。
- 断点调试 TypeScript。

如果读者是从 C/C++、Java、C#、Objective-C 等语言转到 TypeScript 开发的话，就会非常熟悉这种开发调试模式。

工欲善其事，必先利其器。有了 TypeScript 这个利器，接下来就来领略 TypeScript 的强类型、面向对象、面向接口和支持泛型的编程模式吧。

第 2 章 使用 TypeScript 实现 Doom 3 词法解析器

本章的目的是想让大家了解 TypeScript 中常用的一些语法及编程方式（如面向对象编程、面向接口编程、泛型编程，以及常用的设计模式等），因此，特别以面向接口的方式编写了一个 Doom3（原 id Software 公司毁灭战士 3 游戏引擎）词法解析器，并且在此基础上实现了工厂模式和迭代器两种设计模式，使其支持接口的生成，以及使用迭代方式进行 Token 的解析输出。

Doom3 引擎中所有的资源都存储在后缀名为.pk4 的资源文件包中。该资源文件包实际上就是一个 zip 压缩文件，因此可以将.pk4 后缀名更改为.zip，然后就可以直接使用，例如 winzip 解压程序进行解压并浏览该文件。

当打开 pk4 文件浏览后会发现一个事实：Doom3 引擎中大部分的资源都是基于文本描述的（除了图片、视频及音频等资源外），并且基于一套简单的、统一的词法规则。如何将文本按照词法规则描述解析成有意义的标记（Token）是本章的另外一个重要目的。

最后对 XMLHttpRequest 对象进行二次封装用于向服务器请求资源。这样就能够利用第 1 章中部署的 lite-server 服务器，将所有存储在 lite-server 服务器上的文本文件、二进制文件、视频文件和音频文件传输到 TypeScript（网页客户端）中进行处理。

实际上，实现的词法解析器不仅仅用于 Doom3 引擎相关资源的解析，通过些许扩展，还可以支持解析各种不同格式的 ASCII 编码文本文件，例如 Wavefront 的 obj 模型文件，以及 mtl 材质文件等。

2.1 Token 与 Tokenizer

为简单起见，使用 JS 代码作为示例来了解一下 JS 的 Token 相关内容，以加深对 Token 的理解。代码如下：

```
if ( b === true )
    alert ( "true" ) ;
```

在浏览器中输入网址 http://esprima.org/demo/parse.html 后，将上述代码粘贴到左侧文本编辑框内，然后选择 Tokens Tab 选项，就会获得如下结果：

```
[
    {
        "type" : "Keyword",
        "value" : "if"
    },
    {
        "type" : "Punctuator",
        "value" : "("
    },
    {
        "type" : "Identifier",
        "value" : "b"
    },
    {
        "type" : "Punctuator",
        "value" : "==="
    },
    {
        "type" : "Boolean",
        "value" : "true"
    },
    {
        "type" : "Punctuator",
        "value" : ")"
    },
    {
        "type" : "Identifier",
        "value" : "alert"
    },
    {
        "type" : "Punctuator",
        "value" : "("
    },
    {
        "type" : "String",
        "value": "\"true\""
    },
    {
        "type" : "Punctuator",
        "value" : ")"
    },
    {
        "type" : "Punctuator",
        "value" : ";"
    }
]
```

通过上述代码可以看到，Esprima（ECMAScript 词法语法解析器）会将 JS 源码解析成 Token 的集合表示。每个 Token 具有 type 属性，表示该 Token 的类型分类，并且具有 value 属性，表示该 Token 的值是什么。

由此可见，Token（标记或记号）就是指一组不可分割的字符或字符串，它能唯一地、没有歧义地标记出一种状态。从本质上来说，就是特殊的字符或字符串（例如 if 和=== 等）。而 Esprima 则是 Tokenizer，其作用是将字符串表示的 JS 源码数据读取进来，按照预先设定的标准进行分类处理，处理的结果就是 Token。

事实上，Esprima 是一个 ECMAScript 解析器，包含词法解析和语法解析，最终会将 JS 源码解析成抽象语法树（Abstract Syntax Tree，AST），而在这里由于演示的原因，仅仅使用了 Esprima 的词法解析功能，并没有使用到语法解析功能。

2.1.1　Doom3 文本文件格式

如果要实现一个特定文件格式的词法解析器，一定要了解该文件的词法特征，根据文件的词法特征抽象出分类规则，然后才能编码实现词法解析功能。因此本节来讲解一下 Doom3 引擎中的文本文件的相关规则，使用如下一段具有普遍性的文本字符串：

```
numMeshes   5

/*
 * joints 关键字定义了骨骼动画的 bindPose
 */
joints {
    "origin"    -1 ( 0 0 0 ) ( -0.5 -0.5 -0.5 )
    "Body"   0  ( -12.1038131714  0  79.004776001 )   ( -0.5 -0.5 -0.5 )
    // origin
}
```

例如，numMeshes 和 joints 等没有双引号的单词，作为关键字处理，也就是 Doom3 引擎预先定义好的具有特定含义的一些词，它们具有唯一性及不可更改性。

例如，"origin""Body"这些具有双引号的单词，作为标识符处理，这些标识符并非由 Doom3 引擎预先定义，而是由美术设计等相关人员或者模型制作动画师定义的名称。

在"/*"和"*/"之间的文字被 Doom3 引擎的词法解析器视为注释，和 TypeScript 一样，表示多行注释。斜杠"//"后的文字则被视为单行注释，这也和 TypeScript 的单行注释保持一致性。

大括号对"{ }"表示一个块状模块，可以将其视为一个区块分组，和 TypeScript 作用域类似。小括号对"( )"内部是使用浮点数表示的向量或矩阵数据，可以将其看成数组，需要注意的一点是，数组元素之间不是使用逗号分隔，而是使用空格符号进行分隔。

Doom3 文本文件中的数据类型其实只有两种：字符串和数字，其中关键字和标识符都可以看成字符串类型，而数字可以分为整数和浮点数两种类型。

上述内容基本囊括了 Doom3 文本文件格式的关键之处，还有一些隐藏在深处的规则，将在源码实现的过程中进行描述。

2.1.2　使用 IDoom3Token 与 IDoom3Tokenizer 接口

先来看一下如何调用 Doom3 的词法解析器，然后再了解如何实现其过程。首先创建一个名为 doom3TokenizerTest.ts 的文件，并导入如下 4 个结构：

```
import { IDoom3Token , IDoom3Tokenizer , Doom3Factory , ETokenType } from
"./src/doom3Tokenizer" ;
```

然后将要解析的字符串赋值给一个 string 类型的变量，需要注意的是，使用了 ES 6 中的模板字符串（使用了开单引号`xxx`，而不是双引号"xxx"或单引号'xxx'的方式来定义字符串字面值）：

```
let str : string = `                         //注意：这是开单引号`，不是单引号'
    numMeshes  5
    /*
     * joints 关键字定义了骨骼动画的 bindPose
     */

    joints {
        "origin" -1 ( 0 0 0 ) ( -0.5 -0.5 -0.5 )
        "Body"    0 ( -12.1038131714 0 79.004776001 ) ( -0.5 -0.5 -0.5 )
        //origin
    }
`;  // 注意：这是开单引号`，不是单引号'
```

最后来看一下如何使用 IDoom3Token 和 IDoom3Tokenizer 的属性和方法。具体代码如下：

```
// 从 Doom3Factory 工厂创建 IDoom3Tokenizer 接口
let tokenizer : IDoom3Tokenizer = Doom3Factory . createDoom3Tokenizer ( ) ;
// IDoom3Tokenizer 接口创建 IDoomToken 接口
let token : IDoom3Token = tokenizer . createDoom3Token ( ) ;

//设置 IDoom3Tokenizer 要解析的数据源
tokenizer . setSource ( str ) ;

// getNextToken 函数返回 ture,说明没有到达字符串的结尾，仍有 Token 需要解析
// 解析的结果以传引用的方式从参数 token 中传出来
// 如果 getNextToken 返回 false,说明已经到达字符串结尾,则停止循环
while ( tokenizer . getNextToken ( token ) ) {
    //如果当前的 Token 的 type 是 Number 类型
    if ( token . type === ETokenType . NUMBER ) {
        console . log ( " NUMBER : " + token . getFloat ( ) ) ;
                                        //输出该数字的浮点值
    } else if ( token . isString ( "joints" ) ) {
        //如果当前 Token 是字符串类型，并且其值为 joints,则输出
        console . log ( " 开始解析 joints 数据 " ) ;
    }
    else { //否则获取当前 Token 的字符串值
        console . log( " STRING : " + token . getString ( ) ) ;
    }
}
```

使用 F5 快捷键启动 VS Code 的调试器，会看到在浏览器的控制台中输出如图 2.1 所示的内容。

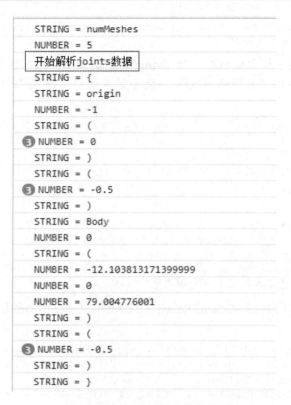

图 2.1　词法解析后 Chrome 输出结果

可以看到，每个关键字（例如 numMeshes）、标识符（例如 origin），以及数字（浮点数、整数及负数）被正确地输出，并且跳过多行注释和单行注释中的内容。代码中对于 joints 关键字则进行了特殊处理，因此并没有输出。除此之外，还将左右小括号和左右大括号都作为单独的一个 Token 输出到控制台。

如果想重新解析整个字符串，那么可以使用 IDoom3Tokenizer 接口的 reset 方法，该方法会将当前索引设置到字符串的首位，这样继续循环调用 getNextToken 就可以重新解析整个字符串。

也可以使用 IDoom3Tokenizer 的 setSource 方法重设要解析的字符串（另外一个字符串），setSource 方法内部也会将当前索引重置到字符串的首位。

2.1.3　ES 6 中的模板字符串

在此强调一下，使用 ES 6 中的模板字符串是很不错的一种体验。笔者认为 ES 6 模板字符串一个最大的优点是：可以一次定义多行字符串，并且保证空格和缩进，而单引号或双引号只能定义单行字符串。下面来看一下笔者经常使用的定义多行字符串的 3 种方式。

（1）在一行字符串中添加\n转义换行符，这种方式耗时又容易出错。具体代码如下：

```
let str1 : string = " numMeshes 5 \n /** \n * joints 关键词定义了骨骼动画的
bindPose \n */ \n joints { \n 'origin'-1 ( 0 0 0 ) ( -0.5 -0.5 -0.5 )
\n 'Body'0 ( -12.1038131714 0 79.004776001 ) ( -0.5 -0.5 -0.5 ) // origin
\n } " ;
```

（2）使用+=符号拼接字符串，代码如下：

```
let str2: string = " numMeshes  5 " ;
        str2 += " /* " ;
        str2 += " * joints 关键词定义了骨骼动画的 bindPose " ;
        str2 += " */ " ;
        str2 += " joints { " ;
        str2 += " 'origin' -1 ( 0 0 0 ) ( -0.5 -0.5 -0.5 ) " ;
        str2 += " 'Body' 0 ( -12.1038131714 0 79.004776001 )
        ( -0.5 -0.5 -0.5 )   // origin " ;
        str2 += " } " ;
```

（3）使用数组方式并调用 join()函数将数组拼接成字符串，这种方式相对来说比较清晰，代码如下：

```
let str3: string = [
  " numMeshes  5 " ,
  " /* " ,
  " * joints 关键词定义了骨骼动画的 bindPose " ,
  " */ " ,
  " joints { " ,
    " 'origin'   -1 ( 0 0 0 ) ( -0.5 -0.5 -0.5 ) " ,
    " 'Body' 0 ( -12.1038131714 0 79.004776001 ) ( -0.5 -0.5 -0.5 )
    // origin " ,
  " } "
] . join( " \n " ) ;
```

需要注意的是，以上 3 种方式中，处理字符串中的子字符串（例如 orgin 和 Body）时，需要使用单引号引起来。

如果不考虑兼容性，并且当前的 JavaScrit 引擎支持 ES 6 标准的话，那么应尽量使用模板字符串。模板字符串还支持以${ }方式定义变量，具体的用法请读者自行查阅 TypeScript 或 ES 6 官方手册。

2.1.4　IDoom3Token 与 IDoom3Tokenizer 接口的定义

可以使用 TypeScript 的 interface 关键字来定义接口。具体代码如下：

```
export interface IDoom3Token {
   reset ( ) : void ;
   isString ( str : string ) : boolean ;
   readonly  type : ETokenType ;
   getString ( ) : string ;
   getFloat ( ) : number ;
   getInt ( ) : number ;
}
```

关于 IDoom3Token 接口中大部分方法的应用，在上一节的代码中有演示，还是比较简单的。这里看一下 type 这个只读属性，该属性使用 readonly 声明，意味着其值只能被读取，不能被更改。同时 type 的数据类型为 ETokenType，是一个枚举类型。在 TypeScript 中，可以使用 enum 关键字来定义枚举类型。具体代码如下：

```
export enum ETokenType {
    NONE ,              // 0 default 情况下，enum 定义的枚举值是以 0 开始的数字类型
    STRING ,                                                    // 1 表示字符串类型
    NUMBER                                                      // 2 表示数字类型
}
```

接下来再看一下 IDoom3Tokenizer 的接口定义，代码如下：

```
export interface IDoom3Tokenizer {
    setSource ( source : string ) : void ;              //设置要解析的字符串
    reset ( ) : void ;                                  // 重置当前索引为 0
    getNextToken ( token : IDoom3Token ) : boolean ;    // 获取下一个 Token
}
```

2.2　IDoom3Token 与 IDoom3Tokenizer 接口的实现

在上一节中声明了 IDoom3Token 接口和 IDoom3Tokenizer 接口，本节来看一下这两个接口的具体实现过程，从中会发现接口和实现类之间的微妙关系，即接口规定了要做什么，接口的实现类则规定了应该怎么做。

2.2.1　Doom3Token 类成员变量的声明

首先来看一下 IDoom3Token 接口的实现类，在 TypeScript 中使用 implements 关键字来实现一个接口，代码如下：

```
class Doom3Token implements IDoom3Token {
    private _type : ETokenType ;
                         // 标识当前 Token 的类型 : NONE / STRING / NUMBER
    private _charArr : string [ ] = [ ] ;              // 字符串数组
    private _val : number ; // 如果当前的 Token 类型是 NUMBER，则会设置该数值，
                         如果是字符串类型，就忽略该变量
}
```

上面的代码很简单，但是也有几个值得关注的地方，如下所述。

（1）在 TypeScript / JavaScript 中，并没有 char 这个数据类型，都是使用 string 类型来表示单个字符，在变量_charArr 中存放的实际是 char（一个字符）类型的数据。

（2）在 TypeScript 中，有两种声明和实列化（内存分配）类型数组的方式，第一种就是上面所使用的方式，另外一种可以使用_charArray : Array < string> = new Array < string > ()的方式，笔者更喜欢第一种方式来声明类型数组变量，简洁明了。

（3）在声明 IDoom3Token 接口时使用了 export 关键字来导出接口，但是在实现类 Doom3Token 中并没有使用 export 关键字。这是接口的一个很棒的特性：只暴露（export）接口（interface），而隐藏类（class）的实现，第三方调用时，只关心接口是怎么使用的，不需要知道具体类是怎么实现的。

（4）TypeScript 支持 public、protected 和 private 这 3 个级别的访问修饰符，如果没有在成员变量前声明访问修饰符，在默认情况下，被定义为 public 级别。3 个访问修饰符的区别如下：

- 被 public 访问修饰符修饰的成员变量或方法能够被所有类访问。
- 被 protected 访问修饰符修饰的成员变量或方法既能被定义它的类访问，也能被继承它的子类访问。
- 被 private 访问修饰符修饰的成员变量或方法只能被定义它的类访问，也就是说不能在声明它的类的外部访问。

2.2.2　Doom3Token 类变量初始化的问题

接下来继续看一下构造函数（constructor 关键字）和 reset()函数，代码如下：

```
public constructor ( ) {
    this . _charArr . length = 0 ;
    this . _type = ETokenType . NONE ;
    this . _val = 0.0 ;
}
public reset ( ) : void {
    this . _charArr . length = 0 ;
    this . _type = ETokenType . NONE ;
    this . _val = 0.0 ;
}
```

可以发现，constructor 中的代码和 reset()函数中的代码一模一样，那么读者可能会问，为什么不在 constructor 中直接调用 reset()函数呢？

其实这里涉及 TypeScript 对成员变量初始化的时机点问题。大家可以试一下，如果在 constructor 中调用 reset()函数，TypeScript 编译器会报"属性 xxx 没有初始化表达式，且未在构造函数中明确赋值。"的错误，如图 2.2 所示。

从上述错误描述中可以知道，TypeScript 对于成员变量的初始化有两个时机点，第一个时机点是在成员变量声明时立即进行赋值（初始化），如 private _charArr : string [ ] = [ ]；这句代码，称为初始化表达式。

如果不在成员变量声明时立即赋值的话，那么就只能在 constructor 构造函数中进行变量赋值（初始化）。但是我们会发现，有时候延迟初始化或重新初始化是很有必要的一种操作。幸运的是，从 TypeScript 2.7 版本开始支持使用!（感叹号）来进行变量的显示断言赋值声明。下面来修改一下代码，看一下效果。具体代码如下：

```typescript
// 使用!操作符来显示断言赋值声明
private _val ! : number ;
private _type ! : ETokenType ;

public constructor ( ) {
    // this . _charArr . length = 0 ;
    // this . _type = ETokenType . NONE ;
    // this . _val = 0.0 ;
    this . reset ( ) ;
}
```

```
class Doom3Token implements IDoom3Token {

    private _charArr: string[] = [];

    private                [ts] 属性"_type"没有初始化表达式，且未在构造函数中明确赋值。
    private _type: (property) Doom3Token._type: ETokenType

    public constructor () {
        /*
        this . _charArr . length = 0 ;
        this . _type = ETokenType . NONE ;
        this . _val = 0.0 ;
        */
        this . reset();
    }

    public reset () : void {
        this . _charArr . length = 0 ;
        this . _type = ETokenType . NONE ;
        this . _val = 0.0 ;
    }
}
```

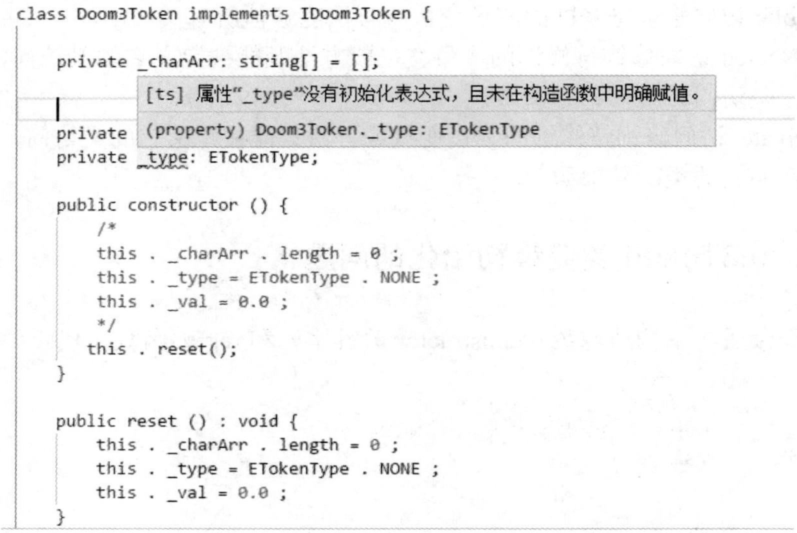

图 2.2　TypeScript 初始化错误

可以发现，TypeScript 不再报初始化的错误了，是不是很棒的感觉？这是一个很有用的功能，可以灵活地处理变量初始化的问题，因此值得在这里花点时间讨论一下。最后还是需要强调一点，在使用该变量前一定要初始化变量。

2.2.3　IDoom3Token 接口方法的实现

接下来看一下 Doom3Token 类的其他几个接口方法的实现。具体代码如下：

```typescript
// 使用 get 关键字来定义属性，get 定义只读属性，set 定义只写属性
public get type ( ) : ETokenType {
    return this . _type ;
}
//获取当前 Token 的字符串值
public getString ( ) : string {
    // _charArr 数组中存放的都是单个字符序列，例如["d","o","o","m","3"]
    // 可以使用数组的 join 方法将字符串联成字符串
    // 下面使用 join 方法后，会返回 doom3 这个字符串
```

```
        return this . _charArr . join ( "" ) ;
}
// 获取当前 Token 的浮点值
public getFloat ( ) : number {
    return this . _val ;
}
// 获取当前 Token 的 int 类型值
public getInt ( ) : number {
    // 使用 parserInt() 函数
    // 第一个参数是一个字符串类型的数字表示
    // 第二个参数是进制,一般用十进制
    return parseInt ( this . _val . toString ( ) , 10 ) ;
}
```

下面来看一个字符串比较的接口方法的实现。具体代码如下:

```
public isString ( str : string ) : boolean {
    let count : number = this . _charArr . length ;
    // 字符串长度不相等,肯定不等
    if ( str . length !== count ) {
        return false ;
    }
    // 遍历每个字符
    for ( let i : number = 0 ; i < count ; i++ ) {
     // _charArr 数组类型中每个 char 和输入的 string 类型中的每个 char 进行严格比较
     (!==操作符而不是!=)
     // 只要任意一个 char 不相等,意味着整个字符串都不相等
        if ( this . _charArr [ i ] !== str [ i ] ) {
            return false ;
        }
    }
    // 完全相等
    return true ;
}
```

2.2.4　Doom3Token 类的非接口方法实现

至此,我们介绍了所有接口方法的实现及涉及的与 TypeScript 相关的语言要点,这些接口方法都是被第三方调用的,还要增加一些方法,这些方法由实现的内部类(例如 IDoom3Tokenizer 的实现类 Doom3Tokenizer)所调用,但是它们并不需要被公开给第三方使用。下面就介绍这些方法。具体代码如下:

```
// 下面 3 个非接口方法被 IDoom3Tokenizer 接口的实现类 Doom3Tokenizer 所使用
// 将一个 char 添加到_charArr 数组的尾部
public addChar ( c : string ) : void {
    this . _charArr . push ( c ) ;
}
// 设置数字,并将类型设置为 NUMBER
public setVal ( num : number ) : void {
    this . _val = num ;
    this . _type = ETokenType . NUMBER ;
```

```
}
//设置类型
public setType ( type : ETokenType ) : void {
    this . _type = type ;
}
```

2.2.5　Doom3Tokenzier 处理数字和空白符

首先声明一下，IDoom3Tokenizer 词法解析器仅支持 ASCII 编码字符串的解析，不支持 UNICODE 编码字符串的解析（换句话说，词法解析器不支持中文解析），实际上 Doom3 引擎文本格式文件也仅支持 ASCII 编码的字符串。

然后再强调一点，像 Java 的 JDK、C#的.NET Framework 或 C 语言的 CRT（C 语言运行库）都内置了强大的 ASCII 字符处理函数，但是在 TypeScript 或 JavaScript 中处理 ASCII 字符的一些操作需要自己来实现。下面就先来实现两个简单但是必需的 ASCII 字符处理函数。具体代码如下：

```
// 接口实现使用 implements 关键字
class Doom3Tokenizer implements IDoom3Tokenizer {
    // 使用了初始化表达式方式初始化字符串数组
    private _digits : string [ ] = [ "0" , "1" , "2" , "3" , "4" , "5" , "6" , "7" , "8" , "9" ] ;
    private _whiteSpaces : string [ ] = [ " " , "\t" , "\v" , "\n" ] ;

    //判断某个字符是不是数字
    private _isDigit ( c : string ) : boolean {
        for ( let i : number = 0 ; i < this . _digits . length ; i++ ) {
            if ( c === this. _digits [ i ] ) {
                return true ;
            }
        }
        return false ;
    }
    //判断某个字符是不是空白符
    //一般将空格符（" "）、水平制表符（"\t"）、垂直制表符（"\v"）及换行符（"\n"）统称为空白符
    private _isWhitespace ( c : string ) : boolean {
        for ( let i : number = 0 ; i < this . _whiteSpaces . length ; i++ ) {
            if ( c === this . _whiteSpaces [ i ] ) {
                return true ;
            }
        }
        return false;
    }
}
```

2.2.6　IDoom3Tokenizer 接口方法实现

接着来看一下解析字符串时所需的一些方法。具体代码如下：

```typescript
//要解析的字符串,使用Doom3Tokenizer字符串来初始化变量
private _source : string = " Doom3Tokenizer " ;
private _currIdx : number = 0 ;
//实现公开的接口方法,设置要解析的字符串,并且重置当前索引
public setSource ( source : string ) : void {
    this . _source = source ;
    this . _currIdx = 0 ;
}
//实现公开的接口方法,不改变要解析的字符串,仅重置当前索引
public reset ( ) : void {
    this . _currIdx = 0 ;
}
```

2.2.7　Doom3Tokenizer 字符处理私有方法

一旦通过 setSource 方法设置好要解析的源字符串或者调用 reset 方法进行重新解析字符串时,则需要一些操作来获取当前字符或探测下一个字符,可以使用这几个成员方法,代码如下:

```typescript
//获得当前的索引指向的char,并且将索引加1,后移一位
//后++特点是返回当前的索引,并将索引加1
//这样的话,_getChar返回的是当前要处理的char,而索引指向的是下一个要处理的char
private _getChar ( ) : string {
    //数组越界检查
    if ( this._currIdx >= 0 && this . _currIdx < this . _source . length ) {
        return this . _source . charAt ( this . _currIdx ++ ) ;
    }
    return "" ;
}

//探测下一个字符是什么
//很微妙的后++操作
private _peekChar ( ): string {
    //数组越界检查,与_getChar的区别是并没移动当前索引
    if ( this . _currIdx >= 0 && this . _currIdx < this . _source.length ) {
        return this . _source . charAt ( this . _currIdx ) ;
    }
    return "" ;
}

private _ungetChar ( ) : void {
    //将索引前移1位,前减操作符
    if ( this . _currIdx > 0 ) {
        -- this . _currIdx ;
    }
}
```

到此为止,我们构建了 IDoom3Tokenizer 词法解析器最小的运行环境,可以设置（setSource）或重置（reset）要解析的数据源,可以正向地获取（_getChar）当前字符,或探测（_peekChar）下一个字符,也可以反向归还（_ungetChar）一个字符,还可以知道当

前字符是数字字符（_isDigit）或者是空白符（_isWhiteSpace）。下一节将进入 Token 解析阶段。

2.2.8 核心的 getNextToken 方法

IDoom3Tokenizer 的 getNextToken 方法是一个相对复杂的实现，其工作原理就是一个有限状态机（Finite State Machine，简称 FSM）。所谓有限状态机就是指状态是有限的，并且根据当前的状态来执行某个操作。那么来看一下 getNextToken 这个有限状态机相关的问题：

- 有哪几个状态（即有限的状态数量）？
- 每个状态的开始条件是什么？
- 每个状态的结束条件是什么？
- 在某个状态下要做什么（操作）？

带着上面的问题来看一下 getNextToken 的源码。具体代码如下：

```
public getNextToken ( tok : IDoom3Token ) : boolean {
    //使用 as 关键字将 IDoom3Token 向下转型为 Doom3Token 类型
    let token : Doom3Token = tok as Doom3Token ;
    //初始化为空字符串
    let c : string = "" ;
    //重用 Token，每次调用 reset()函数时，将 Token 的索引重置为 0
    //避免发生内存重新分配
    token . reset ( ) ;
    do {
        // 第一步：跳过所有的空白字符，返回第一个可显示的字符
        //开始条件：当前字符是空白符
        c = this . _skipWhitespace ( );
        // 第二步：判断非空白字符的第一个字符是什么
        if ( c === '/' && this . _peekChar ( ) === '/' ) {
            // 开始条件：如果是//开头，则跳过单行注释中的所有字符
            c = this . _skipComments0 ( ) ;
        } else if ( c === '/' && this . _peekChar ( ) === '*' ) {
            //开始条件：如果是/*开头的字符，则跳过多行注释中的所有字符
            c = this . _skipComments1 ( ) ;
        } else if ( this . _isDigit( c ) || c === '-' || ( c === '.' && this .
 _isDigit( this . _peekChar ( ) ) ) ) {
            //开始条件：如果当前字符是数字、符号或者以点号且数字开头
            //则返回到上一个字符索引处，因为第一个字符被读取并处理过了，而_getNumber
            会重新处理数字情况，这样需要恢复到数字解析的原始状态
            this . _ungetChar ( ) ;
            this . _getNumber ( token ) ;
            return true ;
        } else if ( c === '\"' || c === '\'' ) {
            //开始条件：如果以\"或\'开头的字符，例如'origin'或'Body'
            this . _getSubstring ( token , c ) ;
            return true ;
```

```
            } else if ( c.length > 0 ) {
                //开始条件：排除上述所有的条件并且在确保数据源没有解析完成的情况下
                //返回到上一个字符索引处，因为_getString会重新处理相关情况
                this . _ungetChar ();
                this . _getString ( token ) ;
                return true ;
            }
        } while ( c . length > 0 ) ;
        return false ;
    }
```

这段代码的关键点都在注释里面，其中状态的开始条件都已标注出来。状态的结束条件都注释在对应的状态处理函数中。

来看一下这段代码中的向下转型相关内容。上面将 IDoom3Token 类型使用 as 操作符向下转型为 Doom3Token，是因为_getNumber / _getSubstring / _getString 这 3 个方法的输出参数类型是 Doom3Token，而不是 IDoom3Token，因此需要从 IDoom3Token 向下转型到 Doom3Token。在 TypeScript 中也可以使用<>来进行类型转换。具体代码如下：

```
let token : Doom3Token = < Doom3Token > tok ;
```

2.2.9 跳过不需处理的空白符和注释

在 getNextToken 函数中，可以看到要处理的、有限的 6 种状态对应的操作：_skipWhitespace、_skipComments0、_skipComments1、_getNumber、_getSubstring 及_getString，并且在注释中都备注出了状态的开始条件。

下面来看一下用来处理跳过无用或空白字符的 3 个方法实现。具体代码如下：

```
//跳过所有的空白字符，将当前索引指向非空白字符
private _skipWhitespace ( ) : string {
    let c : string = "" ;
    do {
        c = this . _getChar ( ) ;              //移动当前索引
        //结束条件：解析全部完成或当前字符不是空白符
    } while ( c . length > 0 && this . _isWhitespace( c ) ) ;

    // 返回的是正常的非空白字符
    return c ;
}
//跳过单行注释中的所有字符
private _skipComments0 ( ) : string {
    let c : string = "" ;
    do {
        c = this . _getChar ( ) ;
        //结束条件：数据源解析全部完成或者遇到换行符
    } while ( c.length > 0 && c !== '\n' ) ;
    //此时返回的是\n 字符
    return c ;
}
//跳过多行注释中的所有字符
```

```typescript
private _skipComments1 ( ) : string {
    //进入本函数时，当前索引是/字符
    let c : string = "" ;
    // 1. 读取*号
    c = this . _getChar ( ) ;
    // 2. 读取所有非* /这两个符号结尾的所有字符
    do {
        c = this . _getChar ( ) ;

        //结束条件：数据源解析全部完成或者当前字符为*且下一个字符是/，也就是以*/结尾
    } while ( c . length > 0 && ( c !== '*' || this . _peekChar ( ) !== '/' ) ) ;
    // 3. 由于上面读取到*字符就停止了，因此要将/也读取并处理掉
    c = this . _getChar ( ) ;
    //此时返回的应该是/字符
    return c ;
}
```

在注释中详细标注了针对每个状态的操作及结束条件。

2.2.10　实现_getNumber 方法解析数字类型

接下来看一下 IDoom3Tokenizer 词法解析器中最复杂的一个解析方法。具体代码如下：

```typescript
private _getNumber ( token: Doom3Token ) : void {
    let val : number = 0.0 ;
    let isFloat : boolean = false ;                         // 是不是浮点数
    let scaleValue : number = 0.1 ;                         // 缩放的倍数

    //获取当前的字符（当前可能的值是[数字，小数点，负号]）
    //目前不支持+3.14 类似的表示
    //如果 - 3.14 这种情况，由于负号和数字之间有空格，所以目前会解析成[ '-' , 3.14 ]
    //这两个 Token
    //目前支持例如：[ 3.14 , -3.14 , .14 , -.14 , 3. , -3. ]的表示
    let c : string = this . _getChar ( ) ;
    //预先判断是不是负数
    let isNegate : boolean = ( c === '-' ) ;                // 是不是负数
    let consumed : boolean = false ;
    //获得 0 的 ASCII 编码，使用了字符串的 charCodeAt 实列方法
    let ascii0 = "0" . charCodeAt ( 0 ) ;
    // 3.14 -3.14 .13 -.13 3. -3.
    // 只能进来 3 种类型的字符 : [ -, ., 数字]
    do {
        // 将当前的字符添加到 Token 中
        token . addChar ( c ) ;
        // 如果当前的字符是.的话，设置为浮点数类型
        if ( c === '.' ) {
            isFloat = true ;
        } else if ( c !== '-' ) {
            // 十进制从字符到浮点数的转换算法
            // 否则如果不是-符号的话，说明是数字（代码运行到这里已经将点和负号操作符都
```

排斥掉了，仅可能是数字）

```
    //这里肯定是数字了，获取当前的数字字符的ASCII编码
    let ascii : number = c . charCodeAt ( 0 ) ;
    //将当前数字的ASCII编码减去0的ASCII编码的算法，其实就是进行字符串-
数字的类型转换算法
    let vc : number = ( ascii - ascii0 ) ;
    if ( ! isFloat )          // 整数部分算法，10倍递增，因为十进制
        val = 10 * val + vc ;
    else {
        // 小数部分算法
        val = val + scaleValue * vc ;
        //10 倍递减
        scaleValue *= 0.1 ;
    }
 } /* else {                     // 运行到这段代码时，当前的变量c肯定为负号
        console.log ( " 运行到此处的只能是 : " + c ) ;
    }*/
    //上面循环中的代码没有读取并处理过字符，之所以使用consumed变量，是为了探测
下一个字符
    if ( consumed === true )
        this . _getChar ( ) ;
    //获得下一个字符后，才设置consumed为true
    c = this . _peekChar() ;
    consumed = true ;
//结束条件：数据源解析全部完成，或下一个字符既不是数字也不是小数点（如果是浮点
数表示的话）
} while (c . length > 0 && ( this . _isDigit ( c ) || ( ! isFloat && c
=== '.' ) ) ) ;
//如果是负数，要取反
if ( isNegate ) {
    val = - val ;
}

//设置数字值和NUMBER类型
token.setVal ( val ) ;
}
```

上面这段代码还是比较复杂的，要理解这段代码，最好的方式就是使用一个具有典型性的例子，来看一下如下代码：

```
let input:string = " [ 3.14 , -3.14 , .14 , -.14 , 3. , -3. , +3.14 ] " ;
//使用setSource重新设置数据源
tokenizer . setSource ( input ) ;
while ( tokenizer . getNextToken ( token ) ) {
    if ( token . type === ETokenType . NUMBER ) {
        console . log ( "NUMBER : " + token . getFloat ( ) ) ;
    }
    else {
        console . log( "STRING : " + token . getString ( ) ) ;
    }
}
```

运行代码后的结果如图 2.3 所示。
- 左右中括号及逗号作为 STRING 类型的 Token 正常地解析出来。
- [3.14 , -3.14 , .14 , -.14 , 3. , -3.]表示方式也正常解析出来。

+3.14 这种形式无法正确解析，如果想要支持正号"+"解析操作也不难，毕竟已经完成了负号"-"解析，处理流程类似，这个问题就交给读者尝试解决。

```
STRING : [
NUMBER : 3.14
STRING : ,
NUMBER : -3.14
STRING : ,
NUMBER : 0.14
STRING : ,
NUMBER : -0.14
STRING : ,
NUMBER : 3
STRING : ,
NUMBER : -3
STRING : ,
STRING : +3
NUMBER : 14
STRING : ]
```

图 2.3　不支持的数字解析格式

2.2.11　实现_getSubstring 方法解析子字符串

在 2.1.1 节中提到过，Doom3 文本文件格式中的标识符是由带一对单引号或双引号的字符串组成的，因此也需要一个方法来解析这种情况。让我们来看一下是如何实现的。具体代码如下：

```
private _getSubstring ( token : Doom3Token, endChar: string ) : void {
    let end : boolean = false ;
    let c : string = "" ;
    token . setType ( ETokenType.STRING ) ;
    do {
        // 获取字符
        c = this . _getChar ( ) ;
        //如果当前字符是结束符(要么是\",要么是\')
        if ( c === endChar ) {
            end = true ;   // 结束符
        }
```

```
        else {
            token . addChar( c ) ;
        }

    //结束条件：数据源解析全部完成或遇到换行符（子串不能多行表示）或是结束符号(要么是
\"，要么是\')
    } while ( c . length > 0 && c !== '\n' && ! end ) ;
}
```

2.2.12 实现_getString 方法解析字符串

本节来看一下正常的字符串是如何解析的，实现代码如下：

```
// 进入该函数，说明肯定不是数字，不是单行注释，不是多行注释，也不是子字符串
// 进入该函数只有两种类型的字符串，即不带双引号或单引号的字符串及 specialChar
private _getString ( token: Doom3Token ) : void {
    // 获取当前字符，因为前面已经判断为字符串了
    let c : string = this . _getChar ( ) ;
    token . setType ( ETokenType . STRING ) ;
    // 进入循环
    do {
        //将当前的 char 添加到 Token 中
        token . addChar ( c ) ;

        if ( ! this . _isSpecialChar ( c ) ) {
            c = this . _getChar ( ) ; // 只有不是特殊操作符号的字符，才调用_getChar
                                      移动当前索引
        }
        //如果 this . _isSpecialChar ( c )为 true，不会调用_getChar()函数，并且
        满足了跳出 while 循环的条件
        //结束条件：数据源解析全部完成，或下一个是空白符或者当前字符是特殊符号
    } while ( c . length > 0 && ! this._isWhitespace ( c ) && !
    this._isSpecialChar ( c ) ) ;
}
```

代码注释比较详细，各位读者可以了解一下。这里会看到，和子字符串不同的一点是，_getString 会将一些特殊的字符（标点符号）作为单独的 Token 返回，具体有哪些特殊的字符，其实依赖于个人的决策。在默认情况下，实现代码如下：

```
// 将左边和右边的大、中、小括号及点号逗号都当作单独的 Token 进行处理
// 如果想要增加更多的标点符号作为 Token，可以在本函数中进行添加
private _isSpecialChar ( c : string ) : boolean {
    switch ( c ) {
        case '(' :
            return true ;
        case ')' :
            return true ;
        case '[' :
            return true ;
```

```
            case ']' :
                return true ;
            case '{' :
                return true ;
            case '}' :
                return true ;
            case ',' :
                return true ;
            case '.' :
                return true ;
        }
        return false ;
    }
```

Doom3 文本文件词法解析器的源码都演示完毕了，最好的研究源码方式是断点调试，大家可以去本书前言中的"本书配套资源获取方式"介绍的网站下载本章的源码进行调试。

2.2.13　IDoom3Tokenizer 词法解析器状态总结

根据上面的源码，总结 IDoom3Tokenizer 词法解析器状态，如表 2.1 所示。

表 2.1　IDoom3Tokenizer词法解析状态表

状　态	共 同 条 件	开 始 条 件	结 束 条 件	动　作	例　子
单行注释	当前正在解析的字符不是空白符	以//开头的字符	以\n结尾	_skipComments0	// 单行跳过
多行注释		以/*开头的字符	以*/结尾	_skipComments1	/* 跳过多行注释 */
解析数字		当前字符是数字或符号或者以点号且下一个是数字开头的字符	数据源解析全部完成或下一个字符既不是数字也不是小数点（如果是浮点数表示的话）	_getNumber	[3.14,-3.14, 14 , -.14 , 3. , -3.]
解析子字符串		以\"或\'开头的字符，例如'origin'或'Body'	数据源解析全部完成或遇到换行符（子串不能多行表示）或是结束符号（要么是\', "要么是\'）	_getSubstring	'origin'或 'Body'
解析字符串		排除上述所有的条件并且在确保数据源没有解析完成的情况下	数据源解析全部完成，或下一个是空白符或者当前字符是特殊符号	_getString	numMeshes

2.3 使用工厂模式和迭代器模式

我们发现，当使用面向接口的编程方式时，可以将接口和实现进行分离，并且使用 TypeScript 的 export 关键字仅仅导出了 IDoom3Token 和 IDoom3Tokenizer 接口，但是没有导出 Doom3Token 和 Doom3Tokenizer 这两个实现类。

这样做的好处是隐藏实现细节，让调用方根本不需要了解具体是如何做的，只需调用接口就可以完成其需求。但是也带来了一个问题：由于没有导出实现类，调用方无法使用例如 new Doom3Tokenizer ( )的方式来初始化实现类，那么调用方该如何初始化接口呢？

2.3.1 微软 COM 中创建接口的方式

在回答这个问题前，我们来看一下微软的 COM（Component Object Modal，组件对象模型）中是如何创建接口的。

以微软的 DirectX 9 为例，在 DirectX 9 SDK 包中提供了一个全局函数 Direct3DCreate9，调用该全局函数后会获得 IDirect3D9 接口的指针，然后可以通过 IDirect3D9 指针的 CreateDevice 接口方法，创建 IDirect3DDevice9 接口指针，接着可以使用 IDirect3DDevice9 接口的 CreateTexture 和 CreateRenderTarget 等接口方法，创建用于渲染的各种资源。

其他的 COM 组件对象创建方式类似，它们的共同点都是精心安排各个接口的层次，通过全局工厂函数 Direct3DCreate9 创建最顶层的接口，然后将上级接口作为工厂，创建下一级接口（使用 CreateXXX 的方法）。

2.3.2 Doom3Factory 工厂类

下面来模拟微软 COM 中的方式，不过笔者更喜欢使用静态方法而不是全局方法，因此增加一个名为 Doom3Factory 的类，用于创建 IDoom3Tokenizer 接口。具体代码如下：

```
// 该工厂需要被调用方使用，因此 export 导出
export class Doom3Factory
{
    // 注意返回的是 IDoom3Tokenizer 接口，而不是 Doom3Tokenizer 实现类
    public static createDoom3Tokenizer ( ) : IDoom3Tokenizer {
       let ret : IDoom3Tokenizer = new Doom3Tokenizer ( ) ;
       return ret;
    }
}
```

可以看到，createDoom3Tokenizer 使用了 static 关键字，说明该方法是静态方法，在调用该方法时不需要使用 new Doom3Factory 方法，可以用类名来直接调用静态方法：

Doom3Factory.createDoom3Tokenizer。需要注意的一点是,在接口中不能声明静态方法或属性。

另外还需要注意的是,在面向对象的语言中(C++、C#、Java 等)对于向下转型都需要使用明确的转换操作符,但是向上转型却不需要。例如上面 let ret : IDoom3Tokenizer = new Doom3Tokenizer ( ); 这句代码,自动将 Doom3Tokenizer 类转型为 IDoom3Tokenizer 接口。

接着在 IDoom3Tokenizer 接口中增加一个创建 IDoom3Token 的接口方法并在 Doom3Tokenizer 中实现该方法。具体代码如下:

```
export interface IDoom3Tokenizer {
    //新增一个创建子接口的方法
    createIDoom3Token ( ) : IDoom3Token;

    setSource ( source : string ) : void ;
    reset () : void ;
    getNextToken ( token : IDoom3Token) : boolean ;
}

// 实现新增的接口方法
class Doom3Tokenizer implements IDoom3Tokenizer {
    //创建 IDoom3Token 接口
    public createIDoom3Token ( ) : IDoom3Token {
        return new Doom3Token ( ) ;
    }
}
```

关于上述使用模式,会在后面的代码中经常用到,因此特别说明一下。

2.3.3 迭代器模式

IDoom3Tokenizer 接口中每次调用 getNextToken 方法会返回下一个可用的 Token,该行为非常符合迭代器模式。迭代器模式是最常用的一种设计模式,每门面向对象的语言在其基础库中都对迭代器模式提供支持,例如 C++ 标准模板库、Java JDK 及微软的.NetFramework 等。

由于迭代器模式是如此常用,特别适合容器对象,因此这些高级语言将迭代器模式升华为语言语法的组成部分。例如 C++ 11 标准、Java 5(及以上版本),以及 C#中都支持 for each 风格的迭代,for each 迭代要求实现各自语言相对应的迭代器接口。

实际上 ES 6 规范也定义了迭代器的接口,并通过使用 for of 语句来支持容器迭代。本书其中的一个定位是"造轮子",因此读者可以自己动手,模拟微软在.NetFramework 中定义的迭代器模式。

2.3.4 模拟微软.NetFramework 中的泛型迭代器

在微软的.NetFramework 中,有两个接口定义了迭代器模式:IEnumerable 可迭代接口

（该接口只有一个 getEnumerator 的方法,不使用该接口,因此不在代码中定义),以及 IEnumerator 迭代器接口。使用 TypeScript 泛型方式来定义 IEnumerator 接口。具体代码如下:

```typescript
export interface IEnumerator < T > {
    // 将迭代器重置为初始位置
    reset ( ) : void ;
    // 如果没越界, moveNext 将 current 设置为下一个元素, 并返回 true
    // 如果已越界, moveNext 返回 false
    moveNext ( ) : boolean ;
    // 只读属性, 用于获取当前的元素, 返回泛型 T
    readonly current : T ;
}
```

上述代码模拟了微软.NetFramework 中的泛型迭代器接口,在 TypeScript 中,可以在声明的类型后面使用< T >方式定义泛型,其中 T 可以替换成其他文字,例如< t >或者< kind >之类的。然后就可以在类中使用 T 进行类型替代,后续章节会广泛使用泛型。

2.3.5　IDoom3Tokenizer 扩展 IEnumerator 接口

为了让 IDoom3Tokenizer 支持 IEnumerator 接口,需要先修改一下 IDoom3Tokenizer 的接口,请看如下代码:

```typescript
//接口扩展和类扩展一样,都是使用 extends 关键字
//类实现接口,则使用 implements 关键字
//注意, IEnumerator 中的泛型在 extends 时替换成了 IDoom3Token 类型
export interface IDoom3Tokenizer extends IEnumerator < IDoom3Token > {
    //只需要保留 setSource 接口方法
    setSource ( source : string ) : void ;
    /*
    // 一但使用迭代器模式,实际上生产 IDoom3Token 的方法就不需要了
    createIDoom3Token ( ) : IDoom3Token;
    // reset 方法已经定义在 IEnumerator 接口中了, 不需要再在子接口中声明
    reset ( ) : void ;
    // getNextToken 被 IEnumerator 的 moveNext 和 current 替代, 因此在此接口中可以取消
    getNextToken ( token : IDoom3Token) : boolean ;
    */
}
```

2.3.6　修改 Doom3Tokenizer 源码

再来修改一下实现类 Doom3Tokenizer 的相关源码,具体修改代码如下:

```typescript
// 增加一个私有变量_current, 并使用 new 进行初始化接口
private _current : IDoom3Token = new Doom3Token ( ) ;
// 实现 moveNext 方法, 实际调用的是 getNextToken 方法
```

```
public moveNext ( ) : boolean {
    return this . getNextToken ( this . _current ) ;
}
// 通过 get 方式实现只读属性 current
public get current ( ) : IDoom3Token {
    return this . _current ;
}
```

2.3.7　使用 VS Code 中的重命名重构方法

一直以来我们遵从的编码风格是：凡是私有方法或成员变量，在命名时都带下划线。而上面将 getNextToken 从 public 改成 private 访问级别后，不符合编码风格。此时正是显示 VS Code 编辑器中重命名重构方法的好时机，重命名重构方法会分析变量名或函数名的依赖关系，正确地进行名字的替换，比全文查询替换要"靠谱"很多。

那么让我们来看一下如何进行重命名。首先将光标定位到要修改名称的函数或变量上，然后单击右键，在弹出的菜单中选择"重命名"命令或者按 F2 键跳出如图 2.4 所示界面。

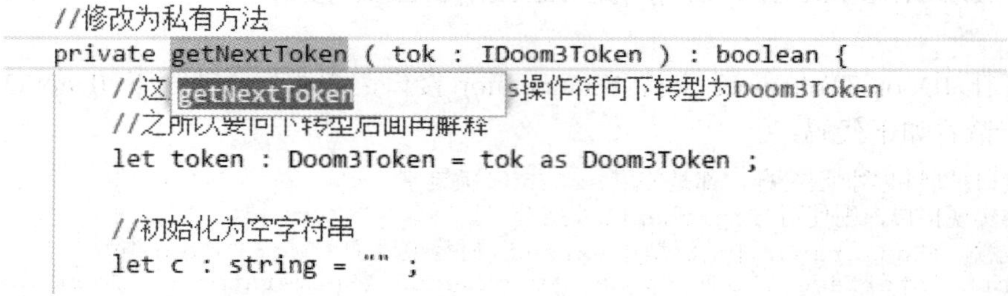

```
//修改为私有方法
private getNextToken ( tok : IDoom3Token ) : boolean {
    //这          getNextToken        s操作符向下转型为Doom3Token
    //之所以要向下转型后再解释
    let token : Doom3Token = tok as Doom3Token ;

    //初始化为空字符串
    let c : string = "" ;
```

图 2.4　重命名重构方法界面

在 getNextToken 方法前添加下划线后按 Enter 键，VS Code 就会自动分析 getNextToken 的所有依赖关系，智能地按需重新命名该方法，这是一个非常有用的重构方法，对开发有着很大的帮助，值得了解。

2.3.8　使用迭代器解析 Token

至此修改了源码，让 IDoom3Tokenizer 支持 IEnumerator < IDoom3Token >迭代器接口，使用该接口也是非常方便的，来看一下下面的代码：

```
let input : string = " [ 3.14 , -3.14 , .14 , -.14 , 3. , -3. ] " ;

tokenizer . setSource ( input ) ;

while ( tokenizer . moveNext ( ) ) {
    if ( tokenizer . current . type === ETokenType . NUMBER ) {
        console . log ( " NUMBER : " + tokenizer . current . getFloat ( ) ) ;
```

```
        }
        else {
            console.log( " STRING : " + tokenizer . current . getString ( ) ) ;
        }
    }
```

之所以花这么多篇幅介绍迭代器，是因为在本书的后面章节中会大量地使用 IEnumerator < T > 这个泛型接口。通过精心设计来确保接口的简单性和一致性，能使用迭代器的地方尽量使用迭代器模式，让调用方尽可能地使用熟悉的模式来进行调用或二次开发。

2.3.9　面向接口与面向对象编程的个人感悟

到此时，来了解面向接口编程的特点是比较适合的时机。面向对象有 3 个要素：继承（Inheritance）、封装（Encapsulation）和多态（Polymorphism）。

继承分为接口继承和实现继承（类继承），Doom3Tokenizer 就是接口继承了 IDoom3Tokenzier 并实现了该接口的所有方法。对于 TypeScript 来说，可以通过关键字来区分实现继承（extends）还是接口继承（implements）。

关于封装，笔者的理解有以下 3 个方面：

- 在接口中声明 readonly 属性，在实现类中使用 get 访问器来提供只读属性，这是对只读属性的封装。
- 将类内部使用的成员变量或成员方法，全部声明为 private 或者 protected 访问级别，决定是使用私有还是受保护级别的访问，依赖于你是否允许自己定义的类被继承，这是第二个封装的体现。
- 使用 export 导出接口，然而并没有导出实现类，让接口与实现相分离。这种情况是最高级别的隐藏，只能看到接口的方法签名，却无法了解具体的成员变量及实现细节。

关于多态，可以用一个最简单的例子来理解。例如，在调用 IDoom3Tokenzier 接口的 moveNext 方法时，实际调用的是实现类 Doom3Tokenizer 的 moveNext 方法。换句话说，就是同一个操作，作用于不同的实列对象，有不同的解释，产生不同的执行结果。其实多态是整个面向对象编程的核心，在后面章节中将会有非常多的例子来演示和了解多态。

而面向接口编程并不是一种独立的编程思想，而是附属于面向对象的编程思想，因此可以将面向接口编程看成是面向对象编程的一个子集。在实现接口时，使用的是接口继承方式，将接口与实现类相分离，从而达到更好的封装效果。对于调用方来说，不需要了解具体的实现细节，只要了解接口的含义就能让程序正常运行。

2.4　从服务器获取资源

到目前为止，在演示 IDoom3Tokenizer 解析文本时，文本来自本地 string 类型变量的定义。但是在实际应用中，文本字符串都是存储在文本文件中的，因此更加方便的方法是从本地或服务器读取文本文件，然后调用 IDoom3Tokenizer 进行解析。本节就来解决这个问题。

2.4.1　HTML 加载本地资源遇到的问题

很不幸的是，当在浏览器中通过 HTML 加载本地资源时你会发现，除了 JS 脚本及 CSS 文件外，其他资源文件（例如图片和视频等）都无法加载。使用如下 HTML 代码进行测试：

```
<html>
<head>
    <link rel = "stylesheet" href = "indexLocal.css" />
    <script src="indexLocal.js"></script>
</head>
<body>
    <image src="/data/test.jpg" ></image>
</body>
</html>
```

其中，indexLocal.js 中仅是弹出对话库的代码，而 indexLocal.css 中是设置背景为红色。当使用第 1 章中的 Open In Broswer 插件本地运行 HTML 文件后会发现，在所有浏览器中 JS 脚本能正常运行，背景也变为红色，但是图片都无法显示。

2.4.2　从服务器加载资源

在第 1 章中，为了热部署及直接在 VS Code 中进行断点调试，引入了 lite-server 服务器，只要在 VS Code 命令行中输入：npm run dev，然后运行 lite-server 服务器，输入 localhost:3000/indexLocal.html，就能很顺利地加载图片，如图 2.5 所示。

图 2.5　从服务器加载资源

2.4.3 使用 XHR 向服务器请求资源文件

我们发现，利用 lite-server 服务器，就能解决 HTML 支持的图片、视频和音频等资源从服务器上自动加载的问题。但是像文本文件和二进制文件等，需要使用 XHR 以编程的方式获取，其中 XHR 是 XMLHttpRequest 对象的简称。

在本书中，访问服务器的需求其实非常简单，只要满足一个要求：通过 GET 方式加载文件。对于这个需求，根本不需要编写服务器端函数来处理 GET 或 POST 请求，也不需要在服务器端处理请求参数问题，只需要在 Web 客户端使用 XMLHttpRequest 以 GET 方式向服务器请求数据即可。让我们来看一下代码，具体如下：

```typescript
export class HttpRequest {
    /**
     *
     * @param url { string } 请求资源的 url
     * @returns HttpResponse
     */
    public static doGet ( url : string ) : HttpResponse {
        // 初始化 XMLHttpRequest 对象
        let xhr : XMLHttpRequest = new XMLHttpRequest ( ) ;
        // XHR 的 open 函数的第 3 个参数 true 表示异步请求，false 表示同步请求
        //本函数是同步请求函数，因此为 false
        xhr . open ( "get" , url , false , null , null ) ;
        // 向服务器发送请求
        xhr . send ( ) ;
        //请求发送成功
        if ( xhr.status === 200 ) {
            //返回自己定义的 HttpResponse 接口对象
            //这里可以看到接口的第二种用法
            //并没有实现该接口，但是可以用大括号及键值对方式来定义接口（其实和 JS 定义对象是一样的方式）
            //可以把这种接口当成纯数据类来使用
            return { success : true , responseType : " text " , response : xhr . response } ;
        } else {
            //请求失败，success 标记为 false, response 返回 null
            return { success : false , responseType : " text " , response : null } ;
        }
    }
}
```

doGet 方法的返回类型是自定义的一个接口，下面来看一下该接口是如何定义的，请参考如下代码：

```typescript
export interface HttpResponse {
    success : boolean ;                //http 请求成功，返回 true, 否则返回 false
    responseType : XMLHttpRequestResponseType ;        //返回请求的资源类型
```

```
            response : any ;// 根据请求的类型不同，可能返回的是字符串、ArrayBuffer 或
                           Blob 对象，因此使用 any 类型
        }
```

在上面的代码的注释中已经提示过，HttpResponse 并没有实现类，使用的是 TypeScript 接口中的第二种用法。接下来看一下 HttpResponse 接口中的 XMLHttpRequestResponseType 这个类型吧。

2.4.4　TypeScript 中的类型别名

XMLHttpRequestResponseType 这个类型用来指示返回的请求资源的类型。下面来看一下该类型是如何定义的。

可以在 VS Code 中，将鼠标指针定位到 doGet 方法的参数类型 XMLHttpRequestResponseType 上，然后按住 Ctrl 键不放，单击鼠标左键，就可以定位到 XMLHttpRequestResponseType 的代码声明处，如图 2.6 所示。

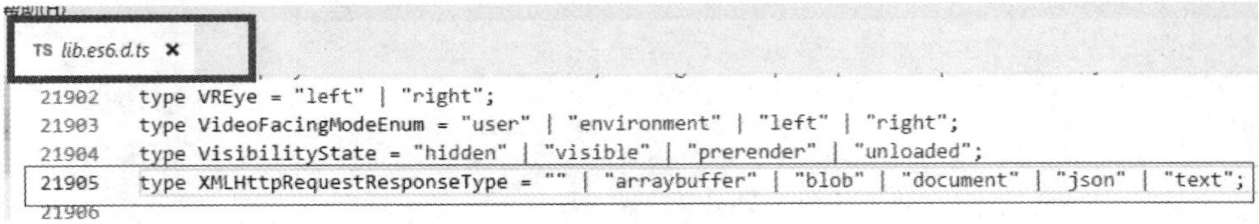

图 2.6　XMLHttpRequestResponseType 的定义

XMLHttpRequestResponseType 的取值中，"arraybuffer"和"blob"返回二进制文件，可以使用 TypeScript / JavaScript 中的 ArrayBuffer 和 Blob 类进行二进制读写。关于 ArrayBuffer 和 Blob 的相关知识点，请大家参考 JavaScript 官方文档。

XMLHttpRequestResponseType 实际上是一个类型别名。在 TypeScript 中可以使用 type 关键字来声明类型别名，类型别名的实际类型不变，它们仅仅是个替代的名字而已。例如 XMLHttpRequesetResponseType 的类型还是 string，其取值范围使用了"|"符号，表示联合类型，在第 1 章中已经了解过了。

type 关键字和 C/C++中的 typedef 作用是一样的，在 C/C++中经常对模板实例化类型进行 typedef 重定义（取别名），目的是减少输入，让代码更清晰且容易理解。在 TypeScript 中也一样，例如可以使用 type 关键字重定义 IEnumerator < IDoom3Token >迭代器，代码如下：

```
type TokenEnumerator = IEnumerator < IDoom3Token > ;
```

再假设，有一个泛型树数据结构，该结构可以挂接例如 number 类型的节点，现在我们想使用迭代器来迭代树节点，来对比一下如下代码：

```
// 不使用 type 关键字来声明类型别名
let nodeEnumerator : IEnumerator < TreeNode < number > > ;
// 使用 type 关键字来声明类型别名
```

```
type NodeEnumerator = IEnumerator < TreeNode < number > > ; // 一次定义，
后续多次使用别名
let nodeEnumerator : NodeEnumerator ;
```

可以看到，代码更加清晰、容易理解了，如果读者看过 C++ 标准模板库的代码，就会知道没有类型别名将是多么痛苦的事情了。

2.4.5 使用 doGet 请求文本文件并解析

下面来测试一下 doGet 静态方法，代码如下：

```
//从服务器请求 level.proc 文件，该文件是 Doom3 的关卡文件，文件大小为 261KB, word 中
字数统计将近 7 万个单词
let response : HttpResponse = HttpRequest . doGet ( "level.proc" ) ;

//如果请求成功，进行文件解析
if ( response . success === true ) {
    //将 response 转换为 string 类型，因为知道是文本文件
    str = response . response as string ;

    //设置要解析的字符串
    tokenizer . setSource ( str ) ;
    while ( tokenizer . moveNext ( ) ) {
        if ( tokenizer . current . type === ETokenType . NUMBER ) {
            console.log ( "NUMBER : " + tokenizer . current . getFloat ( ) ) ;
        }
        else {
            console.log ( "STRING : " + tokenizer . current . getString ( ) ) ;
        }
    }
}
```

运行程序后，会在浏览器的控制台输出所有解析后的 Token，部分输出的截图如图 2.7 所示。

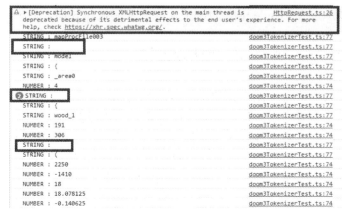

图 2.7 doGet 请求文本文件并解析

从图 2.7 中会发现存在两个问题：
- STRING 输出部分有空白字符（实际上该问题仅出现在 Windows 系统下）。
- doGet 使用同步请求，可能导致主线程堵塞，因此被建议不要使用同步请求。

下面来解决这两个问题。

2.4.6 解决仍有空白字符输出问题

通过使用 VS Code 强大的断点调试功能，很容易地找到了仍有空白字符输出的原因，如图 2.8 所示。

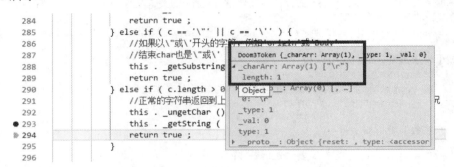

图 2.8 定位空白字符输出问题

原来，Windows 系统下换行是由\r\n 两个字符组成的，而 Linux 和 Mac OS 系统中不存在这个问题。修正 Windows 下的解析 bug 很容易，代码如下：

```
class Doom3Tokenizer implements IDoom3Tokenizer {
    private _whiteSpaces : string [ ] = [ " " , "\t" , "\v" , "\n" , "\r" ] ;
}
```

在_whiteSpaces 列表中增加"\r"后就解决了此问题，Windows 下是使用回车换行方式。这个例子显示了 VS Code 强大的调试功能，掌握调试技能，就能避免浪费大量时间。

2.4.7 实现 doGetAsync 异步请求方法

现在来解决同步请求可能导致的主线程阻塞问题。其实一开始用同步请求的实现，是因为笔者认为同步请求使用方便，不需要编写请求完成后的相关回调代码，不过缺点是如果文件很大的话，会导致主线程堵塞，浏览器失去响应，用户体验很差。

什么是异步请求呢？通俗地讲就是开辟一个新的线程来进行文件传输，等传输完成后通知主线程"我已经做完我该做的，轮到你来处理了"，因此异步回调通知是经典的处理模式。

下面来实现一个新的、使用异步请求的方法。具体代码如下：

```typescript
//异步 Get 请求
   /**
    *
    * @param url { string } 要请求的 URL 地址
    * @param callback { ( response : HttpResponse ) => void 函数类型 }
    请求成功后的回调函数
    * @param responseType { XMLHttpRequestResponseType } 可以取"text"、
    "json"、"arraybuffer"等
    * @returns { HttpResponse } 自己定义的返回结果的接口
    */
   public static doGetAsync ( url : string , callback : ( response :
   HttpResponse ) => void , responseType : XMLHttpRequestResponseType =
   "text" ) : void {
       let xhr : XMLHttpRequest = new XMLHttpRequest ( ) ;
       // 很关键的一句代码,在 doGet 同步请求中,不允许设置要请求的资源是文本、JSON
       还是二进制类型的
       // 但是在异步请求中,却可以自己决定请求 text, json, arraybuffer, blob 和
       document 等类型
       // 大家可以尝试将这句代码粘贴到 doGet 函数中,然后运行 doGet 请求
       // 你会发现浏览器会报错
       xhr . responseType = responseType ;
       // onreadystatechange : ( ( this : XMLHttpRequest , ev : Event ) =>
       any ) | null
       xhr . onreadystatechange = ( ev : Event ) : void => {
           if ( xhr . readyState === 4 && xhr .status === 200 ) {
               //异步请求成功,返回标记成功的 HttpResponse 对象,response 为请求
               资源数据
               let response : HttpResponse = { success : true , responseType :
               responseType , response : xhr . response } ;
               //并且调用回调函数
               callback( response ) ;
           }
           else {
               //异步请求失败,返回标记失败的 HttpResponse 对象,response 为 null
               let response : HttpResponse = { success : false , responseType :
               responseType , response : null } ;
               // 失败也调用回调函数
               callback ( response ) ;
           }
       }
       // open 第三个参数用来设置是同步还是异步请求,本函数设置为 true,表示异步请求
       xhr . open( "get" , url , true , null , null ) ;
       xhr . send ( ) ;
   }
}
```

doGetAsync 与 doGet 的区别主要有以下两点:

(1) 异步请求,当请求完成后,不管成功与否,都会调用 callback()函数,因此使用方必须要实现完成回调操作函数(或方法),后面 Demo 中会看到如何使用。

(2) 可以设置 xhr . responseType 请求为 text、json、arraybuffer 和 blob 等选项。

因此可以使用如 arraybuffer 选项来请求二进制资源，例如经典的 quake3 bsp 关卡文件。二进制文件的特点就是比文本文件小得多，缺点是我们要明确无误地知道数据是如何存储的，每个数据代表什么含义。

在此强调一下，如果在同步请求（doGet）方法中设置 xhr . responseType 属性，虽然不会报错，但是一旦运行浏览器，就会得到如图 2.9 所示的错误。

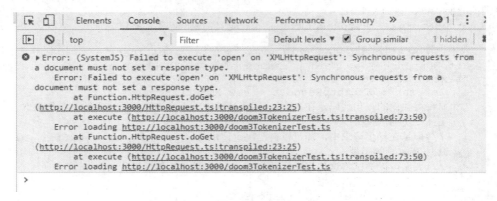

图 2.9　同步请求不允许设置 response type 属性

接下来看一下 TypeScript 中的回调函数（或方法）的相关内容。

2.4.8　声明 TypeScript 中的回调函数

在 2.4.7 节中的 doGetAsync 方法中，声明了回调函数（或方法）类型为（response : HttpResponse）=> void，下面来了解一下其语法组成：

- 没有声明函数（或方法）名，可以将回调函数（或方法）看成匿名函数（或方法）。
- 在( )对中声明了参数名 response 和参数类型 HttpResponse。
- 在箭头=>后声明了返回类型，代码中是返回 void 类型。

然后找出一些隐藏的内容，例如参数的个数，以及如果有多个参数的话，参数声明的顺序，这些要素组成了函数（或方法）的签名（Signature）。

回顾一下，函数（或方法）的签名是由：参数类型、参数个数、参数声明顺序及返回类型组成。函数（或方法）签名确定了函数（或方法）的类型，只要符合该签名的所有函数（或方法），都可以在运行时进行替换（运行时动态绑定），其实这就是面向对象 3 要素中多态的底层原理。

回调函数（或方法）是面向对象编程或写框架的一项重要技术，也是一种思想，即封装一切不变（需求明确）的代码，将可变（需求不明确）的代码通过回调函数（或方法）或虚函数的方式公开给第三方实现。

虚方法就是子类可以覆写（override）的基类方法，在 TypeScript 中，可以将类的所有非静态方法都看作虚方法，在子类中都可以覆写（override）。

关于回调函数（或方法）与基于虚函数覆写（override）的技术选型，依赖于个人的需求或喜好。它们之间的区别是：回调函数（或方法）不需要继承，而虚函数覆写（override）则需要继承基类。

回过头来看一下 doGetAsync 的实现依据，不变的部分（需求明确）是通过 XMLHttpRequest 请求文本或二进制数据，不管成功与否，都要让调用方知道。可变的部分（需求不明确）是，框架或代码编写者根本不知道调用方成功拿到服务器上的数据后要干什么，如果数据请求失败了要干什么，所以通过回调函数（或方法），让第三方来决策想要干什么。

2.4.9　调用回调函数

下面来看一下如何使用 doGetAsync，以及如何使用回调函数（或方法）。具体代码如下：

```
HttpRequest.doGetAsync( "level.proc", ( response : HttpResponse ) :
void => {
    //请求成功或失败都是在回调函数中处理
    if ( response . success === true ) {
        //将 response 转换为 string 类型，因为知道是文本文件
        str = response . response as string ;
        //设置要解析的字符串
        tokenizer . setSource( str ) ;
        while ( tokenizer . moveNext ( ) ) {
            if ( tokenizer . current . type === ETokenType . NUMBER ) {
                console . log ( " NUMBER : " + tokenizer . current . getFloat
                ( ) ) ;
            }
            else {
                console . log ( " STRING : " + tokenizer . current . getString
                ( ) ) ;
            }
        }
    } else {
        console . log( " 请求失败！！！" ) ;
    }
} ) ;
```

从上述代码中我们可以看到，请求成功或失败的处理都是在回调函数（或方法）中。下面来对比一下回调函数（或方法）的使用和声明代码：

```
// 使用回调函数，其中返回类型用冒号，=>符号后面是函数体代码
( response : HttpResponse ) : void => { 处理代码 } ;

// 声明回调函数类型，=>符号后面是返回类型
( response : HttpResponse ) => void ;
```

也可以使用函数对象来赋值调用，实现一个函数名为 processHttpResponse，代码如下：

```typescript
function processHttpResponse ( response : HttpResponse ) : void {
    if ( response . success === true ) {
        //将 response 转换为 string 类型，因为知道是文本文件
        str = response . response as string ;
        //设置要解析的字符串
        tokenizer . setSource( str ) ;
        while ( tokenizer . moveNext ( ) ) {
            if ( tokenizer . current . type === ETokenType . NUMBER ) {
                console . log ( " NUMBER : " + tokenizer . current . getFloat
                ( ) ) ;
            }
            else {
                console . log ( " STRING : " + tokenizer . current . getString
                ( ) ) ;
            }
        }
    } else {
        console . log ( " 请求失败！！！" ) ;
    }
}
```

然后调用函数对象，代码如下：

```typescript
HttpRequest.doGetAsync( "level.proc" , processHttpResponse ) ;
```

最后一种方式是使用 type 来重新定义回调函数，然后再看看如何使用。在 Http Request.ts 中输入如下代码：

```typescript
export type RequestCB = ( ( response : HttpResponse ) => void ) ;
```

看一下如何调用 RequestCB。具体代码如下：

```typescript
//使用 RequestCB 作为回调函数的类型，使用=> { }进行调用
//此时你会发现没有 response 形参，这时候可以使用 RequestCB.response 获取 response
参数
HttpRequest . doGetAsync ( "level.proc" , RequestCB => {
    if ( RequestCB . success === true ) {
        //将 response 转换为 string 类型，因为知道是文本文件
        str = RequestCB . response as string ;
        //设置要解析的字符串
        tokenizer . setSource ( str ) ;
        while ( tokenizer . moveNext ( ) ) {
            if ( tokenizer . current . type === ETokenType . NUMBER ) {
                console . log ( " NUMBER : " + tokenizer . current . getFloat
                ( ) ) ;
            }
            else {
                console . log ( " STRING : " + tokenizer . current . getString
                ( ) ) ;
            }
        }
    }
} ) ;
```

这种使用 type 来重定义回调函数的方式，更加清晰易读。如上述代码所示，需要注意的是，原来的回调函数形参 response 变成了属性。

2.5 本章总结

本章主要通过实现一个解析 Doom3 文本文件格式的词法解析器，介绍了 TypeScript 的一些常用知识点，分为 4 个方面，如下所述。

（1）TypeScript 语言本身的内容非常多，本书中用到的 TypeScript 语法上的知识点如下：
- 模板字符串；
- 接口的定义和扩展；
- 类的定义、继承及类成员的访问级别；
- 显示断言赋值；
- 枚举的定义；
- 类型别名；
- 函数（方法）签名和回调函数（方法）；
- 泛型编程；
- 联合类型；
- 模块化编程（export / import）。

（2）3 个设计模式：
- 工厂设计模式；
- 迭代器设计模式；
- 模板方法设计模式。

（3）了解 Doom3 文本文件的词法规则，以及从头开始实现一个解析这些规则的 Doom3 词法解析器。

（4）封装了 XMLHttpRequest 类，使用 GET 方式从服务器端获取文本或二进制文件。

以上是本章的主要内容总结。

第 2 篇
Canvas2D 篇

- 第 3 章　动画与 Application 类
- 第 4 章　使用 Canvas2D 绘图

第 3 章　动画与 Application 类

本章将从程序实现的角度来了解一下动画的原理，以及 HTML 5 提供的一些基础且必要的方法。

可以将动画的相关功能都封装到一个名为 Application 的类中，该类主要是作为应用程序的入口类。它能启动或关闭动画循环，抽象更新与重绘流程，提供事件分发和处理功能，并且具有一个允许以不同帧率运行的计时器。

在开始本章内容前先声明一下，本书所有的 Demo 以 Chrome 浏览器为主要测试环境，为简单起见，本书并不关注各浏览器之间的兼容性。

3.1　requestAnimationFrame 方法与动画

从程序的角度来描述，笔者认为，动画就是不间断地、基于时间的更新与重绘。可以说这句话贯穿了本节要讲的所有内容。

3.1.1　HTML 中不间断的循环

所谓不间断的，是指动画需要一个不停地重复循环机制，从程序实现的角度来说，一般有两种选择：

一种是类似于 while (true) { }之类的死循环，除非满足退出死循环的条件，否则就一直不停地重复相同的行为。在 Windows 下的 D3D / OpenGL 开发中，经常使用这种模式来驱动动画不断运行。作为知识的延伸点，下面来看一段经典的 Windows 下基于 C / C++语言的动画循环演示代码。若不感兴趣可直接跳过。具体代码如下：

```
MSG msg ;
ZeroMemory ( & msg , sizeof ( msg ) ) ;
// 只有明确地收到 WM_QUIT 消息，才跳出 while 循环，退出应用程序
// 否则一直循环重复相同的行为
// Windows 下经典的 runLoop 操作
while ( msg . message != WM_QUIT )
{
    // 如果当前线程消息队列中有消息，则取出该消息
    if ( PeekMessage ( & msg , NULL , 0U , 0U , PM_REMOVE ) ) {
```

```
        //将键盘的虚拟键消息转换为 WM_CHAR 消息,并将 WM_CHAR 消息再次放入当前线程
        消息队列中,下次还是可以被 PeekMessage 读取并处理
        TranslateMessage ( & msg ) ;
        //将当前的 WM_开头的消息分发到 Window 窗口过程处理回调函数中进行处理
        DispatchMessage ( & msg ) ;
        //上面的代码实际就是处理鼠标、键盘、WM_PAINT,或者计时器等队列消息
    } else {
        //如果当前线程消息队列中(上面的代码处理消息队列)没有消息可处理,就一直更新并重绘
        Update ( ) ;                               //更新
        Render ( ) ;                               //重绘
    }
}
```

另外一种是类似于定时器的回调,例如使用 HTML DOM(Document Object Model, 文档对象模型)中 Window 对象的 setTimeout、setInterval 及 requestAnimationFrame 方法。关于 setTimeout 和 setInterval 的用法,请各位读者自行查阅相关资料(在 3.4 节中将实现类似 setTimeout 和 setInterval 的功能)。下面主要来看一下 requestAnimationFrame 方法的用法。具体代码如下:

```
// start 记录的是第一次调用 step 函数的时间点,用于计算与第一次调用 step 函数的时间差,
以毫秒为单位
let start : number = 0 ;
//lastTime 记录的是上一次调用 step 函数的时间点,用于计算两帧之间的时间差,以毫秒为单位
let lastTime : number = 0 ;
// count 用于记录 step 函数运行的次数
let count : number = 0 ;
// step 函数用于计算:
// 1.获取当前时间点与 HTML 程序启动时的时间差 : timestamp
// 2.获取当前时间点与第一次调用 step 时的时间差 : elapsedMsec
// 3.获取当前时间点与上一次调用 step 时的时间差 : intervalMsec
// step 函数是作为 requestAnimationFrame 方法的回调函数使用的
// 因此 step 函数的签名必须是 ( timestamp : number ) => void
function step ( timestamp : number ) : void {
    // 第一次调用本函数时,设置 start 和 lastTime 为 timestamp
    if ( ! start ) start = timestamp ;
    if ( ! lastTime ) lastTime = timestamp ;
    // 计算当前时间点与第一次调用 step 时间点的差
    let elapsedMsec : number = timestamp - start ;
    // 计算当前时间点与上一次调用 step 时间点的差(可以理解为两帧之间的时间差)
    let intervalMsec : number = timestamp - lastTime ;
    // 记录上一次的时间戳
    lastTime = timestamp ;
    // 计数器,用于记录 step 函数被调用的次数
    count ++ ;
    console . log ( " " + count + " timestamp = " + timestamp ) ;
    console . log ( " " + count + " elapsedMsec = " + elapsedMsec ) ;
    console . log ( " " + count + " intervalMsec = " + intervalMsec) ;
    // 使用 requestAnimationFrame 调用 step 函数
```

```
        window . requestAnimationFrame ( step ) ;
}
// 使用 requestAnimationFrame 启动 step
// 而 step 函数中又会调用 requestAnimationFrame 来回调 step 函数
// 从而形成不间断地递归调用，驱动动画不停地运行
window . requestAnimationFrame ( step ) ;
```

上述代码每次调用 step 函数会在浏览器的 console 控制台窗口中输出当前函数的调用次数，以及 3 个时间差的数值。Chrome 浏览器中 console 控制台窗口输出的结果，如图 3.1 所示。

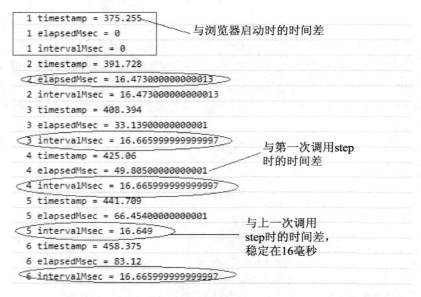

图 3.1　Chrome 浏览器中的 requestAnimationFrame 输出时间差

3.1.2　requestAnimationFrame 与监视器刷新频率

根据图 3.1 所示，通过 requestAnimationFrame 方法启动 step 回调函数后，每次调用 step 函数的时间间隔固定在 16.66 毫秒左右，基本上每秒调用 60 次 step 函数（1000 / 60 约等于 16.66 毫秒，其中 1 秒等于 1000 毫秒）。

每秒 step 函数调用的次数（频率）实际上是和监视器屏幕刷新次数（频率）保持一致。笔者在自己的 Windows 笔记本电脑上做了个实验，看一下监视器的屏幕刷新频率对 requestAnimationFrame 方法的影响。默认情况下，笔者的 Windows 系统电脑的监视器屏幕刷新频率是 60 赫兹，可以人为认定监视器每秒刷新重绘 60 次，如图 3.2 所示，将屏幕刷新频率 60 赫兹更改为 48 赫兹，来看看结果会怎样。

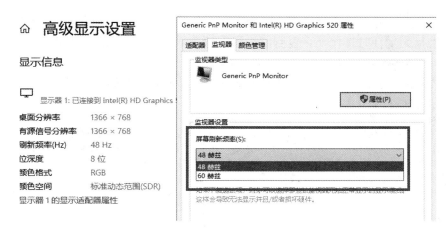

图 3.2　更改监视器屏幕刷新频率

然后运行 3.1 节所示的代码，会得到如图 3.3 所示的结果。

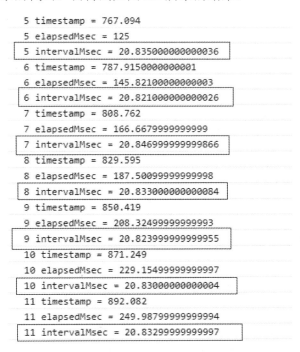

图 3.3　监视器屏幕刷新频率 48 赫兹下 Chrome 浏览器中的运行效果

从图 3.3 中会看到，每次调用 step 函数的时间间隔固定在 20.83 毫秒左右，符合当前监视器屏幕刷新频率 48 赫兹的设置（1000 / 48 约等于 20.83 毫秒），由此可见 requestAnimation Frame 是一个与硬件相关的方法，该方法会保持与监视器刷新频率一致的状态。

再来看一下 3.1 节中的 step 函数，会发现它很简单，只是输出 3 个时间差，这种操作本身花不了多少时间，因此能保持 16 毫秒的频率一直稳定运行是很正常的。那么，如果在 step 中进行大量耗时操作（恢复到监视器屏幕刷新频率 60 赫兹的情况下），结果会如何呢？

继续做个实验来看看，在 step 函数的 count ++ 代码后面添加如下代码：

```
// 每次调用 step 就做累加操作
let sum : number = 0 ;
// 每次调用 step 后，随机生成一个区间位于 [5 百万, 6 百万] 之间的数 num
// 然后从 0 累加到 num, 这样每次操作都需要耗费一定的时间，并且具有不同的结果
// 目的是不让 JS 解释器优化，看看每次暴力穷举所耗费的时间
// 其中 Math . random ( ) 函数返回 [ 0 , 1 ] 之间的浮点数
// 通过 a + Math . random ( ) * b 公式可以生成区间位于 [ a , b ] 之间的随机数
let num : number = 5000000 + Math . random ( ) * 1000000 ;
for ( let i : number = 0 ; i < num ; i++ ) {
    sum ++ ;
}
```

运行上述程序后，会得到如图 3.4 所示的结果。

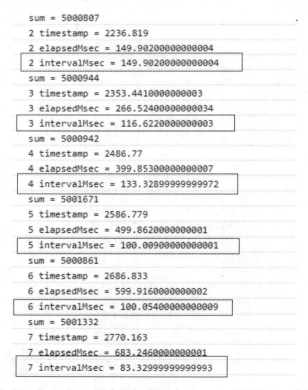

图 3.4　Chrome 浏览器中 step 函数在 500～600 万级别累加计时测试

通过图 3.4 会看到，两次 step 调用间隔所耗时间有很明显的波动，但是通过统计，会发现一个很明显的规律，具体如表 3.1 所示。

表 3.1　Chrome浏览器中两帧时间间隔和监视器屏幕刷新频率之间的关系统计表

帧　号	两 帧 间 隔	计算公式（两帧间隔/16.66）	计 算 结 果
2	149.902	149.902 / 16.66	8.994（9）
3	116.622	116.622 / 16.66	7.000（7）
4	133.329	133.329 / 16.66	8.002（8）
5	100.009	100.009 / 16.66	6.008（6）
6	100.009	100.009 / 16.66	6.008（6）
7	83.330	83.330 / 16.66	5.001（5）

根据表 3.1 统计数据会发现，两帧之间的时间间隔（intervalMsec）总是 16.66 毫秒的倍数（当前监视器屏幕刷新频率 60 赫兹，折算成每帧需要 16.66 毫秒刷新一次）。

由此可见，requestAnimationFrame 方法会稳定间隔时间：

- 如果当前的回调操作（step 函数）在 16.66 毫米内能完成，那么 requestAnimationFrame 会等到 16.66 毫秒时继续下一次 step 回调函数的调用。
- 如果当前的回调操作（step 函数）大于 16.66 毫秒，则会以 16.66 毫秒为倍数的时间间隔进行下一次 step 回调函数的调用。
- 当将监视器屏幕刷新频率 60 赫兹设置成 48 赫兹时，结果也类似，两帧之间间隔时间总是 20.83 毫秒的倍数。

3.1.3　基于时间的更新与重绘

现在有一个很简单的需求，假设想让一个物体（例如一个矩形或圆球）沿着水平轴（x 轴），以每秒 10 个像素的速度进行移动，那么应该怎么做呢？

其实这里涉及两个基本步骤，让物体进行基于时间（每秒 10 个像素）的更新，以及更新完后进行显示。

让我们来搭建一个最简单的更新与重绘框架。上一节中，在 step 回调函数中已经计算出了 3 个以毫秒表示的时间间隔（时间差），即：

- 当前时间点与当前 HTML 应用启动时的时间差 timestamp。
- 当前时间点与第一次调用 step 回调函数时的时间差 elapsedMsec。
- 当前时间点与上一次调用 step 回调函数时的时间差 intervalMsec。

以上 3 个时间间隔或时间差中最有用的是第 3 个时间差，特别适合做基于时间的更新。那么声明一个函数用于更新操作，演示代码如下：

```
let posX : number = 0 ;
let speedX : number = 10 ;              //单位为秒，以每秒10个像素的速度进行位移
function update ( timestamp : number , elapsedMsec : number , intervalMsec : number ) : void {
    // 参数都是使用毫秒为单位，而现在的速度都是以秒为单位
    // 因此需要将毫秒转换为秒来表示
```

```
    let t : number = intervalMsec / 1000.0 ;
    // 线性速度公式 : posX = posX + speedX * t ;
    posX += speedX * t ;
    console . log ( " current posX : " + posX ) ;
}
```

关于渲染，输入如下代码：

```
// 使用 CanvasRenderingContext2D 绘图上下文渲染对象进行物体的绘制
function render ( ctx : CanvasRenderingContext2D | null ) : void {
    // 简单起见，仅仅输出 render 字符串
    console . log ( " render " ) ;
}
```

最后，将这两个函数由 step 函数进行调用，就可以形成一个基本的框架。具体代码如下：

```
// start 记录的是第一次调用 step 函数的时间点，用于计算与第一次调用 step 函数的时间差，以毫秒为单位
let start : number = 0 ;
//lastTime 记录的是上一次调用 step 函数的时间点，用于计算两帧之间的时间差，以毫秒为单位
let lastTime : number = 0 ;
// count 用于记录 step 函数运行的次数
let count : number = 0 ;
// step 函数用于计算
// 1.获取当前时间点与 HTML 程序启动时的时间差 : timestamp
// 2.获取当前时间点与第一次调用 step 时的时间差 : elapsedMsec
// 3.获取当前时间点与上一次调用 step 时的时间差 : intervalMsec
function step ( timestamp : number ) {
    //第一次调用本函数时，设置 start 和 lastTime 为 timestamp
    if ( ! start ) start = timestamp ;
    if( ! lastTime) lastTime = timestamp ;
    //计算当前时间点与第一次调用 step 时间点的差
    let elapsedMsec = timestamp - start ;
    //计算当前时间点与上一次调用 step 时间点的差（可以理解为两帧之间的时间差）
    let intervalMsec = timestamp - lastTime ;
    //记录上一次的时间戳
    lastTime = timestamp ;
    // 进行基于时间的更新
    update ( timestamp , elapsedMsec , intervalMsec ) ;

    // 调用渲染函数，目前并没有使用 CanvasRenderingContext2D 类，因此设置为 null
    render ( null ) ;
    // 使用 requestAnimationFrame 调用 step 函数
    window . requestAnimationFrame ( step ) ;
}
// 使用 requestAnimationFrame 启动 step
// 而 step 函数中通过调用 requestAnimationFrame 来回调 step 函数
// 从而形成不间断地递归调用，驱动动画不停地运行
window . requestAnimationFrame ( step ) ;
```

当运行上述代码后，会发现每次调用 step 回调函数后，posX 总是根据传入的时间差进行更新，这样的好处不管在哪个浏览器中或是不同运行速度的计算机上，每秒钟的运动

距离都是恒定的（每秒 10 像素）。如果不使用基于时间的更新，那么在不同浏览器中不同的 CPU 计算机上，会有明显的快慢差别。

实际上这是一个很固定的流程，将这个动画流程封装起来形成一个类，以后所有的程序都可以使用该类的子类。下一节就来封装这个流程。

3.2 Application 类及其子类

上一节中演示了动画的基本流程，本节将上一节的演示内容封装成一个类，该类的作用主要有以下几点：

- 可以启动动画循环和结束动画循环。
- 可以进行基于时间的更新与重绘。
- 可以对输入事件（例如鼠标或键盘事件）进行分发和响应。
- 可以被继承扩展，用于 Canvas2D 和 WebGL 渲染。

本节将介绍如何实现上述 4 个基本需求。

3.2.1 Application 类体系结构

千言万语，不如一张结构图来得更清晰，先来看一下如图 3.5 所示的 Application 框架静态类结构图。

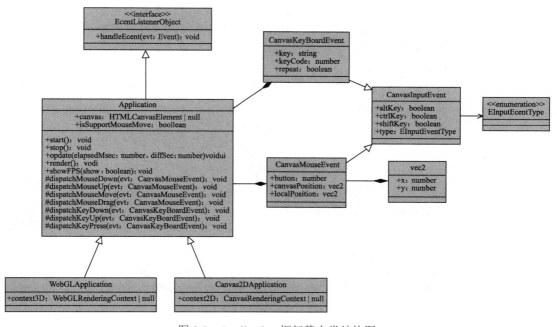

图 3.5　Application 框架静态类结构图

接下来根据图 3.5 所示的结构来实现 Application 这个架构。

3.2.2　启动动画循环和停止动画循环

声明一个 Application 类，具有如下成员变量：

```
export class Application {
    // _start 成员变量用于标记当前 Application 是否进入不间断地循环状态
    protected _start : boolean = false ;
    // 由 Window 对象的 requestAnimationFrame 返回的大于 0 的 id 号
    // 可以使用 cancelAnimationFrame ( this . _requestId )来取消动画循环
    protected _requestId : number = -1 ;
    // 用于基于时间的物理更新，这些成员变量类型前面使用了!，可以进行延迟赋值操作
    protected _lastTime ! : number ;
    protected _startTime ! : number ;
}
```

可以看到，上面的代码中的所有成员变量都使用了 protected 访问修饰符，意味着 Application 本类，以及继承自 Application 的子类都能访问这些成员变量。作为一个经验，如果我们设计的类能够被继承、被扩展，那么最好将成员变量声明为 protected，这样子类也能访问这些成员变量。

接着来看一下 start / stop / isRunning 成员方法。具体代码如下：

```
public start ( ) : void {
    if ( ! this . _start ) {
        this . _start = true ;
        this . _requestId = -1 ;      // 将_requestId 设置为-1
        // 在 start 和 stop 函数中，_lastTime 和 _startTime 都设置为-1
        this . _lastTime = -1 ;
        this . _startTime = -1 ;
        // 启动更新渲染循环
        this . _requestId = requestAnimationFrame ( ( elapsedMsec : number ) : void => {
            // 启动 step 方法
            this . step ( elapsedMsec ) ;
        } ) ;
    }
}
public stop ( ) : void {
    if ( this . _start ) {
        // cancelAnimationFrame 函数用于：
        //取消一个先前通过调用 window.requestAnimationFrame()方法添加到计划中的
        动画帧请求
        cancelAnimationFrame ( this . _requestId ) ;
        this . _requestId = -1 ;      // 将_requestId 设置为-1
        // 在 start 和 stop 函数中，_lastTime 和 _startTime 都设置为-1
        this . _lastTime = -1 ;
        this . _startTime = -1 ;
        this . _start = false ;
```

```
    }
}
// 用于查询当前是否处于动画循环状态
public isRunning ( ) : boolean {
    return this . _start ;
}
```

上述代码很简单,主要是启动动画循环、查询当前是否处于动画循环,以及结束动画循环这 3 种操作,接下来看一下 step 这个成员方法。

3.2.3 Application 类中的更新和重绘

Application 中的 step 成员方法和 3.1 节中的 step 函数仅仅有一些小的差别。下面来看一下代码是如何实现的。

```
// 不停地周而复始运动
protected step ( timeStamp : number ) : void {
    // 第一次调用本函数时,设置 start 和 lastTime 为 timeStamp
    if ( this . _startTime === -1 ) this . _startTime = timeStamp ;
    if( this . _lastTime === -1 ) this . _lastTime = timeStamp ;
    //计算当前时间点与第一次调用 step 时间点的差,以毫秒为单位
    let elapsedMsec : number = timeStamp - this . _startTime ;
    //计算当前时间点与上一次调用 step 时间点的差(可以理解为两帧之间的时间差)
    // 注意:intervalSec 是以秒为单位,因此要除以 1000.0
    let intervalSec : number = ( timeStamp - this . _lastTime ) / 1000.0 ;
    //记录上一次的时间戳
    this . _lastTime = timeStamp ;

    console . log (" elapsedTime = " + elapsedMsec + " intervalSec = " +
    intervalSec ) ;
    // 先更新
    this . update ( elapsedMsec , intervalSec ) ;
    // 后渲染
    this . render ( ) ;
    // 递归调用,形成周而复始地循环操作
    requestAnimationFrame ( ( elapsedMsec : number ) : void => {
        this . step ( elapsedMsec ) ;
    } ) ;
}
```

最后来看一下在 Application 类中的 update 和 render 函数。具体代码如下:

```
//虚方法,子类能覆写(override)
//注意:intervalSec 是以秒为单位的,而 elapsedMsec 是以毫秒为单位
public update ( elapsedMsec : number , intervalSec : number ) : void { }

//虚方法,子类能覆写(override)
public render ( ) : void { }
```

会发现,update 和 render 方法都是空实现,需要继承自 Application 的子类根据实际的

需求来覆写（override）这两个方法，这是面向对象三要素的经典体现，如下所述。
- 封装：将不变的部分（更新和渲染的流程）封装起来放在基类中，也就是基类固定了整个行为规范。
- 多态：将可变部分以虚函数的方式公开给具体实现者，基类并不知道每个子类要如何更新，也不知道每个子类如何渲染（例如，Canvas2DApplication 子类使用 CanvasRenderingContext2D 来进行各种二维图形的绘制，而 WebGLApplication 子类则使用 WebGLRenderingContext 进行二维或三维图形的绘制），让具体的子类来实现具体的行为，运行时动态绑定到实际调用的类成员方法上。
- 继承：很明显，虚函数多态机制依赖于继承，没有继承就没有多态。

3.2.4　回调函数的 this 指向问题

如果仔细看，会发现在 Application 类中的 start 和 step 成员方法中，在调用 requestAnimationFrame 时，其参数都是使用箭头函数的方式：requestAnimationFrame ((elapsedMsec : number) : void => { })；而在 3.1.3 节中，调用 requestAnimationFrame 函数时使用的是函数名作为参数：window . requestAnimationFrame (step)。那么能否在 Application 类中的 start 和 step 方法中像 3.1.3 节的代码一样，使用函数名或方法名作为 requestAnimationFrame 方法的参数呢？

答案是：不能。下面修改一下 start 方法中的代码：

```
public start ( ) : void {
    if ( ! this . _start ) {
        this . _start = true ;
        this . _requestId = -1 ;                // 将_requestId 设置为-1
        // 在 start 和 stop 函数中，_lastTime 和 _startTime 都设置为-1
        this . _lastTime = -1 ;
        this . _startTime = -1 ;
        // 启动更新渲染循环

        /* 注释掉箭头函数
        this . _requestId = requestAnimationFrame ( ( msec : number ) : void => {
            // 启动 step 方法
            this . step ( msec ) ;
        } ) ;
        */

        //注释掉上述代码，使用下面的代码来启动 this . step 方法
        this . _requestId = requestAnimationFrame ( this . step ) ;
    }
}
```

然后如图 3.6 所示，在 step 成员方法体内第一行设置一个断点，按 F5 键启动调试后来看一下运行效果。

第 3 章 动画与 Application 类

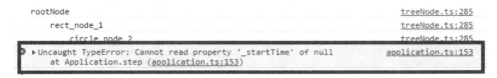

图 3.6 step 方法中 this 指针为 null 值

会看到，当运行到如图 3.6 所示的 153 行断点处，程序中断，此时将鼠标指针放在 this 关键字上，会显示 this 当前值为 null 值，这意味着如果代码继续往下运行（F10），程序肯定会报错，如图 3.7 所示。

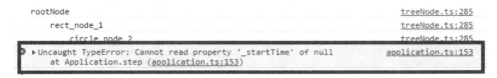

图 3.7 Chrome 浏览器报 null 值错误

之所以报 null 值错误，是因为在编写 step 成员方法代码时，所有的 this 都是假设指向 Application 类的实例对象。但是将 step 成员方法作为参数传递给 requestAnimationFrame 方法时，step 中的 this 实际上被绑定为 null 值了。为什么会这样呢？还是举个简单例子来更加详细地了解一下吧。

3.2.5 函数调用时 this 指向的 Demo 演示

为了更好地了解 TypeScript / JavaScript 在函数调用时 this 到底指向哪里，使用手动编译 TS 代码方式来演示一个 Demo，新建一个 testThis.ts 的文件，输入如下代码：

```typescript
class TestThis {
    // 编写一个打印 this 的方法
    public printCB ( msec : number ) : void {
        // 直接将 this 打印到 console 控制台
        console.log(this);
    }
    // 构造函数
    public constructor ( ) {
        // 将 this . printCB 作为参数传递给 requestAnimationFrame 方法
        window . requestAnimationFrame ( this . printCB ) ;
    }
}
```

```
// 生成TestThis的一个实例对象，该对象的构造函数会调用
// window.requestAnimationFrame方法
new TestThis();
```

在 VS Code 中打开集成终端，输入：tsc ./testThis.ts，将 testThis.ts 编译（转译）成非严格模式下的 testThis.js 源码，然后使用第 1 章中安装的 Open In Browser 插件运行 HTML 页面（该 HTML 文件使用 testThis.js）后，会发现 printCB 打印出来的 this 是 Window 对象，如图 3.8 所示。

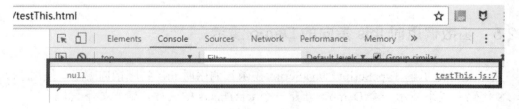

图 3.8　非严格模式下 this 指向 HTML DOM Window 对象

在编译（转译）出来的 testThis.js 的第一行加入" use strict "后，代码如下：

```
"use strict";                    // 使用严格模式，来测试一下this指向问题
var TestThis = /** @class */ (function () {
    function TestThis() {
        window.requestAnimationFrame(this.printCB);
    }
    TestThis.prototype.printCB = function (msec) {
        console.log(this);
    };
    return TestThis;
}());
new TestThis();
```

运行严格模式下的 JS 代码会发现，此时 this 打印出来的是 null 值，如图 3.9 所示。

图 3.9　严格模式下 this 指向 null 值

如果不使用箭头函数，并且要确保 this 指向正确的对象（例如在本 Demo 中，想让 this 指向 TestThis 类的实例对象），可以使用函数对象的 bind 方法。具体代码如下：

```
        // 构造函数
        public constructor() {
            // printCB是一个成员方法，也是一个函数对象
            // 因此printCB也具有bind方法
            // 使用bind(this)来绑定this对象
            // 由于在constructor函数中，此时的this是指向TestThis类的实例对象的
```

```
            // 这样printCB方法中的this指向的是printCB . bind ( this )中的this对象
            window . requestAnimationFrame ( this . printCB . bind ( this ) ) ;
    }
```
运行上述代码后，printCB 中 this 指向的是 TestThis 类的实例对象，如图 3.10 所示。

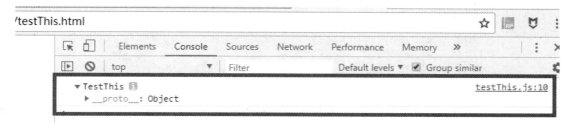

图 3.10　使用函数对象的 bind 方法绑定 this

由此可见，也可以在 Application 类的成员方法 start 和 step 中使用 bind 进行绑定，就能让 this 正确地指向 Application 类的实例对象。

最后以 Application 和 TestThis Demo 为例子来总结一下，当一个回调函数或方法被绑定和调用时，可以按照如下顺序来确定 this 指向的问题：
- 如果使用了箭头函数，this 指向的是当前类的实例对象，例如 Application 中代码所示。
- 如果使用函数对象的 bind 方法并绑定当前类的实例，那么 this 也指向当前类的实例对象，例如 TestThis 中 bind 函数调用所示。
- 如果没有使用上述两种方式，并且在非严格模式下，则 this 指向的是 Window 对象，例如 TestThis 中非严格模式 Demo 所示。
- 如果没有使用上述两种方式，并且在严格模式下（"use strict"），则 this 指向 null 值。

3.2.6　CanvasInputEvent 及其子类

Application 类除了不间断地基于时间的更新与重绘封装外，还有一个很重要的功能就是对输入事件的分发和响应机制。输入事件包括鼠标事件和键盘事件，将这两种事件的共有部分抽象成 CanvasInputEvent 基类，该类的源码如下：

```
// CanvasKeyboardEvent 和 CanvasMouseEvent 都继承自本类
// 基类定义了共同的属性，keyboard 或 mouse 事件都能使用组合键
// 例如可以按住 Ctrl 键的同时单击鼠标左键做某些事情
// 当然也可以按住 Alt + A 键做另外一些事情
export class CanvasInputEvent {
    // 3个boolean 变量，用来指示 Alt、Ctrl、Shift 键是否被按下
    public altKey : boolean ;
    public ctrlKey : boolean ;
    public shiftKey : boolean ;
```

```
    // type 是一个枚举对象，用来表示当前的事件类型，枚举类型定义在下面的代码中
    public type : EInputEventType ;
    // 构造函数，使用了 default 参数，初始化时 3 个组合键都是 false 状态
    public constructor ( altKey : boolean = false , ctrlKey : boolean = false ,
    shiftKey : boolean = false , type : EInputEventType = EInputEventType .
    MOUSEEVENT ) {
        this . altKey = altKey ;
        this . ctrlKey = ctrlKey ;
        this . shiftKey = shiftKey ;
        this . type = type ;
    }
}
```

再来看一下 EInputEventType 这个枚举，该枚举罗列出目前支持的各个输入事件，包括鼠标和键盘事件。具体代码如下：

```
export enum EInputEventType {
    MOUSEEVENT ,                    //总类，表示鼠标事件
    MOUSEDOWN ,                     //鼠标按下事件
    MOUSEUP ,                       //鼠标弹起事件
    MOUSEMOVE ,                     //鼠标移动事件
    MOUSEDRAG ,                     //鼠标拖动事件
    KEYBOARDEVENT ,                 //总类，表示键盘事件
    KEYUP ,                         //键按下事件
    KEYDOWN ,                       //键弹起事件
    KEYPRESS                        //按键事件
} ;
```

接下来看一下子类 CanvasMouseEvent 和 CanvasKeyBoardEvent。具体代码如下：

```
export class CanvasMouseEvent extends CanvasInputEvent {
    // button 表示当前按下鼠标哪个键
    // [ 0 : 鼠标左键 , 1 : 鼠标中键 , 2 : 鼠标右键 ]
    public button : number ;
    // 基于 canvas 坐标系的表示
    public canvasPosition : vec2 ;

    public localPosition : vec2 ;
    public constructor ( canvasPos : vec2 , button : number , altKey : boolean
    = false , ctrlKey : boolean = false , shiftKey : boolean = false ) {
        super ( altKey , ctrlKey , shiftKey ) ;
        this . canvasPosition = canvasPos ;
        this . button = button ;

        // 暂时创建一个 vec2 对象
        this . localPosition = vec2 . create ( ) ;
    }
}
export class CanvasKeyBoardEvent extends CanvasInputEvent {
    // 当前按下的键的 ASCII 字符
    public key : string ;
    // 当前按下的键的 ASCII 码（数字）
    public keyCode : number ;
```

```
            // 当前按下的键是否不停地触发事件
            public repeat : boolean ;
            public constructor ( key : string , keyCode : number , repeat : boolean ,
            altKey : boolean = false , ctrlKey : boolean = false , shiftKey : boolean
            = false ) {
                super ( altKey , ctrlKey , shiftKey , EInputEventType .
                KEYBOARDEVENT ) ;
                this . key = key ;
                this . keyCode = keyCode ;
                this . repeat = repeat ;
            }
        }
```

这些代码比较简单，并且注释也比较详细。相对有点难度的是 CanvasMouseEvent 类中的 canvasPosition 和 localPosition 这两个成员变量，涉及一些坐标变换相关的知识，将会在 3.2.7 节中详解。关于这两个变量的类型是 vec2，表示 2D 向量，目前仅需要了解的是 vec2 是 2D 坐标{ x , y }的表示，关于 vec2 和 mat2d 这两个数学类，后面章节也会重点提示，是比较重要的内容。

3.2.7　使用 getBoundingRect 方法变换坐标系

上一节中提到了 CanvasMouseEvent 中的 canvasPosition 这个成员变量，它表示的是鼠标指针位置相对于 HTMLCanvasElement 原点的偏移向量。如何理解这句话呢？最好的方式是画张图，直观且易于理解，如图 3.11 所示。

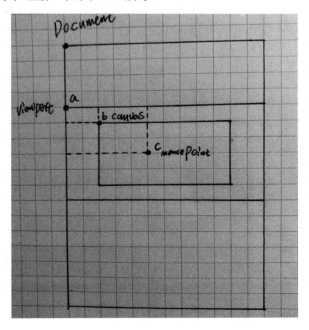

图 3.11　getBoundingRect 方法示例图

根据图 3.11 所示，整个 HTML 结构分为 3 部分：Document 区域、viewport 可视区，以及 canvas 元素区。其中：
- a 点为浏览器的 viewport 原点。
- b 点为 HTMLCanvasElement 元素的原点。
- c 点为当前的鼠标指针的位置点。

如果能够知道 b 点相对于 a 点的偏移坐标，c 点相对于 a 点的偏移坐标，则 b 点和 c 点处于同一个坐标系（也就是浏览器的 viewport 坐标系），那么很容易就能计算出 c 点相对于 b 点的偏移坐标（c - b），这个坐标就是 CanvasMouseEvent 中成员变量 canvasPosition（相对于 canvas 坐标系的偏移向量）。

幸运地是，可以使用现成的类和方法来获取 b 点处的坐标和 c 点处的坐标，让我们来看一下具体的类和方法：
- 可以使用 Element 类的 getBoundingRect 方法获取当前元素的大小及相对于浏览器 viewport 的偏移量，也就是 b 点处的向量。在这里，当前元素是 HTMLCanvasElement，该元素是 Element 的子类。
- 可以使用 MouseEvent 的 clientX / clientY 只读属性获取鼠标事件发生时相对客户区左上角原点处的偏移量，也就是相对于浏览器 viewport 可视区左上角 a 点处的偏移量，即 c 点坐标。

有了上述类的方法，就可以为 Application 类实现一个重要的坐标转换方法。具体代码如下：

```
// 将鼠标事件发生时鼠标指针的位置变换为相对当前 canvas 元素的偏移表示
// 这是一个私有方法，意味着只能在本类中使用，子类和其他类都无法调用本方法
// 只要是鼠标事件（down / up / move / drag .....）都需要调用本方法
// 将相对于浏览器 viewport 表示的点变换到相对于 canvas 表示的点
private _viewportToCanvasCoordinate ( evt : MouseEvent ) : vec2 {
    if ( this . canvas ) {
        let rect : ClientRect = this . canvas . getBoundingClientRect ( ) ;
        // 作为测试，每次 mousedown 时，打印出当前 canvas 的 boundClientRect
        的位置和尺寸
        // 同时打印出 MouseEvent 的 clientX / clientY 属性
        if ( evt . type === "mousedown" ) {
            console . log ( " boundingClientRect : " + JSON . stringify 
              ( rect ) ) ;
            console . log ( " clientX : " + evt . clientX + " clientY : " + 
              evt.clientY ) ;
        }
        let x : number = evt . clientX -  rect . left ;
        let y : number = evt . clientY - rect . top ;
        return vec2 . create ( x , y ) ;
    }

    //到这里，说明 canvas 为 null，直接报错
```

```
        alert ( " canvas 为 null " ) ;
        throw new Error ( " canvas 为 null " ) ;
    }
```

理解了原理后，代码还是很简单的，接下来看一下如何调用该方法。

3.2.8　将 DOM Event 事件转换为 CanvasInputEvent 事件

使用了自己定义的 CanvasInputEvent 事件体系，因此需要将 DOM Event 事件转换为 CanvasInputEvent 对应的子类。具体代码如下：

```
// 将 DOM Event 对象信息转换为自己定义的 CanvasMouseEvent 事件
private _toCanvasMouseEvent ( evt : Event ) : CanvasMouseEvent {
    // 向下转型，将 Event 转换为 MouseEvent
    let event : MouseEvent = evt as MouseEvent ;
    // 将客户区的鼠标 pos 变换到 Canvas 坐标系中表示
    let mousePosition : vec2 = this . _viewportToCanvasCoordinate ( event ) ;
    // 将 Event 一些要用到的信息传递给 CanvasMouseEvent 并返回
    let canvasMouseEvent : CanvasMouseEvent = new CanvasMouseEvent
      ( mousePosition , event . button , event . altKey , event . ctrlKey ,
      event .shiftKey ) ;
    return canvasMouseEvent ;
}
// 将 DOM Event 对象信息转换为自己定义的 Keyboard 事件
private _toCanvasKeyBoardEvent ( evt : Event ) : CanvasKeyBoardEvent {
    let event : KeyboardEvent = evt as KeyboardEvent ;
    // 将 Event 一些要用到的信息传递给 CanvasKeyBoardEvent 并返回
    let canvasKeyboardEvent : CanvasKeyBoardEvent = new CanvasKeyBoardEvent
      ( event . key , event . keyCode , event . repeat , event . altKey , event .
      ctrlKey , event . shiftKey ) ;
    return canvasKeyboardEvent ;
}
```

3.2.9　EventListenerObject 与事件分发

EventListenerObject 是 TypeScript 中预先定义的一个接口，该接口具有如下声明：

```
interface EventListenerObject {
    handleEvent ( evt : Event ) : void ;
}
```

该接口作为回调接口，我们只要实现签名方法 handleEvent 就可以了。在必要时，浏览器系统会自动调用该接口的 handleEvent 方法。因此需要在 Application 类中接口继承 EventListenerObject 并实现 handleEvent 方法。具体代码如下：

```
// 接口继承 EventListenerObject 并实现 handleEvent 签名方法
export class Application implements EventListenerObject {
    // 本书中的 Demo 以浏览器为主
    // 对于 mousemove 事件提供一个开关变量
```

```
// 如果下面变量设置为 true，则每次鼠标移动都会触发 mousemove 事件
// 否则就不会触发
public isSupportMouseMove : boolean ;
// 使用下面变量来标记当前鼠标是否为按下状态
// 目的是提供 mousedrag 事件
protected _isMouseDown : boolean ;
// 调用 dispatchXXXX 虚方法进行事件分发
// handleEvent 是接口 EventListenerObject 定义的接口方法，必须要实现
public handleEvent ( evt : Event ) : void {
    // 根据事件的类型，调用对应的 dispatchXXX 虚方法
    switch ( evt . type ) {
        case "mousedown" :
            this . _isMouseDown = true ;
            this . dispatchMouseDown ( this . _toCanvasMouseEvent
            ( evt ) ) ;
            break ;
        case "mouseup" :
            this . _isMouseDown = false ;
            this . dispatchMouseUp ( this . _toCanvasMouseEvent ( evt ) ) ;
            break ;
        case "mousemove" :
            //如果 isSupportMouseMove 为 true，则每次鼠标移动会触发 mouseMove 事件
            if ( this . isSupportMouseMove ) {
                this . dispatchMouseMove ( this . _toCanvasMouseEvent
                ( evt ) ) ;
            }
            // 同时，如果当前鼠标任意一个键处于按下状态并拖动时，触发 drag 事件
            if ( this . _isMouseDown ) {
                this . dispatchMouseDrag ( this . _toCanvasMouseEvent
                ( evt ) ) ;
            }
            break ;
        case "keypress" :
            this . dispatchKeyPress ( this . _toCanvasKeyBoardEvent
            ( evt ) ) ;
            break ;
        case "keydown" :
            this . dispatchKeyDown ( this . _toCanvasKeyBoardEvent
            ( evt ) ) ;
            break ;
        case "keyup" :
            this . dispatchKeyUp ( this . _toCanvasKeyBoardEvent ( evt ) ) ;
            break ;
    }
}
```

可以看到，对于鼠标事件，在 handleEvent 中先调用_toCanvasXXXEvent 方法将类型为 Event 的参数转换为自己定义的事件，然后调用对应的 dispatchXXX 虚方法，这些方法需要由子类覆写（override）。dispatchXXX 虚方法有多个，在 Application 基类中都是空实现，因此只要知道名字即可，具体实现在子类中详细了解。

3.2.10 让事件起作用

到目前为止，整个事件分发处理流程都打通了，就缺添加监听器监听感兴趣的事件了。将事件监听的操作放在 Application 类的构造函数中，这样就能在内存分配初始化后监听相关事件。具体代码如下：

```
public constructor ( canvas : HTMLCanvasElement ) {
    this . canvas = canvas ;
    // canvas 元素能够监听鼠标事件
    this . canvas . addEventListener ( "mousedown" , this , false ) ;
    this . canvas . addEventListener ( "mouseup" , this , false ) ;
    this . canvas . addEventListener ( "mousemove" , this , false ) ;

    // 很重要的一点，键盘事件不能在 canvas 中触发，但是能在全局的 window 对象中触发
    // 因此能在 window 对象中监听键盘事件
    window . addEventListener ( "keydown" , this , false ) ;
    window . addEventListener ( "keyup" , this , false ) ;
    window . addEventListener ( "keypress" , this , false ) ;

    // 初始化时，mouseDown 为 false
    this . _isMouseDown = false ;
    // 默认状态下，不支持 mousemove 事件
    this . isSupportMouseMove = false ;
}
```

上述代码要强调的一点是：HTMLCanvasElement 元素无法监听键盘事件，幸运的是可以在全局的 Window 对象中监听键盘事件，然后对相应事件进行分发或处理。至此，Application 的基本功能都实现了，接下来需要两个继承自 Application 的子类，分别使用 Canvas2D 和 WebGL API 进行渲染。

3.2.11 Canvas2DApplication 子类和 WebGLApplication 子类

Canvas2DApplication 子类和 WebGLApplication 子类其实很简单，可以从 Application 中的 public 成员变量 canvas 中调用 getContext 方法，根据不同的实参获取 CanvasRenderingContext2D 对象或 WebGLRenderingContext 对象。具体代码如下：

```
export class Canvas2DApplication extends Application {
    public context2D : CanvasRenderingContext2D | null ;
    public constructor ( canvas : HTMLCanvasElement , contextAttributes ? :
    Canvas2DContextAttributes ) {
        super( canvas ) ;
        this . context2D = this . canvas . getContext( "2d" , context
        Attributes ) ;
    }
}
export class WebGLApplication extends Application {
    public context3D : WebGLRenderingContext | null ;
```

```typescript
public constructor ( canvas : HTMLCanvasElement , contextAttributes ? : 
WebGLContextAttributes ) {
    super( canvas ) ;
    this . context3D = this . canvas . getContext( "webgl" , context
    Attributes ) ;
    if ( this . context3D === null ) {
        this . context3D = this . canvas . getContext("experimental-
        webgl" , contextAttributes ) ;
        if ( this . context3D === null ) {
            alert ( " 无法创建 WebGLRenderingContext 上下文对象 " ) ;
            throw new Error ( "无法创建 WebGLRenderingContext 上下文对象" ) ;
        }
    }
}
```

这两个类实际上仅仅是获取 2D/3D 渲染上下文对象，如果要进行具体的渲染逻辑，那么可以继续继承，然后进行相关业务逻辑的扩展，后面的章节会使用 Canvas2D Application 类，到时候再详细地了解相关细节。

至此，Application 及子类都介绍完毕，下一节来测试一下 Application 类是否有效。

3.3 测试及修正 Application 类

在前面章节中将 Application 类的主要功能都实现了，接下来来测试 Application 类是否有效。先创建一个名为 applicationTest.ts 的文件，所有的测试代码都在 applicationTest.ts 中实现。

3.3.1 继承并覆写 Application 基类的虚方法

首先在 applicationTest.ts 中实现一个名为 ApplicationTest 的类，该类继承自 Application 基类，并且覆写（override）4 个基类方法。具体代码如下：

```typescript
import { Application } from "./src/application";
import { CanvasKeyBoardEvent, CanvasMouseEvent } from "./src/application";
// ApplicationTest 继承并扩展了 Application 基类
class ApplicationTest extends Application {
    // 覆写（override）基类的受保护方法 dispatchKeyDown
    protected dispatchKeyDown ( evt : CanvasKeyBoardEvent) : void {
        // 当发生 keydown 事件时，将哪个键按下信息输出到 console 控制台
        console . log ( " key : " + evt.key + " is down " ) ;
    }
    // 覆写（override）基类的受保护方法 dispatchMouseDown
    protected dispatchMouseDown ( evt : CanvasMouseEvent ) : void {
        //当发生 mousedown 事件时,将 canvasPosition 坐标信息输出到 console 控制台
        console . log ( " canvasPosition : " + evt . canvasPosition ) ;
    }
```

```typescript
// 覆写（override）基类公开方法 update
public update ( elapsedMsec : number , intervalSec : number ) : void {
    console . log ( " elapsedMsec : " + elapsedMsec + " intervalSec : 
    " + intervalSec ) ;
}
// 覆写（override）基类公开方法 render
public render ( ) : void {
    console . log ( " 调用 render 方法 " ) ;
}
}
```

3.3.2 测试 ApplicationTest 类

现在来测试一下 ApplicationTest 类，需要一个 HTML 文件，其 css 代码如下：

```html
<style>
    body {
        background : #eeeeee ;
    }
    /*
    css 选择器: # 表示 id 选择器
    */
    #canvas {
        background : #ffffff ;
        margin : 0px; /* 将 canvas 的 margin 设置为 0px */
        /* border : 30px solid ; */
        /* 如果没有 solid ，则 boarder 无效 */
        /* padding: 20px ; */
        /* 给 canvas 增加点阴影 */
        -webkit-box-shadow : 4px 4px 8px rgba ( 0 , 0 , 0 , 0.5 ) ;
        -moz-box-shadow : 4px 4px 8px rgba ( 0 , 0 , 0 , 0.5 ) ;
        box-shadow : 4px 4px 8px rgba ( 0 , 0 , 0 , 0.5 ) ;
    }
</style>
```

HTML 的控件层次树代码如下：

```html
<body id = "body" >
    <div>
        <button id = "start" > start </button >
        <button id = "stop" > stop </button >
    </div>
    <canvas id = "canvas" width = "300" height = "300" > </canvas >
</body>
```

如上述 HTML 源码所示，<button>元素是行内元素，为了让两个<button>元素与 canvas 元素不在同一行，使用<div>这个块级元素来进行换行操作。

接下来在 applicationTest.ts 文件中继续输入测试 ApplicationTest 类的代码，具体源码如下：

```typescript
// 获取 id 为 canvas 的 HTMLCanvasElement 对象
let canvas : HTMLCanvasElement | null = document . getElementById ( 'canvas' )
```

```typescript
as HTMLCanvasElement ;
// 生成一个 Application 对象，这样就能监听 keydown 和 mouseDown 事件了
let app : Application = new ApplicationTest ( canvas ) ;
// 由于 update 和 render 是 public 方法，因此可以在 Application 类外部调用
// 先调用一次，而一些事件分发处理方法是 protected，所以无法外部调用
app . update ( 0 , 0 ) ;
app . render ( ) ;
// 通过 id 号获取 startButton 和 stopButton 两个 button 元素
let startButton : HTMLButtonElement | null = document . getElementById
( 'start' ) as HTMLButtonElement ;
let stopButton : HTMLButtonElement | null = document . getElementById
( 'stop' ) as HTMLButtonElement ;
// 单击 startButton 后，启动 Appliation 启动动画循环
// 在每次循环中会调用 update 和 render 方法
startButton . onclick = ( ev : MouseEvent ) : void => {
    app . start ( ) ;
}
// 单击 stopButton 后，会停止 Application 动画循环
stopButton . onclick = ( ev : MouseEvent ) : void => {
    app . stop ( ) ;
}
```

使用 F5 快捷方式调试运行代码，会出现如图 3.12 所示的结果。

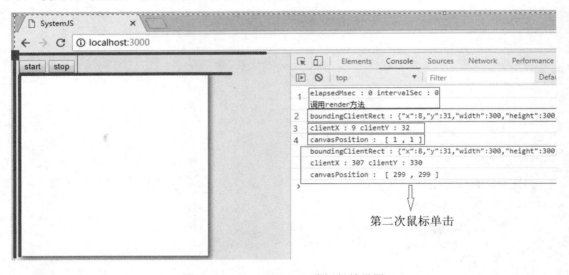

图 3.12　ApplicationTest 类运行效果图

3.3.3　多态（虚函数动态绑定）

此时来了解面向对象三要素中的多态行为是一个好的时机点。下面来看一下 ApplicationTest 是如何创建和调用的。代码在上一节中，为了更好地了解，此处复制具体代码如下：

```
let app : Application = new ApplicationTest ( canvas ) ;
// 手动调用 update 和 render 成员方法
app . update ( 0 , 0 ) ;
app . render ( ) ;
```

要关注的一个事实是，变量 app 声明的类型是 Application 基类，而不是 ApplicationTest 子类，但是 new 的时候却是 ApplicationTest 子类。

调用上述代码后，根据图 3.12 的结果，会发现调用 update 和 render 后在 console 控制台中输出了相关信息。而 Application 基类实现中，update 和 render 都是空实现。

当用 app 调用 update 方法和 render 方法时，就会发生多态行为，也就是运行时函数地址动态绑定。此时会发现，虽然使用基类来调用 update 和 render 方法，但是程序却自动调用了 ApplicationTest 覆写（override）的 update 和 render 方法。这就是多态，基类自动调用子类的同名虚方法。当然，笔者是以 C++ 的方式来理解 TypeScript 的面向对象相关内容，但是从原理上是说得通的。

多态这种行为特别适合面向接口和面向抽象编程。例如在 Application 的 step 中会调用 update 和 render 方法，这个流程很固定，子类不会修改。而 update 和 render 由于不同的子类有不同操作，因此作为虚函数由子类自己来决定如何实现。

当设计框架时，这是一种常用的方法，将不变的部分封装在基类，将可变部分以虚函数（使用多态）或回调函数方式公开给用户自己实现。

事实上，上述的行为从设计模式角度来说，就是经典的模板方法设计模式，在 Application 基类中设计好了一个模板流程，并定义好 update 和 render 这两个模板虚方法，规定好这两个方法的调用流程，子类则负责具体实现 update 和 render 方法。

下面总结以下知识点：
- 如果一门编程语言支持类，那么可以认为该语言是一门基于对象的语言。
- 如果一门编程语言支持类继承的同时还支持多态，那么该语言是一门面向对象的语言。
- 面向对象三要素是：继承、封装和多态。
- 多态就是使用基类来调用实现类（派生类）的虚方法（TypeScript 中，类的成员方法都是虚方法）。

3.3.4 鼠标单击事件测试

如图 3.12 所示，分别在 canvas 元素的左上角和右下角单击两次，输出了两次单击后的结果，通过输出结果可以了解以下几点：
- 在左上角处使用一个方块标记出原点，该原点就是 viewport 的原点。canvas . getBoundingRect ( ) 方法和 evt . ClientX / ClientY 都是相对该原点的偏移量。
- canvas . getBoundingRect ( ) 方法调用后，获得 canvas 相对 viewport 的偏移量为 [8 , 31]，其尺寸是 [300 , 300]。而当前鼠标指针相对 viewport 的偏移量是 [9 , 32]，因此转换到相对 canvas 坐标系中表示为 [1 , 1]（[9 - 8 , 32 - 31]），这是正确的。

- 在 3.2.2 节中，在 CSS 中明确地规定了 canvas 的 margin 为 0px，但是我们会发现在 Chrome 浏览器中 canvas 仍旧与 viewport 之间有[8，8]个像素的偏移（如果去掉<div>元素及两个<button>子元素，y 轴也是偏移 8 个像素）。

从上面最后一条内容，又可以引申出两个问题，如下所述。

（1）为什么明确地设置了 canvas 的 margin 为 0px，却仍然与 viewport 有 8 个像素的偏移？

关于这个问题，首先要说明的一点是，笔者在 Chrome、Firefox、Opera、Safari、Edge，以及 IE 11 中测试了一下，发现 canvas 与 viewport 之间都是[8，8]像素的偏移。

其次在测试过程中发现，Microsoft 公司的 Edge 和 IE 11 中 MouseEvent．clientX / clientY 返回的是浮点数，而其他浏览器都是整数。笔者查阅了 MDN 关于 MouseEvent ClientXs 属性的文档（https ://developer.mozilla.org/en-US/docs/Web/API/MouseEvent/clientX），里面有相关说明：Originally, this property was defined as a long integer. The CSSOM View Module redefined it as a double float.（原本，ClientX 这个属性被定义为长整型，而在 CSSOM View Module 中被重新定义为双精度浮点数），而 Microsoft 公司的实现显然是使用了双精度浮点数。

最后如果想取消 canvas 和 viewport 之间 8 个像素的偏移，可以输入如下代码：

```
body {
    background : #eeeeee ;
    margin : 0px ;
}
```

这也是笔者经过测试后找到的解决方案（笔者的 CSS 水平有限），不是 canvas 和 viewport 之间有 8 个像素的 margin，而是 body 元素与 viewport 之间有 8 个像素的 margin。当然还可以*选择器，选择所有的元素，使其 margin 都为 0px，代码如下：

```
* {
    margin : 0px ;
}
```

（2）margin、border 和 padding 等这些 CSS 盒模型属性对 application 的坐标变换私有方法_viewportToCanvasCoordinate 是否有影响？

这个问题稍微有点复杂，使用已经实现的一个 Demo 来更好地理解盒模型与变换之间的关系，具体请看下一节的内容。

3.3.5　CSS 盒模型对_viewportToCanvasCoordinate 的影响

在 Application 类中，坐标转换是在_viewportToCanvasCoordinate 私有方法中完成的。在上一节鼠标单击测试中对 canvas 的 margin 设置为 0px，并且没有设置 border 和 padding 属性。这些属性都属于 CSS 盒模型属性，那么如果设置这些属性，会不会对原本实现的

_viewportToCanvasCoordinate 函数有影响呢？

首先，设置 canvas 的 margin 为 10px，并使后面章节实现的精灵系统 Demo 截图来看一下效果，效果如图 3.13 所示。

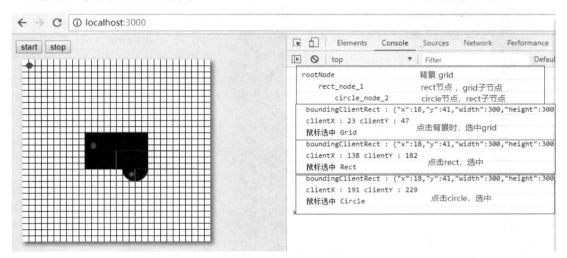

图 3.13　设置 margin 后的测试效果

设置 canvas 的 margin 为 10px，其 boundingClientRect 相对 viewport 的偏移坐标是 [18，41]。其层次关系如图 3.13 所示。由三个节点组成，其中 grid 是整个背景节点，而 rect 节点是 grid 节点的子节点，circle 节点是 rect 的子节点，也就是 grid 节点的孙节点。

当分别单击 grid、rect 和 circle 时，碰撞检测结果是正确的。这说明 margin 的设置并不影响_viewportToCanvasCoordinate 方法。

接下来，看看 border 是否会影响_viewportToCanvasCoordinate 方法。其 CSS 代码如下：

```
#canvas {
    margin : 10px ;   /*将 canvas 的 margin 设置为 10px*/
    border : 80px solid ;
}
```

根据上面的代码所示，将 border 设置为 80px，默认情况下，border 的颜色为黑色，来看一下运行后的效果，如图 3.14 所示。

如图 3.14 所示，当分别单击 grid、rect 和 circle 时，鼠标选中的都是 Grid，碰撞检测结果是不正确的。由此可以得出结论，即 border 的设置影响_viewportToCanvasCoordinate 方法。

这里还要注意一点：设置 border 为某个数值后，必须要标记为 solid，否则 border 设置不会起作用。

以同样的方式设置了 padding 后，同 border 一样，会严重影响到_viewportToCanvasCoordinate 方法。

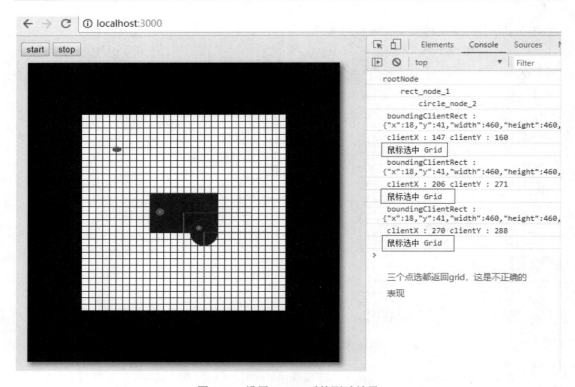

图 3.14　设置 border 后的测试效果

3.3.6　正确的_viewportToCanvasCoordinate 方法实现

通过上节的测试知道，_viewportToCanvasCoordinate 的实现没有考虑 border 和 padding 设置后的情况，那么本节就来修正这个问题。为了更好地显示效果，先设置一个完整的盒模型，如下 CSS 代码所示。

```
/*
 css 选择器：# 表示 id 选择器
*/
#canvas {
    background : #ffffff ;         /* canvas 背景为白色 */
    margin : 10px ;                /* 将 canvas 的 margin 设置为 0px */
    border : 80px solid red ;      /* 如果没有 solid，则 boarder 无效，设置 border
    为红色 */
    padding : 20px ;               /* 设置 padding 为 20 个像素 */
}
```

然后来看一下修正_viewportToCanvasCoordinate 方法后的正确效果，如图 3.15 所示。

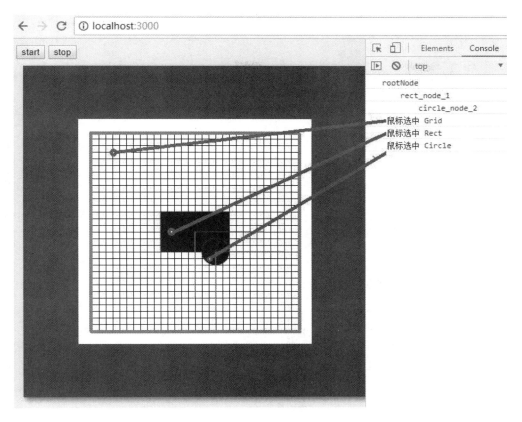

图 3.15　修正_viewportToCanvasCoordinate 方法后的正确效果图

参考图 3.15 会发现，即使有 80px 红色的 border，以及 20px 白色的 padding，鼠标点选也不受影响，正确选中鼠标单击处的节点（或精灵）。那么来看一下实现过程。具体代码如下：

```
// 将鼠标事件发生时鼠标指针的位置变换为相对当前 canvas 元素的偏移表示
// 这是一个私有方法，意味着只能在本类中使用，子类和其他类都无法调用本方法
// 只要是鼠标事件（down / up / move / drag .....）都需要调用本方法
// 将相对于浏览器 viewport 表示的点变换到相对于 canvas 表示的点
private _viewportToCanvasCoordinate ( evt : MouseEvent ) : vec2 {
    if ( this . canvas ) {
        // 切记,很重要一点：getBoundingClientRect ( )方法返回的 ClientRect
        let rect : ClientRect = this . canvas . getBoundingClientRect ( ) ;

        // 作为测试，每次 mousedown 时，打印出当前 canvas 的 boundClientRect 的位置
        和尺寸
        if ( evt . type === "mousedown" ) {
            console . log ( " boundingClientRect : " + JSON . stringify ( rect ) ) ;
            // 测试使用输出相关信息，打印出 MouseEvent 的 clientX / clientY 属性
```

```typescript
        console.log("clientX : " + evt.clientX + " clientY : " +
        evt.clientY);
    }

    // 获取触发鼠标事件的 target 元素,这里总是 HTMLCanvasElement
    if (evt.target)
    {
        let borderLeftWidth : number = 0;
                    //返回 border 左侧离 margin 的宽度
        let borderTopWidth : number = 0;
                    //返回 border 上侧离 margin 的宽度
        let paddingLeft : number = 0;
                    //返回 padding 相对 border (border 存在的话) 左偏移
        let paddingTop : number = 0;
                    //返回 padding 相对 border (border 存在的话) 上偏移
        // 调用 getComputedStyle 方法,这个方法比较有用
        let decl : CSSStyleDeclaration = window.getComputedStyle
        (evt.target as HTMLElement);

        // 需要注意,CSSStyleDeclaration 中的数值都是字符串表示,而且有可能
        返回 null
        // 所以需要进行 null 值判断
        // 并且返回的坐标都是以像素表示的,所以是整数类型
        // 使用 parseInt 转换为十进制整数表示
        let strNumber : string | null = decl.borderLeftWidth;
        if (strNumber !== null) {
            borderLeftWidth = parseInt(strNumber, 10);
        }
        if (strNumber !== null) {
            borderTopWidth = parseInt(strNumber, 10);
        }
        strNumber = decl.paddingLeft;
        if (strNumber !== null) {
            paddingLeft = parseInt(strNumber, 10);
        }
        strNumber = decl.paddingTop;
        if (strNumber !== null) {
            paddingTop = parseInt(strNumber, 10);
        }

        // a = evt.clientX - rect.left,将鼠标点从 viewport 坐标系变换
        到 border 坐标系
        // b = a - borderLeftWidth,将 border 坐标系变换到 padding 坐标系
        // x = b - paddingLeft,将 padding 坐标系变换到 context 坐标系,也就
        是 canvas 元素坐标系
        let x : number = evt.clientX - rect.left - borderLeftWidth
        - paddingLeft;
        let y : number = evt.clientY - rect.top - borderTopWidth -
        paddingTop;
        // 变成向量表示
        let pos : vec2 = vec2.create(x, y);
```

```
        // 测试使用输出相关信息
        if ( evt . type === "mousedown" ) {
            console . log ( " borderLeftWidth : " + borderLeftWidth + "
            borderTopWidth : " + borderTopWidth ) ;
            console . log ( " paddingLeft : " + paddingLeft + " paddingTop :
            " + paddingTop ) ;
            console . log ( " 变换后的canvasPosition : " + pos .
            toString ( ) ) ;
        }
        return pos ;
    }
    // 对于错误,直接报错
    alert ( " canvas 为 null " ) ;
    throw new Error ( " canvas 为 null " ) ;
    }

    // 对于错误,直接报错
    alert ( " evt . target 为 null " ) ;
    throw new Error ( " evt . target 为 null " ) ;
}
```

关于关键的坐标转换算法,在上面的代码中有详细注释。除了图 3.15 的效果外,再看看 border 及 padding 更加通用设置时的效果,CSS 代码如下设置:

```
/*
    css 选择器:# 表示 id 选择器
*/
#canvas {
    background : #ffffff ; /* 背景白色 */
    margin : 10px 15px 20px 25px ; /* 将 canvas 的 margin 设置为 4 个数值 */

    border-style : solid ;
    border-width : 25px 35px 20px 30px ; /* top - right - bottom - left 顺序 */
    border-color : gray ;
    /* 使用各种单位表示 padding */
    padding-top : 10px ;
    padding-righ t: 0.25em ;
    padding-bottom : 2ex ;
    padding-left : 20% ;
    -webkit-box-shadow : 4px 4px 8px rgba ( 0 , 0 , 0 , 0.5 ) ;
    -moz-box-shadow : 4px 4px 8px rgba ( 0 , 0 , 0 , 0.5 ) ;
    box-shadow : 4px 4px 8px rgba ( 0 , 0 , 0 , 0.5 ) ;
}
```

运行后的效果如图 3.16 所示。

根据图 3.16 所示,会看到更新后的_viewportToCanvasCoordinate 能够自适应盒模型的各种参数,精确地进行鼠标点选的碰撞检测,并且即使节点(或精灵)发生位移、缩放或旋转等仿射变换,也能精确地进行点选。而这些变换及碰撞检测,都是建立在_viewportToCanvasCoordinate 基础上的,后续章节会介绍如何实现精灵系统。

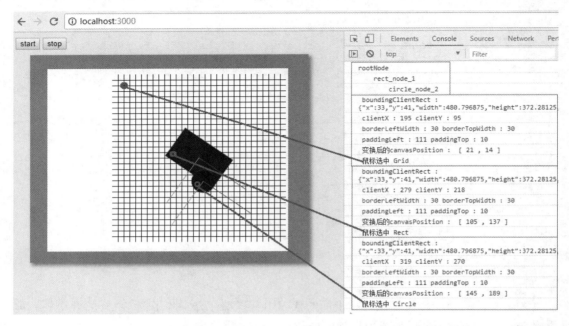

图 3.16 _viewportToCanvasCoordinate 终极效果图

3.4 为 Application 类增加计时器功能

在 Application 类中使用 requestAnimationFrame 来驱动动画不停更新和重绘，该函数与屏幕刷新频率一致（例如 16.66 毫秒）的速度不停重复循环。但有时候，可能还要执行其他任务，比如每秒定时地输出一些信息，或者在某个时间点仅仅执行一次任务。像这些任务并不需要以每秒 60 帧的速度执行，也就是说需要既可以不同帧率来执行同一个任务，也可以倒计时方式执行一次任务。

要实现上述需求，可以使用 HTML DOM 中 Window 对象的 setTimeout 方法或 setInterval 方法。但是既然已经实现了 Application 类基于时间的更新，不如在此基础上自己来实现与 setTimeout / setInterval 相同的功能，这也符合本书的其中的一个定位：造轮子。

3.4.1 Timer 类与 TimeCallback 回调函数

先来总结一下计时器的设计目标，主要有以下 4 个：
- Application 类能够同时触发多个计时器。
- 每个计时器可以以不同帧率来重复执行任务。
- 每个计时器可以倒计时方式执行一次任务。

- 尽量让内存使用与运行效率达到相对平衡。

要实现上述 4 个目标，需要 Timer 类和 Application 类相互配合。下面先来看一下 Timer 的相关代码吧。具体代码如下：

```
// 回调函数类型别名
// 回调函数需要第三方实现和设置，所以导出该回调函数
export type TimerCallback = ( id : number , data : any ) => void ;
// 纯数据类
// 不需要导出 Timer 类，因为只是作为内部类使用
class Timer {
    public id : number = -1 ;                       //计时器的 id 号
    // 标记当前计时器是否有效，很重要的一个变量，具体看后续代码
    public enabled : boolean = false ;
    public callback : TimerCallback;                //回调函数，到时间会自动调用
    public callbackData: any = undefined;           //用作回调函数的参数
    public countdown : number = 0 ;                 //倒计时器，每次 update 时会倒计时
    public timeout : number = 0;    //
    public onlyOnce : boolean = false ;
    constructor ( callback : TimerCallback ) {
        this . callback = callback ;
    }
}
```

这里要注意的一点是：TimerCallback 是 export（被导出）给外部使用的，而 Timer 类没有。这是因为在 Application 类提供的公开方法都是以 Timer 的 id 号来操作 Timer，因此不需要导出 Timer 类。

Timer 类是纯数据类，所有的对 Timer 类的操作都封装在 Application 类中，接下来需要为 Apllication 类增加对 Timer 类的支持。

3.4.2 添加和删除 Timer（计时器）

接下来看一下如何在 Application 类中使用 Timer 类，实现上一节所设定的目标。其中有一条是 Application 类能够同时触发多个 Timer（计时器），这意味着 Application 类是 Timer 的容器。并且要用 id 号来操作计时器，因此需要 Application 类能够自动生成 id 号，为 Application 类增加如下两个成员变量，如下面的代码所示。

```
public timers : Timer[ ] = [ ] ;
private _timeId : number = -1 ; // id 从 0 开始是有效 id，负数是无效 id 值
```

既然 Application 类是 Timer 的容器，需要提供在容器中添加和删除某个 Timer 的方法，先来看一下删除某个 Timer 的源码。具体代码如下：

```
// 根据 id 在 timers 列表中查找
// 如果找到，则设置 timer 的 enabled 为 false，并返回 true
// 如没找到，返回 false
public removeTimer ( id : number ) : boolean {
    let found : boolean = false ;
```

```
        for ( let i = 0 ; i < this . timers . length ; i ++ ) {
            if ( this . timers [ i ] . id === id ) {
                let timer : Timer = this . timers [ i ] ;
                timer . enabled = false ;
                                    // 只是 enabled 设置为 false,并没有从数组中删除掉
                found = true ;
                break ;
            }
        }
        return found ;
    }
```

需要注意的一点是,上述代码并没有真正删除 Timer,而是将要删除的 Timer 的 enabled 标记为 false,这样避免了析构 Timer 的内存,并且不会调整数组的内容,可以称之为逻辑删除。如果下次又要增加一个新的 Timer,会先查找 enabled 为 false 的 Timer,如果存在,可以重用该 Timer,这样又避免了 new 一个新的 Timer 对象。

接下来看一下 addTimer 的代码,具体源码如下:

```
// 初始化时,timers 是空列表
// 为了减少内存析构,在 removeTimer 时,并不从 timers 中删除 timer,而是设置 enabled 为 false
// 这样让内存使用量和析构达到相对平衡状态
// 每次添加一个计时器时,先查看 timers 列表中是否存在可用的 timer,有的话,返回该 timer 的 id 号
// 如果没有可用的 timer,就重新 new 一个 timer,并设置其 id 号及其他属性
public addTimer ( callback : TimerCallback , timeout : number = 1.0 , onlyOnce : boolean = false ,data : any = undefined ) : number {
    let timer : Timer
    let found : boolean = false ;
    for ( let i = 0 ; i < this . timers . length ; i ++ ) {
        let timer : Timer = this . timers [ i ] ;
        if ( timer . enabled === false ) {
            timer . callback = callback ;
            timer . callbackData = data ;
            timer . timeout = timeout ;
            timer . countdown = timeout ;
            timer . enabled = true ;
            timer . onlyOnce = onlyOnce ;
            return timer . id ;
        }
    }
    // 不存在,就 new 一个新的 Timer,并设置所有相关属性
    timer = new Timer ( callback ) ;
    timer . callbackData = data ;
    timer . timeout = timeout ;
    timer . countdown = timeout ;
    timer . enabled = true ;
    timer . id = ++ this . _timeId ;          // 由于初始化时 id 为-1,所以前++
    timer . onlyOnce = onlyOnce ;             //设置是一次回调,还是重复回调
    // 添加到 timers 列表中
    this . timers . push ( timer ) ;
```

```
        // 返回新添加的 timer 的 id 号
        return timer.id ;
}
```

addTimer 的代码注释比较详细，可以看到 Timer 容器管理，使用的策略是初始化容器为空，按需增加新的 Timer，重用原来的 Timer，做到只增不减，重复使用，以达到设计目标的第四条要求：尽量让内存使用与运行效率达到相对平衡。

3.4.3 触发多个定时任务的操作

定时器处理的关键源码封装在 Application 类的私有方法 _handleTimers 中，_handleTimers 实现细节已经注释得很详细了。具体代码如下：

```
// _handleTimers 私有方法被 Application 的 update 函数调用
// update 函数第二个参数是以秒表示的前后帧时间差
// 正符合 _handleTimers 参数要求
// 计时器依赖于 requestAnimationFrame 回调
// 如果当前 Application 没有调用 start 的话
// 则计时器不会生效
private _handleTimers ( intervalSec : number ) : void {
    // 遍历整个 timers 列表
    for ( let i = 0 ; i < this.timers.length ; i ++ ) {
        let timer : Timer = this.timers [ i ] ;
        // 如果当前 timer enabled 为 false，那么继续循环
        // 这句也是重用 Timer 对象的一个关键实现
        if ( timer.enabled === false ) {
            continue ;
        }
        // countdown 初始化时 = timeout
        // 每次调用本函数，会减少上下帧的时间间隔，也就是 update 第二个参数传来的值
        // 从而形成倒计时的效果
        timer.countdown -= intervalSec ;
        // 如果 countdown 小于 0.0，那么说明时间到了
        // 要触发回调了
        // 从这里看到，实际上 timer 并不是很精确的
        // 举个例子，假设 update 每次 0.16 秒
        // timer 设置 0.3 秒回调一次
        // 那么实际上是 ( 0.3 - 0.32 ) < 0，触发回调
        if ( timer.countdown < 0.0 ) {
            // 调用回调函数
            timer.callback ( timer.id , timer.callbackData ) ;
            // 下面的代码两个分支分别处理触发一次和重复触发的操作
            // 如果该计时器需要重复触发
            if ( timer.onlyOnce === false ) {
                // 重新将 countdown 设置为 timeout
                // 由此可见，timeout 不会更改，它规定了触发的时间间隔
                // 每次更新的是 countdown 倒计时器
                timer.countdown = timer.timeout ;       //很精妙的一个技巧
```

```
            } else {  // 如果该计时器只需要触发一次，那么就删除该计时器
               this . removeTimer ( timer . id ) ;
            }
         }
      }
}
```

这段代码实现了设定的 3 个目标：
- Application 类能够同时触发多个计时器。
- 每个计时器可以以不同帧率来重复执行任务。
- 每个计时器可以以倒计时方式执行一次任务。

  _handleTimer 私有方法可以在 Application 类的 step 方法中被调用，也可以在 update 方法中调用。这里就将其放在 step 方法中，并且顺便在 step 方法中添加计算 FPS（Frame Per Second）的源码。具体代码如下：

```
// Application 中声明私有成员变量
private _fps : number = 0 ;
// 提供一个只读函数，用于获得当前帧率
public get fps ( ) {
    return this . _fps ;
}
// 周而复始地运动
protected step ( timeStamp : number ) : void {
    // 第一次调用本函数时，设置 start 和 lastTime 为 timestamp
    if ( this . _startTime === -1 ) this . _startTime = timeStamp ;
    if ( this . _lastTime === -1 ) this . _lastTime = timeStamp ;
    //计算当前时间点与第一次调用 step 时间点的差
    let elapsedMsec = timeStamp - this . _startTime ;
    // 下面的代码和前几节的代码有更改：
    // 1.增加 FPS 计算
    // 2.增加调用_updateTimer 私方法
    //计算当前时间点与上一次调用 step 时间点的差（可以理解为两帧之间的时间差）
    // 此时 intervalSec 实际是毫秒表示
    let intervalSec = ( timeStamp - this . _lastTime ) ;
    // 第一帧的时候, intervalSec 为 0，防止 0 作为分母
    if ( intervalSec !== 0 ) {
       // 计算 fps
       this . _fps = 1000.0 / intervalSec ;
    }
    // update 使用的是以秒为单位，因此转换为秒表示
    intervalSec /= 1000.0 ;
    //记录上一次的时间戳
    this . _lastTime = timeStamp ;

     this . _handleTimers ( intervalSec ) ;
    // console . log ( " elapsedTime = " + elapsedMsec + " diffTime = " + intervalSec);
    // 先更新
    this . update ( elapsedMsec , intervalSec ) ;
    // 后渲染
```

```
      this . render ( ) ;
      // 递归调用，形成周而复始地前进
      requestAnimationFrame ( ( elapsedMsec : number ) : void => {
        this . step ( elapsedMsec ) ;
      } ) ;

      // requestAnimationFrame ( this . step . bind ( this ) ) ;
    }
```

完整地了解了整个 Timer 的实现过程，知道 Timer 定时触发任务操作都是依赖 requestAnimationFrame 方法的，这意味着如果 Application 类不调用 start 方法进入动画循环，是无法触发自定义的 Timer 定时任务回调的，这一点与 Window 对象的 setTimerout 和 setInterval 有所区别，其他的功能都类似。

3.4.4 测试 Timer 功能

来测试一下 Timer 相关功能，测试代码如下：

```
import { Application } from "./src/application";
//获取 canvas 元素，并创建 Application 对象
let canvas: HTMLCanvasElement | null = document.getElementById('canvas') as HTMLCanvasElement;
let app : Application = new Application ( canvas ) ;
// 实现一个 TimerCallback 签名的回调函数，打印当前 Timer 的 id 号，以及传给回调函数的参数 data
function timerCallback ( id : number , data : string ) : void {
    console . log ( "当前调用的 Timer 的 id : " + id + " data : " + data ) ;
}
// 3 秒钟后触发回调函数，仅回调一次
// addTimer 后返回的 id = 0
let timer0 : number = app . addTimer ( timerCallback ,3 ,true , " data 是 timeCallback 的数据 " ) ;
// 每 5 秒钟触发回调函数，回调 n 次
// addTimer 后返回的 id = 1
let timer1 : number = app . addTimer ( timerCallback , 5 ,false , " data 是 timeCallback 的数据 " ) ;
// 获取 stop Button 元素
let button : HTMLButtonElement = document . getElementById ( 'stop') as HTMLButtonElement ;
// 当单击 stop button 后
button.onclick = (evt : MouseEvent ) : void => {
    // remove 掉 timer1 计时器 ，并不实际删除 id=1 的计时器
    // 删除掉 1 号计时器，就不会有重复信息输出了
    app.removeTimer ( timer1 ) ;
    // 这时打印出来的应该是两个计时器
    console . log ( app . timers . length ) ;
    // 重用 0 号计时器，并且 10 秒后回调一次，然后删除
    let id : number = app.addTimer(timerCallback , 10 ,true , " data 是 timeCallback 的数据 ");
    // 返回的应该是 0 号计时器，因为重用了现有的计时器
```

```
            console . log ( id === 0 ) ;
            // 10 秒后打印出 0 号计时器的相关信息,仅打印 1 次

}
// 一开始就启动动画循环
app . start ( ) ;
```

上述源码注释得非常详细了,主要是测试增加、删除,以及一次触发或多次触发等主要的功能。来看一下代码运行结果,如图 3.17 所示。

图 3.17 Timer 测试效果图

当 Application 启动动画循环后,console 控制台输出 4 条消息,这些输出符合测试要求:
第 1 条是 3 秒后,id 为 0 的计时器触发回调函数,该计时器是只运行一次的计时器。
第 2 条是 5 秒后,id 为 1 的计时器触发回调函数,该计时器是 n 次触发计时器。
当按下 stop 按钮后,关闭(removeTimer)了 id 为 1 的计时器,此时该计时器触发过 4 次,以后不再触发。
当按下 stop 按钮后,关闭(removeTimer)了 id 为 1 计时器后,Application 中的 timers 列表仍旧有两个 timer,所以输出为 2。实际上,此时 timers 中的计时器 enabled 都为 false。
当继续 addTimer 一个 10 秒后触发一次的计时器时,它会重用 id 为 0 的计时器,因此输出 id === 0 为 true。
最后,10 秒后,0 号计时器被调用一次后不再触发回调。

3.5 本章总结

通过本章可以知道,如果想使用 Canvas2D 或 WebGL 来绘制动态场景,必须要有一

个不停刷新的机制,让要显示的物体做状态更新,以及重新绘制。幸运的是,HTML 5 提供了强大的 requestAnimationFrame 方法,能够实现帧率稳定的刷新,在 3.1 节中重点讲解了一下该函数的用法,以及与屏幕刷新频率之间的隐秘关系。

我们发现,requestAnimationFrame 不间断的刷新方式是如此地通用,而且刷新的流程也可以预先设计好,这不就是设计模式中的模板方法模式吗?因此提供 Application 类的实现,该类使用模板方法设计模式将固定的流程封装起来,将可变的操作以虚函数的形式提供给第三方,这就是 3.2 节中讲解的内容。

实际上,很多游戏引擎或类库的入口类大部分都命名为 Application,也遵循该规则。Application 类不仅规定了刷新流程,而且还有一个很重要的功能是与用户的交互,因此在 3.2 节中还封装了一个抽象的事件分发和处理系统,并且提供了 Canvas2DApplication 和 WebGLApplication 两个子类。

在 3.3 节中测试 Application 类,并且修正了测试中发现的一些问题,此外更是花了很大篇幅来解释当使用鼠标事件时,CSS 盒模型对坐标转换的影响及其解决方案。

使用 requestAnimationFrame 方法可以提供稳定的刷新操作,而且很多时候它足够快,能保持每秒 60 帧的刷新频率。但是有时候希望以不同的时间频率来操作某些任务,这些任务有的要不停地重复,有的却只需要到达指定时间后只运行一次,因此在 3.4 节中实现了 Timer 类与 Application 类的配合,用于处理上述需求。

以上就是本章的主要内容。

第 4 章 使用 Canvas2D 绘图

第 3 章中我们搭建了一个 Application 框架，并且从其继承得到了 Canvas2DApplicaiton 类。本章将使用 Canvas2DApplication 类来演示与 Canvas2D 相关的内容。

Canvas2D 是一个 2D 渲染 API 集，从可渲染的对象来分类，可以将 Canvas2D 分成 4 类绘制操作，即基本几何图形的绘制、文字的绘制、图像和视频（本书不涉及）的绘制、阴影绘制。除了绘制操作外，Canvas2D 还支持如裁剪、碰撞检测及空间变换等功能。本章主要讲解 Canvas2D 渲染绘制的相关内容。

4.1 绘制基本几何体

先来看一下如何使用 Canvas2DApplication 框架来绘制一个 rect 形状，从而整体地了解一下 Canvas2D 的基本使用步骤。

首先需要知道的是，基于 Canvas2D 的绘图并不是直接在 canvas 标签（或元素）所创建的绘图表面上进行绘制的，而是依赖于 canvas 标签（或元素）所提供的渲染上下文（CanvasRenderingContext2D）对象，该对象包含了所有渲染所需要的属性和绘制命令。

关于如何从 canvas 元素中获取 CanvasRenderingContext2D 对象，在第 3 章中的 Canvas2D Application 类的实现过程中有代码演示。我们已经搭建好了一切所需的绘图环境，本节将介绍如何在 Canvas2DApplication 中使用 CanvasRenderingContext2D 上下文对象。

4.1.1 Canvas2DApplication 的绘制流程

为了演示 Demo，需要从 Canvas2DApplication 类继承，实现一个名为 TestApplication 的子类，并覆写（override）基类的 render 方法。具体代码如下：

```
class TestApplication extends Canvas2DApplication {
    public render ( ) : void {
        //由于canvas.getContext方法返回的CanvasRenderingContext2D可能会是null
        //因此 VS Code 会强制要求 null 值检查，否则报错（tsconfig.json 中设置为类
        型严格检查模式）
        if ( this . context2D !== null )
        {
            // 可以通过CanvasRenderingContext2D获取Canvas元素,然后获取Canvas
```

的尺寸大小
```
        // 流程 1, 总是在渲染其他形体前, 先调用 clearRect 清屏
        this . context2D .clearRect( 0 , 0 , this . context2D . canvas .
        width , this . context2D . canvas .height ) ;
        // 下面是具体的绘制代码
        ..........;
        ..........;
    }
  }
}
```

可以看到，作为 Canvas2D 渲染流程的第一步，通常总是先调用 clearRect 方法来清除渲染上下文（CanvasRenderingContext2D），然后再进行具体的绘图工作。对于渲染阶段的操作来说，所有的代码都会放在 clearRect 方法后面。

4.1.2 绘制矩形 Demo

下面来看一段代码，了解一下如何使用 CanvasRenderingContext2D 对象绘制一个矩形。

```
private _drawRect( x : number, y : number, w : number, h : number ) : void {
    if ( this . context2D !== null ) {
        // 流程 2, 每次要绘制时总是使用 save / restore 对
        this . context2D . save ( ) ;
        // 流程 3, 在渲染状态 save 后, 设置当前的渲染状态
        this . context2D . fillStyle = "grey" ;
        this . context2D . strokeStyle = 'blue' ;
        this . context2D . lineWidth = 20 ;
        // 流程 4, 使用 beginPath 产生一个子路径
        this . context2D . beginPath ( ) ;
        // 左手系, 左手顺时针方向定义顶点坐标
        // 流程 5, 在子路径中添加向量点
        this . context2D . moveTo ( x , y ) ;                //moveTo 到左上角
        this . context2D . lineTo ( x + w , y ) ;            //绘制左上到右上线段
        this . context2D . lineTo ( x + w , y + h ) ;        //绘制右上到右下线段
        this . context2D . lineTo ( x , y + h ) ;            //绘制右下到左下线段
        // 流程 6, 如果是封闭形体, 调用 closePath 方法
        // 要绘制封闭的轮廓边, 因此调用 closePath 后会自动在起始点和结束点之间产生一
        条连线
        this . context2D . closePath ( ) ;
        // 流程 7, 如果是填充, 请使用 fill 方法
        this . context2D . fill ( ) ;
        // 流程 8, 如果是描边, 请使用 stroke 方法
        this . context2D . stroke ( ) ;
        //恢复渲染状态
        this . context2D . restore ( ) ;
    }
}
```

当在 4.1.1 节所示的 clearRect 后面调用本函数：
```
this._drawRect(10, 10, this.canvas.width - 20, this.canvas.height - 20);
```
就会获得如图 4.1 所示的矩形。

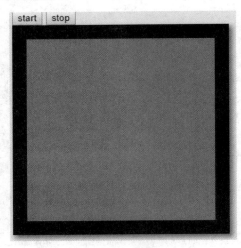

图 4.1　绘制矩形

接下来，通过例子来详细地了解 Canvas2D 相关知识。

4.1.3　模拟 Canvas2D 中渲染状态堆栈

在 4.1.2 节中的代码中，首先设置了 CanvasRenderingContext2D 渲染上下文的渲染状态，然后生成显示使用的路径对象，最后调用 stroke 进行描边、fill 进行填充，这是很经典的渲染状态机模式。

在渲染状态机模式中，一般都使用状态堆栈来实现渲染状态的管理。笔者认为，渲染状态的管理是一个非常重要的操作，为了更好地理解，下面用笔者自己实现的源码来模拟并演示渲染状态堆栈的原理和相关操作。

首先需要一个渲染状态类，用来存储要修改的各个状态，就用上一节中的 lineWidth、strokeStyle 和 fillStyle 这 3 个渲染属性作为例子来演示一下，具体源码如下：

```
class RenderState {
    public lineWidth : number = 1;              //默认情况下，lineWidth 为 1
    public strokeStyle : string = 'red';        //默认情况下，描边状态为红色
    public fillStyle : string = 'green';        //默认情况下，填充状态为绿色
    // 克隆当前的 RenderState 并返回
    public clone() : RenderState {
        let state : RenderState = new RenderState();
        state.lineWidth = this.lineWidth;
        state.strokeStyle = this.strokeStyle;
```

```typescript
        state . fillStyle = this . fillStyle ;
        return state ;
    }
    // 调用 JOSN 的静态方法 stringify，将 this 对象序列化成 JSON 字符串
    // 实现 toString 方法，用来 debug 打印相关信息
    public toString ( ) : string {
        return JSON . stringify ( this , null , ' ' ) ;
    }
}
```

然后需要一个栈数据结构作为 RenderState 的容器，而 TypeScript 和 JavaScript 的 Array 对象支持 push 和 pop 这对经典的栈操作命令，正符合需求。具体代码如下：

```typescript
class RenderStateStack {
    // 初始化情况下，堆栈中有一个渲染状态对象，并且所有状态值都是默认值
    private _stack : RenderState [ ] = [ new RenderState ( ) ] ;

    // 关键的私有 get 辅助属性，获取堆栈栈顶的渲染状态
    private get _currentState ( ) : RenderState {
        // 栈顶就是数组的最后一个元素
        return this . _stack [ this . _stack . length - 1 ] ;
    }
    // save 其实就是克隆栈顶的元素，然后将克隆返回的元素进栈操作
    public save ( ) : void {
        this . _stack . push ( this . _currentState . clone ( ) ) ;
    }
    // restore 就是将栈顶元素丢弃，此时状态会恢复到上一个状态
    public restore ( ) : void {
        this . _stack . pop ( ) ;
    }
    // 下面几个读写属性，都是操作的栈顶元素
    public get lineWidth ( ) : number {
        return this . _currentState .lineWidth ;
    }
    public set lineWidth ( value : number ) {
        this . _currentState . lineWidth = value ;
    }
    public get strokeStyle ( ) : string {
        return this . _currentState . strokeStyle ;
    }
    public set strokeStyle ( value : string ) {
        this . _currentState . strokeStyle = value ;
    }
    public get fillStyle ( ) : string {
        return this . _currentState . strokeStyle ;
    }
    public set fillStyle ( value : string ) {
        this . _currentState . strokeStyle = value ;
    }
    // 辅助方法，用来打印栈顶元素的状态值
    public printCurrentStateInfo ( ) : void {
        console . log ( this . _currentState . toString ( ) ) ;
    }
}
```

上述代码有以下 4 个关键点：
- 在渲染状态堆栈初始化时，在栈顶有一个默认渲染状态对象，这是一个全局状态对象，也就是说，如果不设置，就会默认使用该全局对象的相关渲染属性值。
- save 方法会克隆栈顶的元素，并将克隆后的元素压栈成为当前元素（也就是成为当前渲染状态）。
- restore 方法会将栈顶元素丢弃，这样前一个元素就成为了栈顶元素，这意味着它又恢复到上一级的渲染状态。
- 所有的读写属性（lineWidth、strokeStyle、fillStyle）都是针对栈顶元素（当前渲染状态）进行操作的。

下面来测试一下笔者自己实现的渲染状态堆栈。具体代码如下：

```
// 测试代码，首先创建一个渲染状态堆栈，此时会有一个栈顶并有个默认的渲染状态
let stack : RenderStateStack = new RenderStateStack ( ) ;
// 1.打印出默认的全局状态
stack . printCurrentStateInfo ( ) ;
// 2.克隆栈顶元素，并且将克隆的状态压栈变成当前状态
stack . save ( ) ;
    // 3.修改当前状态（栈顶元素）的值
    stack . lineWidth = 10 ;
    stack . fillStyle = 'black' ;
    // 4.此时打印出当前的状态值
    stack . printCurrentStateInfo ( ) ;
stack . restore ( ) ; // 丢弃当前状态

stack . printCurrentStateInfo ( ) ; // 5.再次打印，应该是和第 1 步一致
```

运行测试代码，会在 Chrome 的 console 控制台中打印出如图 4.2 所示内容。

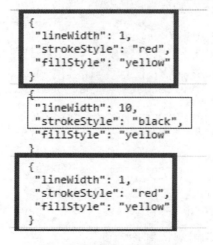

图 4.2　渲染状态堆栈测试

从图中会看到，结果是正确无误的。通过这个代码演示，我们可以了解 Canvas

RenderingContext2D 渲染上下文对象是如何管理渲染状态的。实际上，基本上所有的渲染 API（例如 D3D、OpenGL、GDI+、Skia 和 Quartz2D 等）都是使用这种模式进行渲染状态的管理，属于主流技术。

> **注意**：从上述代码中，还可以得出一个结论：save 和 restore 可以嵌套使用，但必须配对，否则堆栈会乱，出现各种预想不到的问题，切记！

4.1.4 线段属性与描边操作（stroke）

从 4.1.2 节的源码，以及运行效果图可以知道，Canvas2D 中的图形绘制支持 stroke（描边）操作和 fill（填充）操作。

可以将矢量图形分为两个部分：轮廓边和由轮廓边围成的内部填充区域。

对于轮廓边的绘制，在 Canvas2D 中使用 stroke（描边）方法，而对于由轮廓边围成的内部填充区域，则使用 fill（填充）方法进行绘制。

大多数 2D 矢量渲染 API 都是基于路径对象的，Canvas2D 也一样。需要先定义好一个路径，设置好各种渲染属性，如果需要的话，可以对该路径的轮廓部分进行描边操作（stroke）或（和）对路径内部进行填充操作（fill）。

实际上，轮廓边及内部区域并不一定需要是封闭状态，Canvas2D 能处理各种复杂的绘制要求。

下面来看一下影响描边操作的一些渲染属性。在 4.1.2 节的例子中，使用 lineWidth 为 20 像素来处理描边操作时的线宽。除了 lineWidth 属性外，还有 lineJoin、lineCap 及 miterLimit 这 3 个属性可以影响到描边操作的外观显示。可以在 TestApplication 中使用如下代码来打印出线段相关属性的默认值（在调用 printLineStates 方法前没设置过任何渲染属性）：

```
public printLineStates ( ) : void {
   if ( this . context2D !== null ) {
      console . log ( " *********LineState********** " ) ;
      console . log ( " lineWidth : " + this . context2D . lineWidth ) ;
      console . log ( " lineCap : " + this . context2D . lineCap ) ;
      console . log ( " lineJoin : " + this . context2D . lineJoin ) ;
      console . log ( " miterLimit : " + this . context2D . miterLimit ) ;
   }
}
```

调用该函数后，会得到如图 4.3 所示的内容。

下面来看一下 lineCap 这个属性，该属性用来告诉 CanvasRenderingContext2D 渲染上下文对象如何渲染一条线段的两个端点。

lineCap 属性有 3 个可选的 string 类型属性值，分别是：butt（默认值）、round 和 square。

```
*********LineState**********
lineWidth : 1
lineCap : butt
lineJoin : miter
miterLimit : 10
```

图 4.3 线段相关属性默认值

在 4.1.2 节矩形绘制 Demo 中修改一下，来看运行效果。首先要明确的一点是，如果要看到 lineCap 的显示效果，请找到 4.1.2 节中的私有方法 _drawRect，并将如下代码注释掉：

```
// 为了演示 lineCap, 需要将 closePath 给注释掉
// this.context2D.closePath();
```

然后设置 lineCap 的值分别为 butt、round 和 square，效果如图 4.4 所示，我们会看到明显的区别。

图 4.4　lineCap 中 butt、round 和 square 效果对比图

通过上述例子演示 lineCap 属性后，可以得到如下几个结论：
- 如果是封闭线段（调用 closePath 方法），lineCap 看不到任何效果。
- 如果 lineWidth 的值太小，例如默认值为 1 像素时，实际上是无任何效果的。
- 如果 lineCap 的值为 butt，一条线段按原来的长度绘制，两个端点没有任何特殊效果。
- 如果 lineCap 的值为 round，则会在一条线段的两个端点处绘制一个半圆，其半径为 lineWidth 的一半。
- 如果 lineCap 的值为 square，则会在一条线段的两个端点处绘制一个矩形，该矩形的长度与 lineWidth 一致，宽度为 lineWidth 的一半。
- 和 butt 相比，round 和 square 会增加线段的长度。
- 通过上述例子还可以发现，Canvas2D 中图形的绘制顺序是左手顺时针方向的。这里的 4 个点顺序是：左上，右上，右下，左下，明显的左手系，顺时针绘制顺序。
- 最后要明确的一点是，fill（填充）允许非封闭图形自行封闭填满，而描边则不是。

接下来看一下 lineJoin 属性，该属性的作用是如何绘制两条线段的连接处。这个属性的值也是字符串类型，有 3 个选项：miter（默认）、round 和 bevel。

来看一下分别设置这三个选项所产生的不同效果图，如图 4.5 所示。

图 4.5　lineJoin 中 miter、round 和 bevel 效果对比图

通过上述例子演示 lineJoin 属性后，可以得到如下几个结果：
- 如果 lineWidth 的值太小的话，例如默认值为 1 像素时，实际上是无任何效果的。
- 当 lineJoin 为 miter（默认值）时，在两条线段连接处（右上角和右下角）绘制一个四边形填充。
- 当 lineJoin 为 round 时，在两条线段连接处（右上角和右下角）绘制填充好的圆弧。
- 当 lineJoin 为 bevel 时，在两条线段连接处（右上角和右下角）绘制填充好的三角形。
- lineJoin 的实现算法比较复杂，有兴趣的读者可以查阅例如 WebKit（Safari、Chrome）或 Gecko（FireFox）浏览器内核实现。

最后是关于 miterLimit 属性，该属性只有在 lineJoin 属性的值为 miter 时才有意义。miterLimit 属性使用并不广泛，而且实现原理和细节比较复杂，若读者有兴趣了解，建议查阅例如 WebKit（Safari、Chrome）或 Gecko（FireFox）浏览器内核实现。

事实上，在实际的线段绘制过程中，lineWidth 属性是线段渲染最常用的属性，其他三个属性使用机会并不多。

4.1.5 虚线绘制（交替绘制线段）

Canvas2D 中，可以通过 CanvasRenderingContext2D 上下文渲染对象的 setLineDash 方法及 lineDashOffset 属性来实现虚线（交替线段）的绘制。这里还是使用 4.1.2 节的 _drawRect 方法，代码如下：

```
this.context2D.lineWidth = 20;                // 线宽为 20 像素
this.context2D.setLineDash([30, 15]);
                                              // setLineDash 的参数是一个 number 类型的数组
```

运行上述代码后，会得到如图 4.6 左图所示的结果。

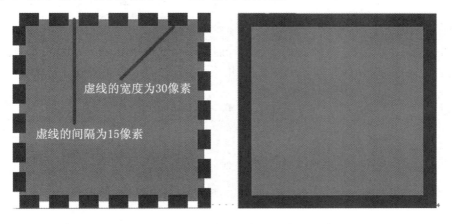

图 4.6　setLineDash 方法绘制与正常线段绘制对比

在上述代码中，并没有使用 lineDashOffset 属性来设置虚线的起始位置偏移量。默认情况下，lineDashOffset 属性的值为 0 像素的偏移。下面来看一下如何通过 lineDashOffset 并结合第 3 章中的 Timer 定时器实现一个虚线转动动画效果。首先实现一个更新 lineDashOffset 的私有成员函数。具体代码如下：

```
// 声明_lineDashOffset 成员变量，初始化为 0
private _lineDashOffset : number = 0 ;

// 实现一个更新 lineDashOffset 的函数
private _updateLineDashOffset ( ) : void {
    //每次计时器回调时，更新 1 像素偏移量
    this . _lineDashOffset ++ ;

    // 如果偏移量操作 10000 像素(可以随意改)，就再从 0 像素偏移量开始
    if ( this . _lineDashOffset > 10000 ) {
        this . _lineDashOffset = 0 ;
    }
}
```

接下来渲染函数使用 4.1.2 节中的_drawRect 私有方法，增加代码如下：

```
// 设置 lineWidth 为 2 像素比较好看
this . context2D . lineWidth = 2 ;
// 长度为 10 像素，中间间隔为 5 像素
this . context2D . setLineDash ( [ 10 , 5 ] ) ;
// 每次重绘修改 lineDashOffset 的偏移值，从而形成动画
this . context2D . lineDashOffset = this . _lineDashOffset ;
```

然后，要使用上一章中自己定义的计时器，就必须要实现一个计时器回调方法，来看一下如下所示的源码：

```
// TimeCallback 函数原型签名为 ( id : number , data : any ) => void
public timeCallback ( id : number , data : any ) : void {
    this . _updateLineDashOffset ( ) ;
    this . _drawRect ( 10 , 10 , this . canvas . width - 20 , this .canvas . height - 20 ) ;
}
```

万事俱备，可以通过两种方式来启动计时器（二选一），第一种是覆写（override）构造函数的方式来启动计时器。具体代码如下：

```
// 覆写（override）基类的构造方法
public constructor ( canvsa : HTMLCanvasElement) {
    // 构造函数中调用 super 方法
    super( canvas ) ;

    // 添加计时器，以每秒 30 帧的速度运行
    // 使用 bind 方法绑定回调函数
    this . addTimer ( this . timeCallback . bind ( this ) , 0.033 ) ;
}
```

或者使用另外一种方法，即覆写（override）基类 start 公开方法。具体代码如下：

```
// 覆写（override）基类的公开方法 start
public start ( ) : void {
```

```
    // 添加计时器，使用箭头方法进行 this 绑定
    // 以每秒 20 帧的速度运行
    this . addTimer ( ( id : number , data : any ) : void => {
        this . timeCallback ( id , data ) ;
    } , 0.05 ) ;

    // 很重要，需要通过 super 属性（super 方法只能在构造函数中调用）调用基类同名方法
    super . start ( ) ;
}
```

最后，需要使用如下代码来启动定时器：

```
//获取 canvas 元素
let canvas : HTMLCanvasElement = document . getElementById ( 'canvas' ) as
HTMLCanvasElement ;
// app 类型是基类 Application 对象
// new 的是 TestApplication 类
let app : Application = new TestApplication ( canvas ) ;
// 虚函数多态调用
app . start ( ) ;
```

当按 F5 键调试运行上述代码后，会发现虚线在不停地运动，如图 4.7 所示。

图 4.7 逆时针运动的虚线动画

关于上述代码，需要注意如下几点：
- 可以覆写（override）构造函数或 start 公开方法。
- 如果不在被覆写（override）的构造函数或 start 方法中使用 bind 或箭头函数来绑定 this 指针，程序就会崩溃。因为不使用 bind 或箭头函数，此时 timeCallback 中的 this 指针指向的是 Timer 对象而不是 TestApplication 对象，要避免这种错误。

注意：作为一个常识，在 TypeScript 或 JavaScript 中，回调函数总是使用 bind 方法或箭头函数方式，切记！

- 在上述代码中，_drawRect 方法中的 this . context2D . lineDashOffset = this . _lineDashOffset 是逆时针运行，如果修改成 - this . _lineDashOffset （负数表示）

- 后，就是顺时针运行了。
- 没有使用覆写（override）基类的 update 和 render 方法，而是使用定时器的回调方法，这样可以自由地以不同帧率来运行动画。如果是使用 update 来更新，使用 render 来渲染，那么就使用正常的刷新频率。

了解了上述内容后，在此再来总结一下 TypeScript 中 super 的用法，可以分为一个前提、两种调用方式，如下所述。

1. 一个前提

只有使用了 extends 的子类才能使用 super 关键字或 super 函数，这里 TestApplication 是 extends 了 Canvas2DApplication 类，因此在 TestApplication 中可以使用 super 指向基类 Canvas2DApplication。

2. 两种调用方式

- super 函数调用方式，在覆写构造函数（override）中使用 super（[基类构造函数参数列表]）函数方式调用基类构造函数，super 函数必须最先调用，然后才能继续在构造函数中调用其他函数或属性。
- super 关键字调用方式，例如在覆写（override）start 方法中使用的是 super 关键字来调用基类同名方法。

4.1.6 使用颜色描边和填充

当对图形进行 stroke（描边）或 fill（填充）操作时，可以使用 strokeStyle 和 fillStyle 属性来设置要呈现的颜色或图案。下面来看一下 CanvasRenderingContext2D 中 strokeStyle 和 fillStyle 相关的属性和方法的原型声明。具体代码如下：

```
interface CanvasRenderingContext2D extends CanvasPathMethods {
    // 描边和填充 style 属性原型声明
    strokeStyle : string | CanvasGradient | CanvasPattern ;
    fillStyle : string | CanvasGradient | CanvasPattern ;
    // 线性和放射性渐变色 CanvasGradient 对象的创建函数
    createLinearGradient ( x0 : number , y0 : number , x1 : number , y1 : number ) : CanvasGradient ;
    createRadialGradient ( x0 : number , y0 : number , r0 : number , x1 : number , y1 : number , r1 : number) : CanvasGradient ;
    // 创建图案 CanvasPattern 对象的方法，可以通过 HTML 的 image、canvas 和 video 元素创建图案
    createPattern ( image : HTMLImageElement | HTMLCanvasElement | HTMLVideoElement , repetition : string ) : CanvasPattern ;
}
```

根据上述代码可以看到，strokeStyle 和 fillStyle 的类型可以是字符串颜色、CanvasGradient 渐变色对象或 CanvasPattern 对象这 3 种类型，对于 CanvasGradien 渐变色对象和

CanvasPattern 对象,可通过 CanvasRenderingContext2D 上下文渲染对象的 3 个专门的方法进行创建。

先来看一下 string 类型的颜色,可以使用任意有效的 CSS 颜色字符串。关于 CSS 颜色字符串,目前一共有如下几种表达式方式(全部是以字符串表示)。

1. 预定义颜色名

根据 HTML 和 CSS 规范,一共预先定义了 147 种颜色名,其中 17 种为标准色,130 种为其他颜色。由于 17 种标准色比较常用,后续会经常用到,因此可以在 TestApplication 中定义一个静态颜色数组容纳这 17 个标准颜色名称。具体代码如下:

```
// 由于 Colors 独一无二,没有多个实例
// 可以声明为公开的静态的数组类型
public static Colors : string [ ] = [
    'aqua' ,             //浅绿色
    'black' ,            //黑色
    'blue' ,             //蓝色
    'fuchsia' ,          //紫红色
    'gray',              //灰色
    'green' ,            //绿色
    'lime' ,             //绿黄色
    'maroon' ,           //褐红色
    'navy' ,             //海军蓝
    'olive' ,            //橄榄色
    'orange' ,           //橙色
    'purple' ,           //紫色
    'red',               //红色
    'silver' ,           //银灰色
    'teal' ,             //蓝绿色
    'white' ,            //白色
    'yellow'             //黄色
] ;
```

2. rgb或rgba表示的颜色

其中 rgb 由 3 个分量(red、green、blue)组成,每个分量取值范围为[0 , 255]或[0% , 100%]。而 rgba 则由 4 个(red、green、blue 和 alpha)分量组成,其中前面 3 个分量和 rgb 表示一致,而 alpha 表示不透明度,取值范围为[0 , 1]之间,0 表示全透明,1 表示全不透明。可以按照如下方式来使用,代码如下:

```
// this . context2D . strokeStyle = 'blue' ;
// 使用每个分量为 [ 0 , 255 ]取值范围
this . context2D . strokeStyle = 'rgb( 0 , 0 , 255 ) ' ;
// 使用每个分量为[ 0% , 100% ]取值范围
// this . context2D . strokeStyle = 'rgb( 0% , 0% , 100% ) ' ;
// 不允许混合使用数字和百分比,如下代码错误
//this . context2D . strokeStyle = 'rgb( 0% , 0% , 255 ) ' ;
```

```
// 使用alpha值，实现半透明效果
// 一定要注意，alpha取值为[ 0 , 1 ]之间，否则错误
this . context2D . fillStyle = "rgba( 0 , 255 , 0 , 0.5 )" ;
```

3．使用十六进制颜色值#rrggbb

以#号开头，表示rgb的3个分量，不支持不透明度（alpha），例如蓝色的rgb为[0，0，255]，而十六进制表示为#0000FF。

4．使用hsl或hsla表示的颜色

关于该颜色坐标系，有兴趣的读者可以网上查询相关资料。

使用预定义颜色名（如black和blue等）能满足大部分需求，若需要一些特殊颜色或需要带alpha值的颜色，则可以使用rgb或rgba颜色值。

4.1.7　使用渐变对象描边和填充

除了颜色外，还可以为strokeStyle和fillStyle指定渐变色。在Canvas2D中，渐变可以分为线性渐变和放射渐变，使用CanvasGradiant对象来表示。通过一个例子来看一下如何使用线性渐变色。具体代码如下：

```
// 在TestApplication中增加一个线性渐变对象
private _linearGradient ! : CanvasGradient ;
public  fillLinearRect ( x : number, y : number, w : number, h : number ) : void {
    if ( this . context2D !== null ) {
        // 每次要绘制时总是使用save / restore对
        this . context2D . save() ;
        // 如果当前的线性渐变对象为undefined，就生成一个
        // 使用延迟创建方式，只有在用到时才创建一次
        if ( this . _linearGradient === undefined ) {
            // 使用createLinearGradient
            // 创建从左到右水平线性渐变，也可以将参数改为( x + w , y , x , y )
            创建从右到左线性渐变
            this . _linearGradient = this . context2D . createLinearGradient 
                ( x , y, x + w , y ) ;
            // 下面的代码创建从上到下线性渐变，也可以将参数改为( x , y + h , x  , y )
            创建从下到上线性渐变
            // this . _linearGradient = this . context2D . createLinear
            Gradient ( x , y , x , y + h ) ;
            // 下面的代码创建从左上角到右下角的线性渐变
            // this . _linearGradient = this . context2D . createLinear
            Gradient ( x , y , x + w , y + h ) ;
            // 下面的代码创建从右下角到左上角的线性渐变
            // this . _linearGradient = this . context2D . createLinear
            Gradient ( x + w , y + h , x , y) ;
```

```
            // 可以使用 CanvasGradiant 对象的 addColorStop 添加 5 个 color stop
            // 第一个参数是[ 0 , 1 ]的浮点数,可以将其当成线性颜色线所占的百分比
            // 第二个参数是一个字符串颜色值,下面使用了上一节中讲到的 CSS 颜色表示法
            this . _linearGradient . addColorStop ( 0.0 , 'grey' ) ;
            this . _linearGradient . addColorStop ( 0.25 , 'rgba( 255 , 0 ,
            0 , 1 ) ' ) ;
            this . _linearGradient . addColorStop ( 0.5 , 'green' ) ;
            this . _linearGradient . addColorStop ( 0.75 , '#0000FF' ) ;
            this . _linearGradient . addColorStop ( 1.0 , 'black' ) ;
        }
        // 设置线性渐变对象
        this . context2D . fillStyle = this . _linearGradient ;
        // 这里使用 rect 方法直接绘制一个封闭的矩形对象
        // 前面代码使用 moveTo 和 lineTo 来实现矩形对象的绘制
        this . context2D . beginPath ( ) ;
        this . context2D . rect ( x , y , w , h ) ;
        this . context2D . fill ( ) ; // 这里只填充,不描边
        //恢复渲染状态
        this . context2D . restore ( ) ;
    }
}
```

在上面的代码中,设计了 4 种不同方向的线性渐变(从左到右线性渐变、从上到下线性渐变、从左上角到右下角线性渐变,以及从右下角到左上角线性渐变),调用该函数后,生成如图 4.8 所示的效果。

图 4.8 各方向线性渐变

关于线性渐变的用法,在上述代码注释中有详细描述,接下来看一下放射渐变的使用。还是以代码先行的方式来介绍。具体代码如下:

```
// 在 TestApplication 中增加一个放射渐变对象
private _radialGradient ! : CanvasGradient ;
public  fillRadialRect ( x : number, y : number, w : number, h : number ) : void {
    if ( this . context2D !== null ) {
        // 每次要绘制时总是使用 save / restore 对
        this . context2D . save() ;
        // 如果当前的放射渐变对象为 undefined,就生成一个
        // 使用延迟创建方式,只有在用到时才创建一次
        if ( this . _radialGradient === undefined ) {
            // 计算矩形的中心点坐标
            let centX : number = x + w * 0.5 ;
            let centY : number = y + h * 0.5 ;
            // 矩形可能不是正方形,此时我们选择矩形的长度和宽度中最小的值作为直径
```

```
        let radius : number = Math . min ( w , h ) ;
        // 计算半径
        radius *= 0.5 ;
        // 放射渐变由内圆和外圆来定义
        // 可以调整一下内圆的半径来看一下效果
        // 提供内圆圈 radius * 0.1 / 0.3 / 0.8 对比图
        this . _radialGradient = this . context2D . createRadialGradient
        ( centX , centY , radius * 0.3 , centX , centY , radius ) ;

        this . _radialGradient . addColorStop ( 0.0 , 'black' ) ;
        this . _radialGradient . addColorStop ( 0.25 , 'rgba( 255 , 0 ,
        0 , 1 ) ' ) ;
        this . _radialGradient . addColorStop ( 0.5 , 'green' ) ;
        this . _radialGradient . addColorStop ( 0.75 , '#0000FF' ) ;
        this . _radialGradient . addColorStop ( 1.0 , 'white' ) ;
    }
    // 如果当前的线性渐变对象为 undefined，就生成一个
    // 使用延迟创建方式，只有在用到时才创建一次

    // 设置放射渐变对象
    this . context2D . fillStyle = this . _radialGradient ;
    // 这里使用 fillRect 方法直接绘制一个封闭的矩形对象
    // 前面代码使用 moveTo 和 lineTo 以及 rect 方法来实现矩形对象的绘制
    this . context2D . fillRect ( x , y , w , h ) ;
    //恢复渲染状态
    this . context2D . restore ( ) ;
    }
}
```

设置不同的内圆半径，代码如下：

```
// 使用 radius * 0.1
this . _radialGradient = this . context2D . createRadialGradient ( centX ,
centY , radius * 0.1 , centX , centY , radius ) ;

// 使用 radius * 0.3
this . _radialGradient = this . context2D . createRadialGradient ( centX ,
centY , radius * 0.3 , centX , centY , radius ) ;

// 使用 radius * 0.8
this . _radialGradient = this . context2D . createRadialGradient ( centX ,
centY , radius * 0.8 , centX , centY , radius ) ;
```

运行代码后来看一下对放射渐变的影响，结果如图 4.9 所示。

图 4.9 内圆不同半径放射渐变效果

从图中会看到，放射渐变发生在内圆和外圆之间的区域。

4.1.8 使用图案对象描边和填充

接下来看一下最后一个可以用于 stroke（描边）和 fill（填充）的对象：CanvasPattern。该对象由 CanvasRenderingContext2D 上下文渲染对象的 createPattern 函数创建，其中第一个参数是 HTMLImageElement、HTMLCanvasElement 或者是 HTMLVideoElement 类型。第二个参数是个字符串，表示当前要填充的图形尺寸在大于 pattern 图像尺寸时该如何处理，可选值可以使用 type 关键字来重新定义。具体代码如下：

```
type Repetition = "repeat" | "repeat-x" | "repeat-y" | "no-repeat" ;
```

使用 type 重新定义的好处是可以利用 VS Code 编辑器的智能感知功能，根据提示选取上面四个值之一，这样既可以少输入字符，又能防止字符串拼写错误，好处多多。

下面来看一下如何使用编程方式从服务器加载图像，然后生成 CanvasPattern 对象，最后绘制到一个矩形上。具体代码如下：

```
public fillPatternRect ( x : number, y : number, w : number, h : number ,
repeat : Repetition = "repeat" ) : void {
  if ( this . context2D !== null ) {
    if ( this . _pattern === undefined ) {
      // 注意，createElement 中 image 类型使用'img'拼写，不能写错
      let img : HTMLImageElement = document . createElement ( 'img' )
      as HTMLImageElement ;
      // 设置要载入的图片 URL 相对路径
      img . src = './data/test.jpg' ;
      // 使用箭头函数后，this 指向 TestApplication 类
      img . onload = ( ev : Event ) : void => {
        if ( this . context2D !== null ) {
          // 调用 createPattern 方法
          this . _pattern = this . context2D . createPattern ( img ,
          repeat ) ;
          // 会看到 onload 是异步调用的,只有整个图片从服务器载入到浏览器后
          // 才会调用下面的代码
          this . context2D . save ( ) ;
          // 设置线性渐变对象
          this . context2D . fillStyle = this . _pattern ;
          // 这里使用 rect 方法直接绘制一个封闭的矩形对象
          // 前面代码使用 moveTo 和 lineTo 来实现矩形对象的绘制
          this . context2D . beginPath ( ) ;
          this . context2D . rect ( x , y , w , h ) ;
          this . context2D . fill ( ) ;         //这里只填充,不描边
            //恢复渲染状态
          this . context2D . restore ( ) ;
        }
      }
    } else {
      // 如果已经存在 pattern 后,会运行这段代码
```

```
            this . context2D . save ( ) ;
                // 设置线性渐变对象
                this . context2D . fillStyle = this . _pattern ;
                // 这里使用 rect 方法直接绘制一个封闭的矩形对象
                // 前面代码使用 moveTo 和 lineTo 来实现矩形对象的绘制
                this . context2D . beginPath ( ) ;
                this . context2D . rect ( x , y , w , h ) ;
                this . context2D . fill ( ) ;                    //这里只填充,不描边
            //恢复渲染状态
            this . context2D . restore ( ) ;
        }
    }

}
```

调用上述方法,输入不同的 Repeatition 的值,可以看到如图 4.10 和图 4.11 所示的不同效果。

图 4.10 CanvasPattern 使用 repeat 和 no-repeat 效果

图 4.11 CanvasPattern 使用 repeat-x 和 repeat-y 的效果

从图 4.11 中会发现 Canvas2D 中没有拉伸缩放(stretch)效果,也就是将一个图像的 4 个顶点与要绘制的矩形的 4 个顶点重合。此时,如果图像的尺寸和要绘制的矩形尺寸不一致,必然导致图像产生形变,从而导致长宽比发生改变(在后面小节中将实现更加通用的 repeat 和 stretch 效果)。

实际上,Canvas2D 中的 Pattern 和 WebGL 中的纹理贴图技术是类似的,只是 WebGL 中的纹理贴图自己可以精确控制纹理坐标,而 Canvas2D 中的 Pattern 则无法精确控制。

事实上，Canvas2D 中的 Pattern 可以应用到任何路径对象上，例如圆、多边形和线段等。但是如何进行图像与图形的映射，则是由 Canvas2D 自行决定，我们无法控制。

关于如何使用 HTMLCanvasElement 和 HTMLVideoElement，就由给大家自行查阅资料，自己动手来验证相关结果。

4.1.9 后续要用到的一些常用绘制方法

为了更好地演示后续的 Demo，在此提供如下几个常用的绘制方法。

1．点或圆的绘制

关于点的绘制，可以使用圆来代替。在 Canvas2D 中，可以使用 CanvasRenderingContext2D 上下文渲染对象中的 arc 方法来绘制圆圈，事实上，圆是圆弧绘制的特殊情况，具体来看一下如下代码：

```
//在坐标[ x , y ]处绘制半径为 radius 的圆,可以使用指定的 style 来填充（颜色、渐变色或图案）
public  fillCircle ( x : number , y : number , radius : number , fillStyle : string | CanvasGradient | CanvasPattern = 'red' ) : void {
    if ( this . context2D !== null ) {
        //流程
        this . context2D . save ( ) ;
            this . context2D .fillStyle = fillStyle ;
            this . context2D . beginPath ( ) ;

            // 圆是圆弧的特殊表现形式[startAngle = 0 , endAngle = 2 * Math . PI ]
            this . context2D . arc ( x , y , radius , 0 , Math . PI * 2 ) ;
            //只是使用 fill,如要用 stroke,请设置 strokeStyle 属性和调用 stroke 方法
            this . context2D . fill ( ) ;
        //流程
        this . context2D . restore ( ) ;
    }
}
```

上述代码使用 arc 方法绘制点或圆。

2．线段绘制

线段的绘制也是一个基础、常用的操作，来看一下如下代码：

```
// 和以前的绘制方法相比,strokeLine 比较特别
// 没有进行状态的 save / restore 操作
// 也没有任何的修改渲染属性
// 纯粹 stroke 操作
// 这是因为这个方法被其他方法多次调用,由调用方进行状态管理和状态设置
// （参考 strokeCoord 和 strokeGrid 方法）
// 要记住本方法并没有进行状态管理和状态修改
public strokeLine ( x0 : number , y0 : number , x1 : number , y1 : number ) :
```

```
void {
    if ( this . context2D !== null ) {
        // 一定要调用 beginPath 方法
        this . context2D . beginPath ( ) ;
        this . context2D . moveTo ( x0 , y0 ) ;
        this . context2D . lineTo ( x1 , y1 ) ;
        this . context2D . stroke ( ) ;
    }
}
```

3. 坐标轴绘制

调用上面的 strokeLine 方法，绘制坐标系，红色为 x 轴，蓝色为 y 轴。具体代码如下：

```
public strokeCoord (orginX : number , orginY : number , width : number ,
height : number ) : void {
    if ( this . context2D !== null ) {
        this . context2D . save ( ) ;
        // 红色为 x 轴
        this . context2D .strokeStyle = 'red' ;
        this .strokeLine ( orginX , orginY , orginX + width , orginY ) ;
        // 蓝色为 y 轴
        this . context2D .strokeStyle = 'blue' ;
        this .strokeLine ( orginX , orginY , orginX , orginY + height ) ;
        this . context2D . restore ( ) ;
    }
}
```

4. 网格背景绘制

下面来看一下网格的绘制代码，该网格大小和 canvas 画布大小一致，并且会标记出画布的坐标系。具体代码如下：

```
// grid 绘制的区域为整个 canvas 的大小
// 其中参数 interval 控制每个网格横向和纵向的间隔大小
public strokeGrid ( color : string = 'grey' , interval : number = 10 ) :
void {
    if ( this . context2D !== null ) {
        this . context2D . save ( ) ;
        this . context2D . strokeStyle = color ;
        this . context2D . lineWidth = 0.5 ;
        // 从左到右每隔 interval 个像素画一条垂直线
        for ( let i : number = interval + 0.5 ; i < this . canvas . width ;
        i += interval ) {
            this . strokeLine ( i , 0 , i , this .canvas . height ) ;
        }

        // 从上到下每隔 interval 个像素画一条水平线
        for ( let i : number = interval + 0.5 ; i < this . canvas . height ;
        i += interval ) {
            this . strokeLine ( 0 , i , this . canvas . width , i ) ;
```

```
            this . context2D . restore ( ) ;

            // 绘制网格背景全局坐标系的原点
            this . fillCircle ( 0 , 0 , 5 , 'green' ) ;
            // 为网格背景绘制全局坐标
            // Canvas 中全局坐标系的原点在左上角,并且 x 轴总是指向右侧,y 轴指向下方
            // 全局坐标系永远不会变换,总是固定的
            this . strokeCoord ( 0 , 0 , this . canvas . width , this . canvas .
            height ) ;
        }
    }
```

关于 strokeCoord 和 strokeGrid 方法,会在后续 Demo 中经常使用,特别适合用来研究和观察 Canvas2D 中的坐标变换操作(Transformation)。

到目前为止,我们了解了关于 stroke(描边)和 fill(填充)相关的属性和操作,接下来看一下另外一个相关的绘制内容:文本的绘制。

4.2 绘制文本

在 CanvasRenderingContext2D 上下文渲染对象中,和文本绘制相关的属性和方法主要有如下几个:

```
font : string ;
textAlign : string ;
textBaseline : string ;
strokeText ( text : string , x : number , y : number , maxWidth ? : number ) :
void ;
fillText ( text : string , x : number , y : number , maxWidth ? : number ) :
void ;
measureText ( text : string ) : TextMetrics ;
// TextMetrics 接口签名
interface TextMetrics {
    readonly width : number ;
}
```

接下来看一下如何使用上述与文本相关的属性和方法。

4.2.1 封装 fillText 方法

和图形一样,文本也可以使用 stroke(描边)和(或)fill(填充)两种方式来绘制,但方法名称改为 strokeText 和 fillText,更加明确化。同样地,也可以使用例如 strokeStyle 和 fillStyle 这两个属性来控制文本使用哪种颜色、渐变色或图案来绘制。

接下来先看一段代码,了解一下文本相关的默认属性的值有哪些。具体代码如下:

```
public printTextStates ( ) : void {
    if ( this . context2D !== null ) {
```

```
        console . log ( " *********TextState********** " ) ;
        console . log ( " font : " + this . context2D . font ) ;
        console . log ( " textAlign : " + this . context2D . textAlign );
        console . log ( " textBaseline : " + this . context2D . textBaseline ) ;
    }
}
```

当运行上述代码后，会得到如图 4.12 所示的结果。

```
*********TextState**********
font : 10px sans-serif
textAlign : start
textBaseline : alphabetic
```

图 4.12 默认文本属性值

接下来看一下如何使用这些文本相关的渲染属性。先封装一个文本绘制函数，用于后续的操作使用。在封装该函数前，先使用 type 关键字来重新定义相关属性值。具体代码如下：

```
// Canvas2D 中 TextAlign 用于设置文字左右如何对齐，默认情况下是 start
type TextAlign = 'start' | 'left' | 'center' | 'right' | 'end' ;
// Canvas2D 中 TextBaseline 用于设置当前绘制文本的基线，默认情况下是 alphabetic
// 可以认为用来设置文字是如何对齐的
type TextBaseline = 'alphabetic' | 'hanging' | 'top' | 'middle' | 'bottom' ;

// 字体大小和字体类型，默认情况下是 10px sans-serif
// 设置 15px 和 20px 及 25px 大小的字体，后续代码会使用
// 利用 VS Code 智能感知功能，减少输入和拼写错误
type FontType = "10px sans-serif" | "15px sans-serif" | "20px sans-serif"
 | "25px sans-serif" ;
```

有了上述 type 定义后，下面来实现文本绘制函数 fillText。具体代码如下：

```
public  fillText(text : string , x : number , y : number , color : string =
'white' , align : TextAlign = 'left' , baseline : TextBaseline = 'top' ,
font : FontType = '10px sans-serif' ) : void {
    if ( this . context2D !== null ) {
        this . context2D . save ( ) ;                    //管理渲染属性经典模式
        this . context2D . textAlign = align ;
                                //文字左右对齐方式，类型为 TextAlign
        this . context2D . textBaseline = baseline ;
                                //文字上下对齐方式，类型为 TextBaseline
        this . context2D . font = font ;          //使用哪种字体，多少大小绘制
        this . context2D . fillStyle = color ;          //文字填充的颜色
        this . context2D . fillText ( text , x , y ) ;
                                //调用 fillText()函数，指定文字要绘制的坐标
        this . context2D . restore ( ) ;                //状态恢复
    }
}
```

上面的 fillText 函数，只是使用填充方式绘制文字，并没有使用描边，如有需要可以自行使用文字描边的相关属性和方法。

4.2.2 文本的对齐方式

下面用一个例子来演示一下 fillText 函数的用法,特别是 TextAlign 和 TextBaseline 的不同属性值对文字绘制的不同效果,具体结果可以参考图 4.13 所示。

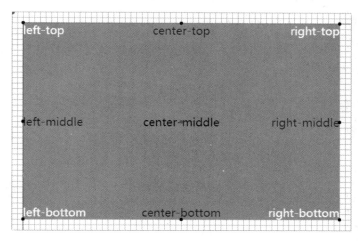

图 4.13 文本对齐演示

先来了解一下这个 Demo:
- 网格背景及坐标系由 strokeGrid 方法绘制。
- 灰色矩形及坐标系由 fillRectWithTitle 方法绘制,该方法比较特别,将在后面章节介绍相关细节。
- 9 个绘制文本显示出相对整个灰色矩形的对齐效果,例如 left-top 和 right-bottom 等,使用 fillText 方法来完成。
- 9 个圆点标记出在绘制 9 个文本时的 x 和 y 坐标,使用 fillCircle 方法。
- 本 Demo 中使用了 TextAlign 中的 left、center 和 right 属性值,用于文本左、中、右对齐。
- 本 Demo 中使用了 TextBaseline 中的 top、middle 和 bottom 属性值,用于文本上、中、下对齐。

下面来看一下实现过程,代码如下:

```
public testCanvas2DTextLayout ( ) : void {

    // 要绘制的矩形离 canvas 的 margin(外边距)分别是 [ 20 , 20 , 20 , 20 ] ;
    let x : number = 20 ;
    let y : number = 20 ;
    let width : number = this . canvas .width - x * 2 ;
    let height : number = this . canvas . height - y * 2 ;
    let drawX : number = x ;
```

```
let drawY : number = y ;
// 原点的半径为 3 像素
let radius : number = 3 ;
// 1.画背景 rect, 该函数在下面一节介绍
this . fillRectWithTitle ( x , y , width , height ) ;
// 使用 20px sans-serif 字体绘制 (默认为 10px sans-serif)
// 每个位置, 先绘制 drawX 和 drawY 的坐标原点, 然后绘制文本
// 2.左上
this . fillText ( "left - top" , drawX , drawY , 'white' , 'left' , 'top' ,
    '20px sans-serif' ) ;
this . fillCircle ( drawX , drawY , radius , 'black' ) ;
// 3.右上
drawX = x + width ;
drawY = y ;
this . fillText ( "right - top" , drawX , drawY , 'white' , 'right' ,
    'top' , '20px sans-serif' ) ;
this . fillCircle ( drawX , drawY , radius , 'black' ) ;
// 4.右下
drawX = x + width ;
drawY = y + height ;
this . fillText ( "right - bottom" , drawX , drawY , 'white' , 'right' ,
    'bottom' , '20px sans-serif' ) ;
this . fillCircle ( drawX , drawY , radius , 'black' ) ;
// 5.左下
drawX = x ;
drawY = y + height ;
this . fillText ( "left - bottom" , drawX , drawY , 'white' , 'left' ,
    'bottom' , '20px sans-serif' ) ;
this . fillCircle ( drawX , drawY , radius , 'black' ) ;
// 6.中心
drawX =  x + width * 0.5 ;
drawY =  y + height * 0.5 ;
this . fillText ( "center - middle" , drawX , drawY , 'black' , 'center' ,
    'middle' , '20px sans-serif' ) ;
this . fillCircle ( drawX , drawY , radius , 'red' ) ;
// 7.中上
drawX = x + width * 0.5 ;
drawY = y ;
this . fillText ( "center - top" , drawX , drawY , 'blue' , 'center' ,
    'top' , '20px sans-serif' ) ;
this . fillCircle ( drawX , drawY , radius , 'black' ) ;
// 8.右中
drawX = x + width ;
drawY = y + height * 0.5 ;
this . fillText ( "right - middle" , drawX , drawY , 'blue' , 'right' ,
    'middle' , '20px sans-serif' ) ;
this . fillCircle ( drawX , drawY , radius , 'black' ) ;
// 9.中下
drawX = x + width * 0.5 ;
drawY = y + height ;
this . fillText ( "center - bottom" , drawX , drawY , 'blue' , 'center' ,
    'bottom' , '20px sans-serif' ) ;
this . fillCircle ( drawX , drawY , radius , 'black' ) ;
```

```
        // 10.左中
        drawX = x ;
        drawY = y + height * 0.5 ;
        this . fillText ( "left - middle" , drawX , drawY , 'blue' , 'left' ,
        'middle' , '20px sans-serif' ) ;
        this . fillCircle ( drawX , drawY , radius , 'black' ) ;
    }
```

4.2.3 自行实现文本对齐效果

本节来增加一下难度，假设 Canvas2D 中只支持文本基于左上角原点定位的绘制效果，那么在这种情况下该如何计算出其他对齐方式的坐标位置，从而实现类似如图 4.13 所示的效果？

为了更好地理解，仍旧先来看看效果，如图 4.14 所示。

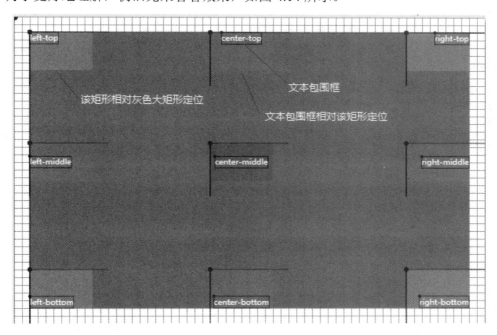

图 4.14　自己实现文本对齐效果图

来了解一下这个 Demo：
- 网格背景及坐标系。
- 灰色的大矩形，相对于网格背景定位，用于容纳子矩形。
- 9 个带有坐标系的子矩形，这些子矩形相对灰色的大矩形定位。
- 9 个包围文本的矩形，这 9 个文本包围框又相对于各自所对应的子矩形定位。
- 要实现上述的需求，实际上关键需解决如下两个问题：

- 如何计算要绘制的文本的尺寸（宽度和高度）？
- 已知两个嵌套的矩形 A 和 B，其中矩形 A 可以完全容纳矩形 B，如何计算矩形 B 相对矩形 A 的 9 种局部偏移量？

4.2.4 计算文本高度算法

关于如何计算文本的尺寸，CanvasRenderingContext2D 渲染上下文对象内置了一个名为 mesaureText 的方法，该方法具有一个字符串类型的参数，表示要计算尺寸的文本内容。调用该方法后返回一个 TextMetrics 接口，该接口所包含的 width 属性表示传入文本的像素宽度。

令人遗憾的是，TextMetrics 只有 width 属性，没有 height 属性，但是可以使用变通的方式来计算 height 值（行高）：

- 计算某一个字符的宽度。
- 在该宽度上增减一定比例的像素。

上述算法精度不高，但是的确行得通。下面来看 showTextRectangle 方法，该方法使用上述算法计算要绘制文本的宽度和行高。具体代码如下：

```
// 笔者测试大小写 26 个英文字母后（10px sans-serif 默认字体）
// 决定使用大写 W 的宽度加上 scale 为 0.5 作为行高计算的要点（默认参数）
// 其他字体或字体尺寸请自行做实验
public calcTextSize ( text : string , char : string = 'W' , scale : number = 0.5 ) : Size {
  if ( this . context2D !== null ) {
    let size : Size = new Size ( ) ;
    size . width = this . context2D .measureText ( text ) . width ;
    let w : number = this . context2D . measureText ( char ) . width ;
    size . height =  w + w * scale ; // 宽度上加 scale 比例
    return size ;
  }
  // 直接报错
  alert ( " context2D 渲染上下文为 null " ) ;
  throw new Error ( " context2D 渲染上下文为 null " ) ;
}
```

笔者测试大小写 26 个英文字母后（10px sans-serif 默认字体），决定使用大写 W 的宽度加上 scale 为 0.5，效果如图 4.15 所示。

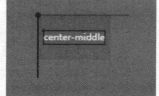

图 4.15　calcTextSize 居中对齐绘制效果

4.2.5 嵌套矩形定位算法

已知两个嵌套的矩形 A 和 B，其中矩形 A 可以完全容纳矩形 B，如何计算矩形 B 相对矩形 A 的 9 种局部偏移量？

以图 4.15 为图示，矩形 A 是带坐标的矩形，矩形 B 则是文本框矩形，矩形 A 和 B 的原点都在左上角，现在要计算矩形 B 在矩形 A 中居中对齐：

```
( A . width - B . width ) * 0.5 ;
( A . height - B . height ) * 0.5 ;
```

上述代码就计算出了矩形 B 在矩形 A 中居中对齐时左上角的坐标位置（图 4.15 中文本包围框左上角的原点）。其他 8 种坐标计算方式类似，下面来看其代码实现。

首先，为了方便，使用 TypeScript 的 enum 关键字来定义一个枚举 ETextLayout，这些枚举值如下代码所示。

```
export enum ETextLayout {
    LEFT_TOP ,
    RIGHT_TOP ,
    RIGHT_BOTTOM ,
    LEFT_BOTTOM ,
    CENTER_MIDDLE ,
    CENTER_TOP ,
    RIGHT_MIDDLE ,
    CENTER_BOTTOM ,
    LEFT_MIDDLE
}
```

然后定义 3 个数学数据结构，将这些数据结构放在 math2D.ts 文件中。这些数据结构在下一章及后续章节中会经常使用。具体代码如下：

```
// 二维向量
export class vec2 {
    public values : Float32Array ; // 使用 float32Array
    public constructor ( x : number = 0 , y : number = 0 ) {
        this . values = new Float32Array ( [ x , y ] ) ;
    }
    public toString ( ) : string {
        return " [ " + this . values [ 0 ] + " , " + this . values [ 1 ] + " ] " ;
    }
    public get x ( ) : number { return this . values [ 0 ] ; }
    public set x ( x : number ) { this . values [ 0 ] = x ; }
    public get y () : number { return this . values[ 1 ] ; }
    public set y ( y : number ) { this . values[ 1 ] = y ; }

    // 静态 create 方法
    public static create ( x : number = 0 , y : number = 0 ) : vec2 {
        return new vec2 ( x , y ) ;
    }
}
// 2D 尺寸
export class Size {
    public values : Float32Array ; // 使用 float32Array
    public constructor ( w : number = 1 , h : number = 1 ) {
        this . values = new Float32Array ( [ w , h ] ) ;
    }
    public set width ( value : number ) { this . values [ 0 ] = value ; }
```

```typescript
        public get width ( ) : number { return this . values [ 0 ] ; }
        public set height ( value : number ) { this . values [ 1 ] = value ; }
        public get height ( ) : number { return this . values [ 1 ] ; }

        // 静态 create 方法
        public static create ( w : number = 1 , h : number = 1 ) : Size {
            return new Size ( w , h ) ;
        }
    }
    // 矩形包围框
    export class Rectangle {
        public origin : vec2 ;
        public size : Size ;
        public constructor ( orign : vec2 = new vec2 ( ) , size : Size = new Size ( 1 , 1 ) ) {
            this . origin = orign ;
            this . size =  size ;
        }
    // 静态 create 方法
        public static create ( x : number = 0 , y : number = 0 , w : number = 1 , h : number = 1 ) : Rectangle {
            let origin : vec2 = new vec2 ( x , y ) ;
            let size : Size = new Size ( w , h ) ;
            return new Rectangle ( origin , size ) ;
        }
    }
```

有了上述这些结构，来看一下 calcLocalTextRectangle 这个方法的实现，该方法计算并返回 9 种局部偏移矩形之一。具体代码如下：

```typescript
// parentWidth / parentHeight 是父矩形的尺寸
// 函数返回类型是 Rectangle，表示 9 个文本子矩形之一
// 这些子矩形是相对父矩形坐标系的表示
// 这意味着父矩形原点为[ 0 , 0]，所以参数是父矩形的 width 和 height，而没有 x 和 y 坐标
public calcLocalTextRectangle ( layout : ETextLayout , text : string , parentWidth : number , parentHeight : number ) : Rectangle {
    // 首先计算出要绘制的文本的尺寸（width / hegiht）
    let s : Size = this . calcTextSize ( text ) ;
    // 创建一个二维向量
    let o : vec2 = vec2 . create ( ) ;
    // 计算出当前文本子矩形左上角相对父矩形空间中的 3 个关键点（左上、中心、右下）坐标
    // 1.当前文本子矩形左上角相对父矩形左上角坐标，由于局部表示，所以为[ 0 , 0 ]
    let left : number = 0 ;
    let top : number = 0 ;
    // 2.当前文本子矩形左上角相对父矩形右下角坐标
    let right : number = parentWidth - s . width ;
    let bottom : number = parentHeight - s . height ;
    // 3.当前文本子矩形左上角相对父矩形中心点坐标
    let center : number = right * 0.5 ;
    let middle : number = bottom * 0.5 ;
    // 根据 ETextLayout 的值来匹配这 3 个点的分量
    // 计算子矩形相对父矩形原点[ 0 , 0 ]偏移量
    switch ( layout ) {
```

```
        case ETextLayout . LEFT_TOP :
            o . x = left ;
            o . y = top ;
            break ;
        case ETextLayout . RIGHT_TOP :
            o.x = right ;
            o.y = top ;
            break ;
        case ETextLayout . RIGHT_BOTTOM :
            o.x = right ;
            o.y = bottom ;
            break ;
        case ETextLayout . LEFT_BOTTOM :
            o.x = left ;
            o.y = bottom ;
            break;
        case ETextLayout . CENTER_MIDDLE:
            o.x = center ;
            o.y = middle ;
            break ;
        case ETextLayout . CENTER_TOP :
            o.x = center ;
            o.y = 0;
            break;
        case ETextLayout . RIGHT_MIDDLE :
            o.x = right ;
            o.y = middle ;
            break;
        case ETextLayout . CENTER_BOTTOM:
            o.x = center ;
            o.y = bottom ;
            break ;
        case ETextLayout . LEFT_MIDDLE :
            o.x = left ;
            o.y = middle ;
            break ;
    }
    // 返回子矩形
    return new Rectangle ( o , s ) ;
}
```

4.2.6 fillRectWithTitle 方法的实现

本节实现一个名为 fillRectWithTitle 的方法,该方法调用上述的 calcLocalTextRectangle 方法,目的是实现如下要求（参考图 4.15 所示效果）：
- 带坐标的父矩形绘制。
- 文本绘制。
- 文本框绘制。
- 绘制文本框左上角原点（相对父矩形表示）。

下面来看一下实现细节。具体代码如下：

```typescript
public fillRectWithTitle ( x : number , y : number , width : number , height :
number , title : string = '' , layout : ETextLayout = ETextLayout .
CENTER_MIDDLE , color : string = 'grey' , showCoord : boolean = true ) :
void {
    if ( this . context2D !== null ) {
        this . context2D . save ( ) ;
        // 1. 绘制矩形
        this . context2D . fillStyle = color ;
        this . context2D . beginPath ( ) ;
        this . context2D . rect ( x , y , width , height ) ;
        this . context2D . fill ( ) ;
        // 如果有文字的话，先根据枚举值计算 x、y 坐标
        if ( title . length !== 0 ) {
            // 2. 绘制文字信息
            // 在矩形的左上角绘制出相关文字信息，使用的是 10px 大小的文字
            // 调用 calcLocalTextRectangle 方法
            let rect : Rectangle = this . calcLocalTextRectangle ( layout ,
            title , width , height ) ;
            // 绘制文本
            this . fillText ( title , x + rect . origin . x , y + rect .
            origin . y , 'white' , 'left' , 'top' , '10px sans-serif' ) ;
            // 绘制文本框
            this . strokeRect ( x + rect . origin . x , y + rect . origin .
            y , rect .size . width , rect . size . height , 'rgba( 0 , 0 ,
            0 , 0.5 ) ' ) ;
            // 绘制文本框左上角坐标（相对父矩形表示）
            this . fillCircle ( x + rect . origin . x , y + rect . origin .
            y , 2 ) ;
        }
        // 3. 绘制变换的局部坐标系
        // 附加一个坐标，x 轴和 y 轴比矩形的 width 和 height 多 20 个像素
        // 并且绘制 3 个像素的原点
        if ( showCoord ) {
            this . strokeCoord ( x , y , width + 20 , height + 20 ) ;
            this . fillCircle ( x , y , 3 ) ;
        }
        this . context2D . restore ( ) ;
    }
}
```

4.2.7 自行文本对齐实现 Demo

所有的代码都准备完毕，那么来组合一下，实现如图 4.14 所示的效果。具体代码如下：

```typescript
public testMyTextLayout ( ) : void {
    let x : number = 20 ;
    let y : number = 20 ;
    let width : number = this . canvas . width - x * 2 ;
    let height : number = this . canvas . height - y * 2 ;
```

```
let right : number = x + width ;
let bottom : number = y + height ;
let drawX : number = x ;
let drawY : number = y ;
let drawWidth : number = 80 ;
let drawHeight : number = 50 ;
// 1. 画背景 rect
this . fillRectWithTitle ( x , y , width , height ) ;
// 2. 左上
this . fillRectWithTitle ( drawX , drawY , drawWidth , drawHeight ,
'left - top' , ETextLayout . LEFT_TOP , 'rgba ( 255 , 255 , 0 , 0.2 )' ) ;
// 3. 右上
drawX = right - drawWidth ;
drawY = y ;
this . fillRectWithTitle ( drawX , drawY , drawWidth , drawHeight ,
'right - top' , ETextLayout . RIGHT_TOP , 'rgba ( 255 , 255 , 0 , 0.2 )' ) ;

// 4. 右下
drawX = right - drawWidth ;
drawY = bottom - drawHeight ;
this . fillRectWithTitle ( drawX , drawY , drawWidth , drawHeight ,
'right - bottom' , ETextLayout . RIGHT_BOTTOM , 'rgba ( 255 , 255 ,
0 , 0.2 )' ) ;
// 5. 左下
drawX = x ;
drawY = bottom - drawHeight ;
this . fillRectWithTitle ( drawX , drawY , drawWidth , drawHeight ,
'left - bottom' , ETextLayout . LEFT_BOTTOM , 'rgba ( 255 , 255 , 0 ,
0.2 )' ) ;
// 6. 中心
drawX = ( right - drawWidth ) * 0.5 ;
drawY = ( bottom - drawHeight ) * 0.5 ;
this . fillRectWithTitle ( drawX , drawY , drawWidth , drawHeight ,
'center - middle' , ETextLayout . CENTER_MIDDLE , 'rgba ( 255 , 0 ,
0 , 0.2 )' ) ;
// 7. 中上
drawX = ( right - drawWidth ) * 0.5 ;
drawY = y ;
this . fillRectWithTitle ( drawX , drawY , drawWidth , drawHeight ,
'center - top' , ETextLayout . CENTER_TOP , 'rgba ( 0 , 255 , 0 , 0.2 )' ) ;
// 8. 右中
drawX = ( right - drawWidth ) ;
drawY = ( bottom - drawHeight ) * 0.5 ;
this . fillRectWithTitle ( drawX , drawY , drawWidth , drawHeight ,
'right - middle' , ETextLayout . RIGHT_MIDDLE , 'rgba ( 0 , 255 , 0 ,
0.2 )' ) ;
// 9. 中下
drawX = ( right - drawWidth ) * 0.5 ;
drawY = ( bottom - drawHeight ) ;
this . fillRectWithTitle ( drawX , drawY , drawWidth , drawHeight ,
```

```
                'center - bottom' , ETextLayout . CENTER_BOTTOM , 'rgba( 0 , 255 ,
                0 , 0.2 )' ) ;
                // 10. 左中
                drawX = x ;
                drawY = ( bottom - drawHeight ) * 0.5 ;
                this . fillRectWithTitle ( drawX , drawY , drawWidth , drawHeight ,
                'left - middle' , ETextLayout . LEFT_MIDDLE , 'rgba( 0 , 255 , 0 ,
                0.2 )' ) ;
}
```

调用上述方法后，会得到如图 4.14 所示的结果。

4.2.8　font 属性

在介绍文字绘制相关章节中我们已经使用了大部分与文字绘制相关的方法和属性，本节中将介绍一个重要的属性：font。

font 属性的值是一个 CSS 格式的字型字符串，用于一次设置元素字体的两个或多个方面的属性，例如 "italic small-caps bold 18px sans-serif"，该 font 字符串具有典型性，具体分析如下：

1．italic 是 font-style 属性值之一，其取值范围如下代码所示。

```
// CSS 规范，应该支持如下 3 个选项
// normal 是默认值，其他两个值表示斜体
type FontStyle = "normal" | "italic" | "oblique" ;
```

2．small-caps 是 font-variant 属性值之一，其取值范围如下代码所示。

```
// CSS 规范，目前只有如下两个选项
type FontVariant = "normal" | "small-caps" ;
```

3．bold 是 font-weight 属性值之一，其取值范围如下代码所示。

```
// CSS 规范，应该支持 normal | bold | bolder | lighter | 100 | 200 | 300 | 400
| 500 | 600 | 700 | 800 | 900
type FontWeight = "normal" | "bold" | "bolder" | "lighter" | "100" | "200"
| "300" | "400" | "500" | "600" | "700" | "800" | "900" ;
```

4．18px 是 font-size 属性值之一，其取值范围如下代码所示。

```
// 前面 4 个 px 预先设定，后面使用名字方式表达字体大小，也可以设置其他的 px 值
// CSS 规范，可以设置像素值 | 百分比值 | 名字值
type FontSize =  "10px" | "12px" | "16px" | "18px" | "24px" | "50%" | "75%"
| "100%" | "125%" | "150%" | "xx-small" | "x-small" | "small" | "medium"
| "large" | "x-large" | "xx-large" ;
```

5．sans-serif 是 font-family 属性值之一，其取值范围如下代码所示。

```
// CSS 规定了如下 5 种通用字体系列
type FontFamily = "sans-serif" | "serif" | "courier" | "fantasy" |
"monospace" ;
```

4.2.9 实现 makeFontString 辅助方法

由于设置 font 属性时，font 字符串必须要按照 font-style font-variant font-weight font-size font-family 的顺序来设置相关的值，如果顺序不正确，会导致 font 属性不起作用。

为了避免设置错误，下面来实现一个名为 makeFontString 的辅助方法，用于合成正确的字符串。具体代码如下：

```
// 1. 本方法并没有使用本类中的任何成员变量或成员方法,因此可以声明为static方法,
当然也可以定义为实例方法
// 2. css font 属性字符串中每个属性都是有先后顺序之分的,因此编写此方法,内部会使用
正确的属性字符串合成顺序,减少错误
// 3. 按照笔者认为最常用的频度来声明参数的顺序,但是内部生成字符串时会按照正确的属性
顺序来合成
public /* static */ makeFontString ( size : FontSize = '10px' ,
                                     weight : FontWeight = 'normal' ,
                                     style : FontStyle = 'normal' ,
                                     variant : FontVariant = 'normal' ,
                                     family : FontFamily = 'sans-serif' ,
                                   ) : string
{
  let strs : string [ ] = [ ] ;
  // 第一个是 fontStyle
  strs . push ( style ) ;
  // 第二个是 fontVariant
  strs . push ( variant ) ;
  // 第三个是 fontWeight
  strs . push ( weight ) ;
  // 第四个是 fontSize
  strs . push ( size ) ;
  // 第五个是 fontFamily
  strs . push ( family ) ;
  // 最后需要将数组中的每个属性字符串以空格键合成
  // 使用 Array 对象的 join 方法,其参数是空格字符串: " "
  let ret : string = strs . join ( " " ) ;
  console . log ( ret ) ;
  return ret ;
}
```

可以修改 4.2.7 节中实现的 testMyTextLayout 方法，让其支持 font 属性的设置。至于如何修改，读者可以查阅相关代码。下面来看一下 testMyTextLayout 的方法声明。具体代码如下：

```
public testMyTextLayout ( font : string = this . makeFontString ( "18px",
"bold" , "italic" , "small-caps" , 'sans-serif' ) ) : void
```

可以看到，testMyTextLayout 方法原型中 font 参数使用了默认参数的形式，在默认参

数中调用了 this．makeFontString 方法生成"italic small-caps bold 18px sans-serif"的字符串值，我们使用这个字体属性设置来演示一下效果，具体如图 4.16 所示。

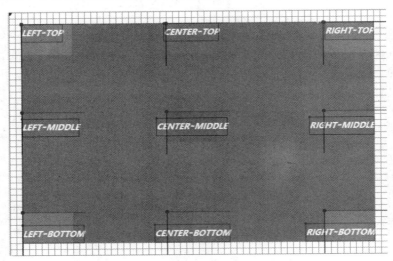

图 4.16　makeFontString 方法调用效果

通过图 4.16，会发现几个重要问题：
- font 属性设定后会影响 Canvas2D 中的 measureText 方法计算文本的宽度值，因为 measureText 方法是基于当前的 font 值来计算文本的宽度。
- 图中文本矩形框的宽度并没有完全容纳整个文字的宽度，由此可见内置方法 measureText 调用后获得的文本宽度并不是精确的值。
- 如果大家在 makeFontString 中用不同的字体（FontFamily）来测试，会发现自己实现的 calcTextSize 方法返回的高度也不是很精确，需要自行测试和调整，以获取最佳的 scale 比例值。
- 在图 4.16 中，文本宽度超过了父矩形的宽度，因此左上角坐标可能为负值，但是计算结果是正确的。可以调整 testMyTextLayout 中的 drawWidth，将原来的 80 调整为 150，父矩形就能完全容纳 9 个文本矩形框了。

至此，除了 Canvas2D 中的 strokeText 方法外，文字的绘制涉及的属性和方法都使用到了。至于 strokeText 方法，则和路径对象的 stroke 方法一样，都用于描边，有了上述相关背景知识后，就很容知道如何使用了。

4.3　绘 制 图 像

在 CanvasRenderingContext2D 上下文渲染对象中，和图像处理相关的方法和接口有如下几个：

```
// 3 个重载的 drawImage 方法，具体使用要点在下面小节中演示
drawImage ( image : HTMLImageElement | HTMLCanvasElement | HTMLVideoElement |
    ImageBitmap , dstX : number , dstY : number ) : void ;
drawImage ( image : HTMLImageElement | HTMLCanvasElement | HTMLVideoElement |
    ImageBitmap , dstX : number , dstY : number , dstW : number , dstH :
    number ) : void ;
drawImage ( image : HTMLImageElement | HTMLCanvasElement | HTMLVideoElement |
    ImageBitmap , srcX : number , srcY : number , srcW : number , srcH :
    number , dstX :       number , dstY : number , dstW : number , dstH :
    number ) : void ;
// 创建 ImageData 对象
createImageData ( imageDataOrSw : number | ImageData , sh ? : number ) :
ImageData ;
// 从 CanvasRenderingContext2D 渲染上下文对象中获取 ImageData 对象
getImageData ( sx : number , sy : number , sw : number , sh : number ) :
ImageData ;
// 设置 CanvasRenderingContext2D 渲染上下文对象中部分或全部像素数据
putImageData ( imagedata : ImageData , dx : number , dy : number , dirtyX ? :
number , dirtyY ? :     number , dirtyWidth ? : number , dirtyHeight ? :
number ) : void ;
// ImageData 接口
interface ImageData {
    readonly data : Uint8ClampedArray ;
    readonly height : number ;
    readonly width : number ;
}
```

接下来看一下如何进行图像的绘制。

4.3.1　drawImage 方法

首先来看一下 drawImage 方法中参数最少的那个重载方法：

```
drawImage ( image : HTMLImageElement | HTMLCanvasElement | HTMLVideoElement |
    ImageBitmap , dstX : number , dstY : number ) : void ;
```

会发现第一个参数 image 事实上支持 4 种类型，但是在本书中，只会使用到前面两种类型，即 HTMLImageElement 和 HTMLCanvasElement。至于视频绘制和 ImageBitmap 对象，请大家自行查询相关资料。

dstX 和 dstY 参数规定了要将 image（源图像）绘制到 Canvas 画布（目标）的哪个位置。这个重载函数最大的特点是将整个源图像保持原样地绘制到目标画布上（源图像和目标图像尺寸一致）。

下面通过实现一个名为 loadAndDrawImage 的方法，演示上述重载方法。具体代码如下：

```
// 参数是要载入的 image 的 URL 路径
public loadAndDrawImage ( url : string ) : void {
    // 调用 document 对象的 createElement 方法
    // 并且提供 tagName 为"img"的字符串生成一个 HTMLElement 对象
    // 要注意 3 点：
```

```
// 1. 必须使用 img 的 tagName，千万别拼错
// 2. ts 是强类型的语言，createEelemet 返回的是 HTMLElement 类型，所以必须要用
as 关键字向下转型成 HTMLImageElement 类型
// 3. HTMLImageElement 的 src 设置后会以异步方式载入数据，所以如果要绘制相关图
像，必须要放在 onload 事件中，否则图像不能正确显示
let img : HTMLImageElement = document . createElement ( 'img' ) as
HTMLImageElement ;
// 设置要载入的图片 URL 路径
img . src = url ;
// 使用箭头函数后，this 指向 TestApplication 类
img . onload = ( ev : Event ) : void => {
    // onload 事件表示图像载入完成
    if ( this . context2D !== null ) {
        // 在 console 控制台输出载入图像的尺寸
        console . log ( url + " 尺寸为 [ " + img . width + " , " + img .
        height + " ] " ) ;
        // 将 srcImage 以保持原样的方式绘制到 Canvas 画布上[ 10 , 10 ]的位置处
        this . context2D . drawImage ( img , 10 , 10 ) ;
    }
}
```

当调用上述方法后，如图 4.17 中的左图所示，会在 Canvas 元素的[10 , 10] 处显示一张原始大小的图片，该图片的尺寸为[256 , 192]，和服务器上原始图像具有一致性。

接下来看一下 drawImage 的第 2 个重载方法：

```
drawImage (image : HTMLImageElement | HTMLCanvasElement | HTMLVideoElement |
    ImageBitmap , dstX : number , dstY : number , dstW : number , dstH :
number ) : void ;
```

这个重载方法在上述方法的基础上增加了 dstW 和 dstH 两个参数，表示要绘制的目标区域的宽度和高度，由此可知，该方法能够将源图像以拉伸缩放（Stretch）的方式绘制到画布的目标区域中，在 loadAndDrawImage 方法的 this . context2D . drawImage (img , 10 , 10) ;语句后再增加如下语句进行测试：

```
// 将 srcImage 以拉伸缩放的方式绘制到 Canvas 画布指定的矩形中
this . context2D . drawImage ( img , img . width + 30 , 10 , 200 , img .
height ) ;
```

如果调用该方法后，如图 4.17 中上右图像所示，会在右侧显示一张经过拉伸压缩（Stretch）的图像。

最后来看一下 drawImage 的第 3 个重载方法，该方法更加强大：

```
drawImage ( image : HTMLImageElement | HTMLCanvasElement | HTMLVideoElement |
    ImageBitmap , srcX : number , srcY : number , srcW : number , srcH :
number , dstX :     number , dstY : number , dstW : number , dstH : number ) :
void ;
```

该方法中的 srcX、srcY、srcW 和 srcH 这 4 个参数规定了要复制的源图像的矩形区域，而 dstX、dstY、dstW 和 dstH 这 4 个参数规定了目标画布的绘制矩形区域。通过参数就可以知道，该方法能够将源图像的部分区域以拉伸缩放（Stretch）的方式绘制到画布的目标

区域中。继续来做个实验,在上面的代码中再输入如下代码:

```
// 将 srcImage 的部分区域[ 44 , 6 , 162 , 175 , 200 ]以拉伸缩放的方式绘制到 Canvas
画布指定的矩形[ 200 , img . height + 30 , 200 , 130 ]中
this . context2D . drawImage ( img , 44 , 6 , 162 , 175 , 200 , img . height
+ 30 , 200 , 130 ) ;
```

当重新调用 loadAndDrawImage 方法后,会获得如图 4.17 中下侧图像所示的效果。

图 4.17　drawImage 的 3 个重载方法调用效果

4.3.2　Repeat 图像填充模式

在上一节中,演示了 drawImage 的 3 个重载方法,可以选择将一幅源图像的全部或部分区域绘制到 Canvas 画布中指定的目标区域内。在绘制过程中,drawImage 会根据目标区域大小的不同,自动应用拉伸缩放(Stretch)效果。

在 4.1.8 节中,曾经使用图案对象(CanvasPattern)来填充一个路径对象(矢量图形)。在填充的过程中使用了 repeat、repeat_x 和 repeat_y 这 3 种不同的图案填充模式来演示相关效果。

在 Canvas2D 中,repeat、repeat_x 和 repeat_y 这种重复填充模式仅支持使用图案对象(CanvasPattern)来填充路径对象(矢量图形),而 drawImage 使用的是图像对象(HTMLImageElement)来填充目标矩形区域,并且仅支持拉伸缩放(Stretch)的模式,本节目的就是让 drawImage 也能支持 repeat、repeat_x 和 repeat_y 这种图像重复填充模式。

首先使用 enum 关键字来定义这些填充枚举值。具体代码如下:

```
export enum EImageFillType {
    STRETCH ,           // 拉伸模式
    REPEAT ,            // xy 重复填充模式
    REPEAT_X ,          // x 方向重复填充模式
    REPEAT_Y            // y 方向重复填充模式
}
```

接着，来看一下 REPEAT、REPEAT_X 和 REPEAT_Y 的实现过程。先来看一种典型情况，具体如图 4.18 所示。

图 4.18　REPEAT 模式实现要点

根据图 4.18 所示，要实现 REPEAT 填充模式的关键点：
- 如何计算 x 轴方向（左右）和 y 轴方向（上下）需要填充的图像的数量。
- 如何计算 x 轴方向（左右）和 y 轴方向（上下）剩余灰色部分的尺寸。
- REPEAT_X 和 REPEAT_Y 是 REPEAT 的一种特殊形式。

4.3.3　加强版 drawImage 方法的实现

加强版的 drawImage 方法的实现过程如下：

```
public drawImage ( img : HTMLImageElement ,  destRect : Rectangle , srcRect :
Rectangle = Rectangle . create ( 0 , 0 , img . width , img . height ) , fillType :
EImageFillType = EImageFillType . STRETCH ) : boolean {
    // 绘制 image 要满足一些条件
    if ( this . context2D === null ) {
      return false;
    }
    if ( srcRect . isEmpty ( ) ) {
      return false;
    }
    if ( destRect . isEmpty ( ) ) {
      return false ;
    }
    // 分为 stretch 和 repeat 两种方式
    if ( fillType === EImageFillType . STRETCH ) {
      this . context2D . drawImage ( img ,
                    srcRect . origin . x ,
                    srcRect . origin . y ,
                    srcRect . size . width ,
                    srcRect . size . height ,
                    destRect . origin . x ,
                    destRect . origin . y ,
                    destRect . size . width ,
                    destRect . size . height
          ) ;
```

```
} else { // 使用 repeat 模式
    // 测试使用，绘制出目标区域的大小
    this . fillRectangleWithColor ( destRect , 'grey' ) ;
    // 调用 Math . ceil 方法，ceil 是天花板的意思，向上升级，例如 1.3 会变成整数
    2，2.1 会变成整数 3
    // 然而 Math . floor 方法，floor 是地板的意思，向下降级，例如 1.3 会变成整数
    1，2.1 会变成整数 2
    // 还有 Math . round 方法，该方法则是四舍五入，例如 1.3 变成 1，而 1.8 会变成 2
    // 计算 x 轴方向（左右）需要填充的图像的数量，使用 ceil 向上升级
    let rows : number = Math . ceil ( destRect . size . width / srcRect . size . width ) ;
    // 计算 y 轴方向（上下）需要填充的图像的数量，使用 ceil 向上升级
    let colums : number = Math . ceil ( destRect . size . height / srcRect . size . height ) ;
    // 下面 6 个变量在行列双重循环中每次都会更新
    // 表示的是当前要绘制的区域的位置与尺寸
    let left : number = 0;
    let top : number = 0 ;
    let right : number = 0 ;
    let bottom : number = 0 ;
    let width : number = 0 ;
    let height : number = 0 ;
    // 计算出目标 Rectangle 的 right 和 bottom 坐标
    let destRight : number = destRect . origin . x + destRect . size . width ;
    let destBottom : number = destRect . origin . y + destRect . size . height ;
    // REPEAT_X 和 REPEAT_Y 是 REPEAT 的一种特殊形式
    if ( fillType === EImageFillType . REPEAT_X ) {
        colums = 1 ;  // 如果是重复填充 x 轴，则让 y 轴列数设置为 1
    } else if ( fillType === EImageFillType . REPEAT_Y ) {
        rows = 1 ;  // 如果是重复填充 y 轴，则让 x 轴行数设置为 1
    }
    for ( let i : number = 0 ; i < rows ; i ++ ) {
        for ( let j : number = 0 ; j < colums ; j ++ )
        {
            // 如何计算第 i 行第 j 列的坐标
            left = destRect . origin . x + i * srcRect . size . width ;
            top =  destRect . origin . y + j * srcRect . size . height ;
            width = srcRect . size . width ;
            height = srcRect . size . height ;

            // 计算出当前要绘制的区域的右下坐标
            right = left + width ;
            bottom = top + height ;
            // 参见图 4.19
            // 计算 x 轴方向（左右）剩余灰色部分的尺寸的算法
            if ( right > destRight ) {
                width = srcRect . size . width - ( right - destRight ) ;
            }
            // 参见图 4.19
            // 计算 y 轴方向（上下）剩余灰色部分的尺寸的算法
```

```
                if ( bottom > destBottom ) {
                    height = srcRect . size . height - ( bottom - destBottom ) ;
                }

                // 调用 Canvas2D 的 drawImage 方法
                this . context2D . drawImage ( img ,
                    srcRect . origin . x ,
                    srcRect . origin . y ,
                    width ,
                    height ,
                    left , top , width , height
                ) ;
            }
        }
    }
    return true ;
}
```

在上述函数的注释中，标明了实现的关键点。

4.3.4　加强版 drawImage 方法效果演示

接下来看一下调用我们自己实现的加强版 drawImage 方法后的几种典型效果。

（1）当 EImageFillType 为 REPEAT 时，如图 4.19 所示效果。

图 4.19　REPEAT 效果图

（2）当 EImageFillType 为 REPEAT_X 时，效果如图 4.20 所示。

图 4.20　REPEAT_X 效果图

（3）当 EImageFillType 为 REPEAT_Y 时，效果如图 4.21 所示。

图 4.21　REPEAT_Y 效果图

（4）当 EImageFillType 为 STRETCH 时，效果如图 4.22 所示。

图 4.22　STRETCH 效果图

（5）当 EImageFillType 是 REPEAT、REPEAT_X 和 REPEAT_Y 之一，并且源图像的尺寸大于要绘制的目标尺寸时，采用的是裁剪绘制的方式，当然也可以选择使用 STRETCH 方式进行绘制，这一切依赖于具体的需求和实现。

4.3.5　离屏 Canvas 的使用

在 CanvasRenderingContext2D 渲染上下文对象中，drawImage 方法第一个参数还可以使用 HTMLCanvasElement 类型。下面用一个例子来介绍。具体代码如下：

```
// 获取 4 * 4 = 16 种基本颜色的离屏画布
public getColorCanvas ( amount : number = 32 ) : HTMLCanvasElement {
    let step : number = 4 ;
    // 第 1 步，使用 createElement 方法，提供 tagName 为"canvas"关键字创建一个离屏
    画布对象
    let canvas : HTMLCanvasElement = document . createElement ( 'canvas' )
    as HTMLCanvasElement ;
    // 第 2 步，设置该画布的尺寸
    canvas . width = amount * step ;
    canvas . height = amount * step ;
    // 第 3 步，从离屏画布中获取渲染上下文对象
```

```
    let context: CanvasRenderingContext2D | null = canvas.getContext("2d");
    if ( context === null ) {
        alert ( "离屏 Canvas 获取渲染上下文失败！" ) ;
        throw new Error ( "离屏 Canvas 获取渲染上下文失败！" ) ;
    }

    for ( let i : number = 0 ; i < step ; i ++ ) {
        for ( let j : number = 0 ; j < step ; j++ ) {
            // 将二维索引转换成一维索引，用来在静态的 Colors 数组中寻址
            let idx : number = step * i + j ;
            // 第 4 步，使用渲染上下文对象绘图
            context . save ( ) ;
                // 在 4.1.6 节中定义了一个静态颜色数组，表示 17 种标准色名
                // 使用其中 16 种颜色（由于背景是白色，17 种颜色包含白色，所以去除白色）
                context . fillStyle = TestApplication . Colors [ idx ] ;
                context . fillRect ( i * amount , j * amount , amount , amount ) ;
            context . restore ( ) ;
        }
    }
    return canvas ;
}
```

上述代码中，使用 document . createElement ('canvas') as HTMLCanvasElement 生成一个离屏画布，并且设置离屏画布的 width 和 height 属性，后续所有的绘制都是在该画布中进行的，最后函数退出时将离屏画布返回给调用者。

接下来要做的是，将 getColorCanvas 方法返回的离屏画布绘制到当前浏览器中，此时需要修改一下我们自己实现的加强版 drawImage 方法，代码如下：

```
public drawImage ( img : HTMLImageElement | HTMLCanvasElement , destRect :
Rectangle , srcRect : Rectangle = Rectangle . create ( 0 , 0 , img . width ,
img .height ) , fillType : EImageFillType = EImageFillType . STRETCH ) :
boolean
```

测试代码如下：

```
public drawColorCanvas ( ) : void {
    let colorCanvas : HTMLCanvasElement = this .getColorCanvas ( ) ;
    this . drawImage ( colorCanvas , ectangle . create ( 100 , 100 ,
    colorCanvas . width , colorCanvas . height ) ) ;
}
```

最后调用 drawColorCanvas 后，可以得到如图 4.23 所示的结果。

图 4.23 显示离屏画布中的内容

4.3.6 操作 Canvas 中的图像数据

在 Canvas2D 中，可以使用 CanvasRenderingContext2D 上下文渲染对象的以下方法：

- createImageData ( ) 方法，用来创建 ImageData 对象。

- getImageData ( ) 方法，用来获取 ImageData 对象。
- putImageData ( ) 方法，用来更新 ImageData 对象。

为了更好地了解这 3 个方法，以及 ImageData 对象的用法，先来看一下要实现的 Demo 的效果及需求描述，然后再来了解实现细节，从而掌握这些方法和对象的用法。具体的图像效果如图 4.24 所示。

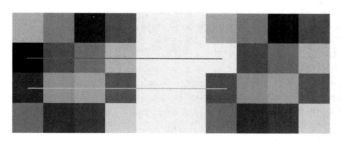

图 4.24 操作图像像素 Demo 演示

请参考图 4.24，来了解以下关键需求：

（1）左侧的 16 色填充矩形是由 4.3.5 节中的 getColorCanvas 方法获取离屏画布后绘制出来的。

（2）要将左侧颜色矩形中第 2 行、第 1 列中的黑色子矩形的颜色反转为如右侧颜色矩形所示的白色。

（3）要将左侧颜色矩形中第 3 行、第 1 列中的蓝色子矩形的颜色替换为如右侧颜色矩形所示的红色。

（4）实现的过程中，会使用两种方式（一重循环和三重循环）来寻址 ImageData 中的像素值，将会在实现代码中注明。

下面来看一下实现过程。先声明一个方法及一些前序操作。具体代码如下：

```
// 参数 rRow / rColum 表示要替换（replace）的颜色的行列索引，默认情况下，将第 3 行，
第 1 列的蓝色子矩形替换为红色
// 参数 cRow / cColum 表示要改变（change）的颜色的行列索引，默认情况下，将第 2 行，
第 1 列的黑色子矩形反转为白色
public testChangePartCanvasImageData(rRow : number = 2 , rColum : number =
0 , cRow : number = 1 , cColum : number = 0 , size : number = 32 ) : void {
    // 调用 getColorCanvas 方法生成 16 种标准色块离屏画布
    let colorCanvas :  HTMLCanvasElement = this .getColorCanvas ( size ) ;
    // 获取离屏画布的上下文渲染对象
    let context : CanvasRenderingContext2D | null = colorCanvas . getContext
( "2d" ) ;
    if( context === null ) {
        alert ( "Canvas 获取渲染上下文失败！" ) ;
        throw new Error ( "Canvas 获取渲染上下文失败！" ) ;
    }
    // 显示未修改时的离屏画布的效果
    this . drawImage ( colorCanvas , Rectangle . create ( 100 , 100 ,
colorCanvas . width , colorCanvas . height ) ) ;
```

然后来实现需求中的第 3 条，使用 createImageData 生成 ImageData 对象，接着将 ImageData 对象中的像素都设置成红色，最后使用 putImageData( )方法来替换离屏画布对应位置的颜色区块。在下面的代码中要注意的是，使用的是第一种像素寻址方式（一维数组表示）。具体代码如下：

```typescript
// 接上面的代码继续往下来替换颜色
//使用 creatImageData 方法，大小为 size * size 个像素
// 每个像素又有 4 个分量 [ r , g , b , a ]
let imgData : ImageData = context . createImageData ( size , size ) ;
// imgData 有 3 个属性，其中 data 属性存储的是一个 Uint8ClampedArray 类型数组对象
// 该数组中存储方式为：[ r , g , b , a , r , g , b , a , ........ ]
// 所以 imgData . data . length = size * size * 4 ;
let data : Uint8ClampedArray = imgData . data ;
// 上面也提到过，imgData . data . length 表示的是所有分量的个数
// 而为了方便寻址，希望使用像素个数进行遍历，因此要除以 4（一个像素由 r、g、b、a 这 4 个分量组成）
let rbgaCount : number = data . length / 4 ;
for ( let i = 0 ; i < rbgaCount ; i ++ ) {
    // 注意下面索引的计算方式
    data [ i * 4 + 0 ] = 255 ;          //红色的 rbga = [ 255 , 0 , 0 , 255 ]
    data [ i * 4 + 1 ] = 0 ;
    data [ i * 4 + 2 ] = 0 ;
    data [ i * 4 + 3 ] = 255 ;          // alpha 这里设置为 255, 全不透明
}

// 一定要调用 putImageData 方法来替换 context 中的像素数据
// 参数 imgData 表示要替换的像素数据
// 参数 [ size * rColum , size * rRow ]表示要绘制到 context 中的哪个位置
// 参数 [ 0 , 0 , size , size ]表示从 imgData 哪个位置获取多少像素
context . putImageData ( imgData , size * rColum , size * rRow , 0 , 0 , size , size ) ;
```

最后来实现上述需求中的第 2 条，获取指定区域中的颜色值，然后将该区域中的颜色值（rgb）倒转过来，但是 alpha 保持不变。由于需要离屏画布指定区域的像素数据，因此不能使用 createImageData 方法，但是可以使用 getImageData 方法。具体代码如下：

```typescript
// 获取离屏画布中位于[ size * cColum , size * cRow ] 处，尺寸为 [ size , size ] 大小的像素数据
imgData = context . getImageData ( size * cColum , size * cRow , size , size ) ;
data = imgData . data ;
let component : number = 0 ;
// 下面使用 imgDate 的 width 和 height 属性，二维方式表示像素
for ( let i : number = 0 ; i < imgData . width ; i ++ ) {
    for ( let j : number = 0 ; j < imgData . height ; j ++ ) {
        // 由于每个像素有包含 4 个分量，[ r g b a ] 因此三重循环
        for ( let k : number = 0 ; k < 4 ; k ++ )
        {
            // 因为 data 是一维数组表示，而使用三重循环，因此需要下面算法
```

```
            // 将三维数组表示的索引转换为一维数组表示的索引，该算法很重要
            let idx : number = ( i * imgData . height + j ) * 4 + k ;
            component = data [ idx ] ;
            // 在 data 数组中，idx % 4 为 3 时，说明是 alpha 值
            // 需求是 alpha 总是保持不变，因此需要下面判断代码，切记
            if ( idx % 4 !== 3 ) {
                data [ idx ] = 255 - component ;  //反转 rgb, 但是 alpha
                    不变, 仍旧是 255
            }
        }
    }
}
// 使用 putImageData 更新像素数据
context . putImageData ( imgData , size * cColum , size * cRow , 0 , 0 ,
    size , size ) ;
// 将修改后的结果绘制显示出来
this . drawImage ( colorCanvas ,
    Rectangle . create ( 300 , 100 , colorCanvas . width , colorCanvas .
    height ) ) ;
}
```

可以以不同的行、列参数及 size 来调用上面实现的函数，观察一下相应的效果。

4.4 绘制阴影

实际上，阴影比较特别，因为它具有叠加效果。可以在绘制图形、图像及文本时，让其产生阴影。绘制阴影所要做的是设置如下 4 个阴影相关的属性。

- shadowColor：CSS 颜色字符串。
- shadowBlur：指定一个数值，参与高斯模糊计算。
- shadowOffsetX：指定一个数值，表示水平偏移像素。
- shadowOffsetY：指定一个数值，表示垂直偏移像素。

先来打印一下上面四个阴影相关属性的默认值。具体代码如下：

```
public printShadowStates ( ) : void {
    if ( this . context2D !== null ) {
        console . log ( " ********* ShadowState ********** " ) ;
        console . log ( " shadowBlur : " + this . context2D . shadowBlur ) ;
        console . log ( " shadowColor : " + this . context2D . shadowColor ) ;
        console . log ( " shadowOffsetX : " + this . context2D .
            shadowOffsetX ) ;
        console . log ( " shadowOffsetY : " + this . context2D .
            shadowOffsetY ) ;
    }
}
```

当调用上述方法后，会得到如图 4.25 所示的结果。

```
*********ShadowState**********
shadowBlur : 0
shadowColor : rgba(0, 0, 0, 0)
shadowOffsetX : 0
shadowOffsetY : 0
```

图 4.25　阴影默认属性值

实现一个设置阴影的方法。具体代码如下：

```
public setShadowState ( shadowBlur: number = 5, shadowColor: string =
"rgba( 127 , 127 , 127 , 0.5 )", shadowOffsetX: number = 10 , shadowOffsetY:
number = 10 ): void {
    if ( this . context2D !== null ) {
        this . context2D . shadowBlur = shadowBlur ;
        this . context2D . shadowColor = shadowColor ;
        this . context2D . shadowOffsetX = shadowOffsetX ;
        this . context2D . shadowOffsetY = shadowOffsetY ;
    }
}
```

在上一节实现的 testChangePartCanvasImageData 方法中调用 setShadowState 后，会得到如图 4.26 所示的效果。

图 4.26　带阴影绘制图像

阴影的使用非常简单，但是阴影的绘制非常耗时，除非必要，否则尽量不要使用阴影技术。

4.5　本 章 总 结

本章主要介绍了 Canvas2D 中绘制部分的相关内容，章节的安排是根据可渲染的对象来分类的。

首先通过继承前面实现的 Canvas2DApplication 类，实现了一个本章及后续章节都要使用的基于 Canvas2D 的演示和测试环境。

然后通过一个绘制矩形的 Demo 介绍了 Canvas2D 中的绘制方式，并且重点讲述了 Canvas2D 中渲染状态管理的关键原理，即渲染状态堆栈，使用 TypeScript 代码模拟了渲染堆栈的实现和应用。

接着介绍了线段和形体绘制的相关内容。先是介绍了与线段绘制相关的 lineWidth 和 lineCap 等属性，并通过一个例子演示了虚线动画效果（如 Photoshop 中选中某个几何形体时虚线滚动效果），然后详细讲解了描边（stroke）和填充（fill）的相关知识，重点讲述了使用颜色、渐变色及图案填充的相关内容。在此强调一点，使用图案对象可以填充任何几何体或线段，但是无法像 WebGL 中那样自行控制纹理（图案）坐标，并且使用图案填充矩形时只能用 repeat 等模式，无法使用拉伸方式。

然后介绍了文本绘制的相关内容。首先通过一个 Demo 来演示 Canvas2D 内置的文本对齐的相关知识点。然后提升了一下难度，在限定条件下，自己实现类似于 Canvas2D 中文本对齐的算法，做到不仅知其然，而且知其所以然。最后实现了一个名为 makeFontString 的辅助方法，通过该方法可以确保 font 对象各个属性设置的正确性。

接下来继续讲解图像绘制相关的知识，通过一些演示代码，讲解了 drawImage 的 3 个重载方法各自的用途及限制。发现 drawImage 支持拉伸模式的绘制，但是却不支持 repeat、repeatX 和 repeatY 这些模式的绘制，而前面的矩形几何体图案填充正好相反，它支持 repeat、repeatX 和 repeatY，却不支持拉伸模式的绘制，于是决定实现加强版的 drawImage 方法，让其支持 repeat、repeatX 和 repeatY 模式，以及拉伸模式并进行相关测试。接着实现了在离屏画布上绘制 16 种不同颜色的正方形，然后通过修改其中两种矩形在不同色块区域中的像素 rgba 值来引申出两个重要的图像数据操作方法，即 getImageDat 和 putImageData 的用法。

最后演示了如何绘制阴影，但是对于阴影绘制，建议除非需要，否则不要使用。

由上述内容可知，Canvas2D 中可绘制的对象可以分为 4 种类型：

- 矢量图形对象（基于路径对象）；
- 文本对象；
- 像素图像对象；
- 阴影。

其中，阴影属于全局绘制对象，它影响到矢量图形、图像及文本的绘制效果。

在上面 4 个主题中，文本、图像及阴影绘制演示的内容在本章中有比较全面的介绍，但是在矢量图形绘制这一主题中，还有很多绘制方法，例如曲线、弧线、椭圆和多边形等内容本章中并没有涉及。像这些基础重要的图形，在后面章节的精灵和形体系统中会更加深入地介绍。

由于渲染状态非常重要，而且必须要清晰地知道各个可选值和默认值，在此提供一段 CanvasRenderingContext2D 对象中所有渲染状态默认值的输出代码，具体如下：

```
public static printAllStates ( ctx: CanvasRenderingContext2D ): void {
    console.log( "*********LineState**********" );
    console.log( " lineWidth : " + ctx.lineWidth );
    console.log( " lineCap : " + ctx.lineCap );
    console.log( " lineJoin : " + ctx.lineJoin );
    console.log( " miterLimit : " + ctx.miterLimit );
    console.log( "*********LineDashState**********" );
    console.log( " lineDashOffset : " + ctx.lineDashOffset );
    console.log( "*********ShadowState**********" );
    console.log( " shadowBlur : " + ctx.shadowBlur );
    console.log( " shadowColor : " + ctx.shadowColor );
    console.log( " shadowOffsetX : " + ctx.shadowOffsetX );
    console.log( " shadowOffsetY : " + ctx.shadowOffsetY );
    console.log( "*********TextState**********" );
    console.log( " font : " + ctx.font );
    console.log( " textAlign : " + ctx.textAlign );
    console.log( " textBaseline : " + ctx.textBaseline );
    console.log( "*********RenderState**********" );
    console.log( " strokeStyle : " + ctx.strokeStyle );
    console.log( " fillStyle : " + ctx.fillStyle );
    console.log( " globalAlpha : " + ctx.globalAlpha );
    console.log( " globalCompositeOperation : " + ctx.globalComposite
Operation );
}
```

通过上述代码，可以得到如下几个结论：

- globalAlpha 和 globalCompositeOperation 这两个全局渲染属性本章没有涉及，如有需要，请自行查阅官方文档。
- 上述渲染状态都是受到渲染状态堆栈管理的（渲染状态堆栈管理参考 4.1.3 节）。
- 图像绘制和渲染状态及渲染堆栈无任何关系。

第 5 章，我们将重点介绍 Canvas2D 中空间变换的相关技术，这是非常重要的一章，也是本书的精华所在。

第 3 篇
图形数学篇

▶▶ 第 5 章　Canvas2D 坐标系变换

▶▶ 第 6 章　向量数学及基本形体的点选

▶▶ 第 7 章　矩阵数学及贝塞尔曲线

第 5 章　Canvas2D 坐标系变换

在第 4 章中，主要以矢量图形绘制、文本字符串绘制、图像绘制及阴影绘制这 4 个主题介绍了 Canvas2D 中渲染显示的相关内容。

本章主要介绍 Canvas2D 中的另外一个重要知识点，即物体的局部坐标系变换。本章是本书的重点章节。

5.1　局部坐标系变换

在 Canvas2D 中，CanvasRenderingContext2D 渲染上下文对象使用如下 5 个方法来进行局部坐标系的变换：

```
translate ( x : number , y : number ) : void ;              // 局部坐标系的平移操作
rotate ( angle : number ) : void ;                          // 局部坐标系的旋转操作
scale ( x : number , y : number ) : void ;                  // 局部坐标系的缩放操作
transform (m11 : number , m12 : number , m21 : number , m22 : number , dx :
number , dy : number ) : void ;                             // 矩阵相乘操作
setTransform ( m11 : number , m12 : number , m21 : number , m22 : number ,
dx : number , dy : number ) : void ;                        // 设置变换矩阵操作
```

接下来介绍这些方法是如何使用的。

5.1.1　准备工作

为了图文并茂地介绍局部坐标系变换，先来做一些基础的准备工作。仍然使用第 4 章中实现的 TestApplication 类，在该类中继续增加如下一些辅助代码，用来绘制相关信息。

1．drawCanvasCoordCenter方法

drawCanvasCoordCenter 方法用来绘制 Canvas 画布的中心点及相交于中心点的 x 和 y 轴。具体代码如下：

```
public drawCanvasCoordCenter ( ) : void {
    // 绘制 image 要满足一些条件
    if ( this . context2D === null ) {
```

```
        return ;
    }
    // 计算出 Canvas 的中心点坐标
    let halfWidth : number = this . canvas . width * 0.5 ;
    let halfHeight : number = this . canvas . height * 0.5 ;
    this . context2D . save ( ) ;
    this . context2D . lineWidth = 2 ;
    this . context2D . strokeStyle = 'rgba( 255 , 0 , 0 , 0.5 ) ' ;
    // 使用 alpha 为 0.5 的红色来绘制 x 轴
    // 调用第 4 章中实现的 strokeLine 方法
    this . strokeLine ( 0 , halfHeight ,  this . canvas . width , halfHeight ) ;
    this . context2D . strokeStyle = 'rgba( 0 , 0 , 255 , 0.5 )' ;
    // 使用 alpha 为 0.5 的蓝色来绘制 y 轴
    this . strokeLine ( halfWidth , 0 , halfWidth , this . canvas . height ) ;
    this . context2D . restore ( ) ;
    // 使用 alpha 为 0.5 的黑色来绘制画布中心点
    // 调用第 4 章中实现的 fillCircle 方法
    this . fillCircle(halfWidth , halfHeight,5 ,'rgba(0 , 0 , 0 , 0.5 ) ');
}
```

2. drawCoordInfo方法

drawCoordInfo 方法用于绘制某个点处的坐标信息。具体代码如下：

```
public drawCoordInfo ( info : string , x : number , y : number ) : void {
    // 调用第 4 章实现的 fillText 方法
    // 使用黑色字体，在（中下）绘制文字
    this . fillText ( info , x , y , 'black' , 'center' , 'bottom' ) ;
}
```

在调用 drawCoordInfo 方法后，会在鼠标指针的上方绘制当前鼠标指针相对于 Canvas 坐标系的位置信息。

3. distance方法

distance 方法用于计算两点间的距离，其公式如下：

$$|AB| = \sqrt{(x_1 - x_0)^2 + (y_1 - y_0)^2}$$

在上述公式中，(x_0, y_0) 与 (x_1, y_1) 分别为 A、B 两个点的坐标信息。

下面实现这个公式。具体代码如下：

```
// 两点间距离公式
public distance ( x0 : number , y0 : number , x1 : number , y1 : number ) : number {
    let diffX : number = x1 - x0 ;
    let diffY : number = y1 - y0 ;
    return Math . sqrt ( diffX * diffX + diffY * diffY ) ;
}
```

可以使用 Math 类的静态方法 sqrt 进行开根号操作。

4. 弧度与角度的互相转换代码

定义成 Math2D 类的静态方法。具体代码如下：

```
// 使用 const 关键字定义常数
const PiBy180 : number = 0.017453292519943295 ; // Math . PI / 180.0
// 将以角度表示的参数转换为弧度表示
public static toRadian ( degree : number ) : number {
    return degree * PiBy180 ;
}

// 将以弧度表示的参数转换为角度表示
public static toDegree ( radian : number ) : number {
    return radian / PiBy180 ;
}
```

5. TestApplication 类

让 TestApplication 类支持 mouseMove 事件并记录当前的鼠标指针位置信息（位于 Canvase 坐标系内）。

首先在 TestApplication 中增加两个私有成员变量，用于记录鼠标指针信息：

```
private _mouseX : number = 0 ;
private _mouseY : number = 0 ;
```

然后为了让 TestApplication 支持 mouseMove 事件，需要在 TestApplication 的构造函数中输入如下代码：

```
// 默认情况下，isSupportMouseMove 为 false，不处理鼠标指针移动事件，所以要设置为
true 让其支持鼠标指针移动事件处理
this . isSupportMouseMove = true ;
```

最后需要覆写（override）基类的受保护方法 dispatchMouseMove。具体代码如下：

```
protected dispatchMouseMove ( evt : CanvasMouseEvent ) : void {
    // 必须要设置 this . isSupportMouseMove = true 才能处理 moveMove 事件
    this . _mouseX = evt . canvasPosition . x ;
    this . _mouseY = evt . canvasPosition . y ;
}
```

6. 覆写（override）基类的 render 方法

具体实现代码如下：

```
public render ( ) : void {
    // 由于 canvas . getContext 方法返回的 CanvasRenderingContext2D 可能会是 null
    // 因此 VSCode 会强制要求 null 值检查，否则报错
    if ( this . context2D !== null ) {
        // 每次重绘都先清屏
        this . context2D .clearRect ( 0 , 0 , this . canvas .width ,this .
        canvas .height ) ;
        // 调用第 4 章实现背景网格绘制方法
        this . strokeGrid ( ) ;
        // 绘制中心原点和 x、y 轴
```

```
            this . drawCanvasCoordCenter ( ) ;
            // 本章后续绘制代码请写在下面
            // .... .... .... ... ...... ..... .... ...
            //.. ...... ... .... .... ... ...... .... ...
            // 坐标信息总是在最后绘制
            this . drawCoordInfo (
                '[' + this . _mouseX + ',' + this . _mouseY + "]" ,
                this . _mouseX ,
                this . _mouseY
            ) ;
        }
    }
```

当使用 TestApplication 类中的 start 方法进入动画循环后，每次移动鼠标时就会实时更新当前鼠标指针位置信息，具体效果可以参考图 5.1。

至此，我们搭建了本章要用到的一些必备方法框架。

5.1.2 平移操作演示

接下来看一下如何使用 translate 方法绘制一个左上角位于画布中心的矩形。具体代码如下：

```
public doTransform ( ) : void {
    if ( this . context2D !== null ) {
        // 要绘制的矩形的尺寸
        let width : number = 100 ;
        let height : number = 60 ;
        // 计算出画布中心点坐标
        let x : number = this . canvas . width * 0.5 ;
        let y : number = this . canvas . height * 0.5 ;
        this . context2D . save ( ) ;
            // 调用 translate 平移到画布中心
            this . context2D . translate ( x , y ) ;
            this . fillRectWithTitle ( 0 , 0 , width , height , ' 0 度旋转 ' ) ;
        this . context2D . restore ( ) ;
    }
}
```

当在 TestApplication 类的 render 方法中调用 doTransform 方法后，得到的结果如图 5.1 所示。

使用 doTransform 方法需要注意 3 点：

- 使用 translate 方法，将物体（要绘制的矩形）的局部坐标系移动到画布中心点。
- 使用 fillRectWithTitle 方法绘制以画布为中心的矩形时，[x，y]坐标参数使用的是[0，0]，而不是以前的[canvas . width * 0.5，canvas . height * 0.5]，这是因为现在使用了 translate 方法将矩形的局部坐标系平移到了画布中心，接下来的所有绘制命令都是以画布中心作为基准点，因此当前填充矩形时与画布中心基准点的偏移量为[0，0]。如果将其设置为[10，0]，则会看到填充的矩形左上角位于画布中心点右侧 10 个单位处。

- 由于 doTransform 方法是每帧都被调用（doTransform 由 render 方法调用），因此，如果不在 doTransform 中调用 save / restore 方法，则会出现绘制不正确的现象。至于导致绘制不正确的原因，会在后续章节中解释。

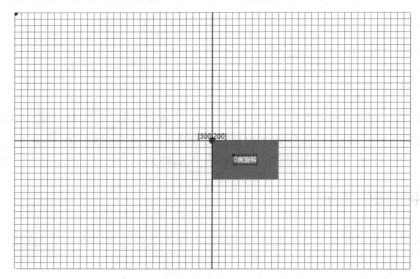

图 5.1 doTransform 调用效果

5.1.3 平移和旋转组合操作演示

在上一节中，仅仅使用 translate 方法来平移矩形，现在要做的是将矩形平移到画布中心点并进行 ±20°的旋转操作。由于平移和旋转操作按照调用顺序的不同，可以先平移后旋转，或者先旋转后平移操作，因此决定在 doTransform 函数中添加一个 rotateFirst 参数来确定先旋转还是先平移。具体代码如下：

```
public doTransform(degree : number , rotateFirst : boolean = true ):void {
    // 将角度转换为弧度，由此可见，本方法的参数 degree 是以角度而不是弧度表示
    let radians : number = Math2D . toRadian ( degree ) ;
    // 顺时针旋转
    this . context2D . save ( ) ;
    // 根据 rotateFirst 进行平移和旋转变换
    if ( rotateFirst ) {
        // 先顺时针旋转 20°
        this . context2D . rotate ( radians ) ;
        // 然后再平移到画布中心
        this . context2D . translate ( this . canvas . width * 0.5 , this . canvas . height * 0.5 ) ;
    } else {
        // 和上面正好相反
        // 先平移到画布中心
```

```
            this . context2D . translate ( this . canvas . width * 0.5 , this .
            canvas . height * 0.5 ) ;
            // 然后再顺时针旋转 20°
            this . context2D . rotate ( radians ) ;
        }
        // 注意是[ 0 , 0 ]坐标
        this . fillRectWithTitle ( 0 , 0 , 100 , 60 , '+' + degree +
        '度旋转' ) ;
    this . context2D . restore ( ) ;
    // 逆时针旋转，代码与上面顺时针旋转一样
    this . context2D . save ( ) ;
        if ( rotateFirst ) {
            this . context2D . rotate ( - radians ) ;
            this . context2D . translate ( this . canvas . width * 0.5 , this .
            canvas . height * 0.5 ) ;
        } else {
            this . context2D . translate ( this . canvas . width * 0.5 , this .
            canvas . height * 0.5 ) ;
            this . context2D . rotate ( - radians ) ;
        }
        // 注意是[ 0 , 0 ]坐标
        this . fillRectWithTitle ( 0 , 0 , 100 , 60 , '-' + degree +
        '度旋转' ) ;
    this . context2D . restore ( ) ;
}
```

当使用 rotateFirst 为 true 的参数来调用 doTransform 方法后，会得到如图 5.2 所示的效果。

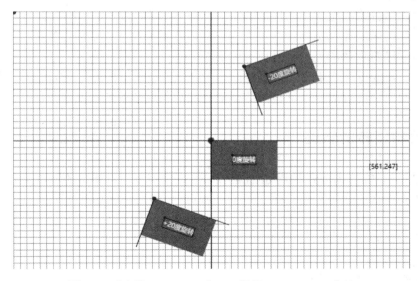

图 5.2　以参数 rotateFirst 为 true 调用 doTransform 方法

当将 rotateFirst 设置为 false 后调用 doTransform 方法，则得到完全不一样的结果，如图 5.3 所示的结果。

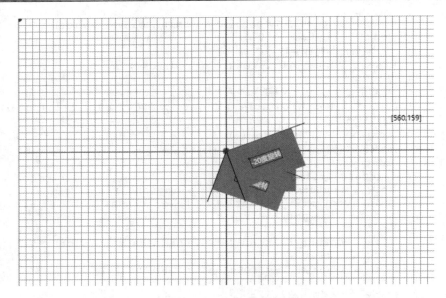

图 5.3 以参数 rotateFirst 为 false 调用 doTransform 方法

接下来,将会以这个简单的例子为核心,来看一下隐藏在背后的一些值得关注的细节,先要牢记以下 4 个要点:

- 物体的局部坐标系就是上面这些图示中依附在物体原点处的,随着物体的运动而运动的那个坐标系。
- 在局部坐标系上定义渲染用的顶点坐标数据(例如通过矩形的左上角及尺寸可以以顺时针的顺序计算出物体的左上、右上、右下和左下 4 个顶点坐标)。
- 变换顺序对物体的显示有着巨大的影响。
- save / restore 与矩阵堆栈之间关系(将在第 7 章介绍自定义实现矩阵堆栈来替换 Canvs2D 中内置的矩阵堆栈,这样能更加深入地了解几何变换)。

5.1.4 绘制旋转的轨迹

在使用 rotateFirst 为 true 调用 doTransoform 的情况下,将旋转的轨迹绘制出来,这样让后续的变换操作更加可视化。在 doTransform 代码最后添加如下代码:

```
// 调用前面实现的两点间距离公式
// 第一个点是原点,第二个点是画布中心点
let radius : number = this . distance ( 0 , 0 , this . canvas . width * 0.5 ,
this . canvas . height * 0.5 ) ;
// 然后绘制一个圆
this . strokeCircle ( 0 , 0 , radius , 'black' ) ;
```

当调用 doTransoform 方法后,具体效果如图 5.4 所示。
通过图 5.4 显示的效果会发现,在先旋转后平移的情况下,3 个矩形原点(左上角)

都在以全局坐标[0 , 0]为圆心，以圆心到 Canvas 画布中心的距离为半径所形成的一个圆的轨迹上。

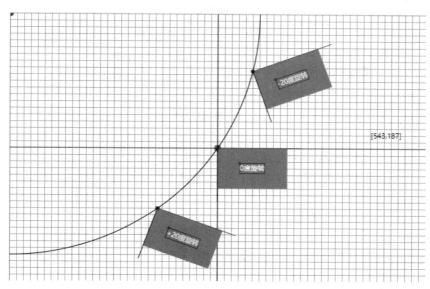

图 5.4　旋转的轨迹

现在增加一点难度，让 3 个矩形的原点从左上角变成中心点，如图 5.5 所示的效果，该如何做？

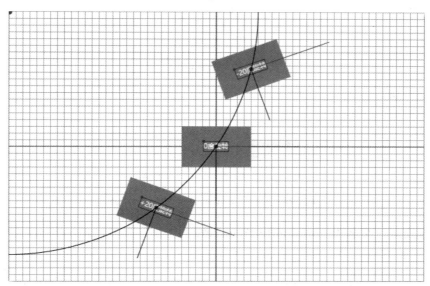

图 5.5　矩形中心为原点效果

5.1.5 变换局部坐标系的原点

通过观察图 5.4 和图 5.5 的效果会发现，依附在这些矩形上的局部坐标系并没有发生任何改变，其原点都是在圆的轨迹上（没改变圆的中心和半径），并且 x 轴和 y 轴的方向也是一致的。改变的是矩形的四个顶点（左上、右上、右下和左下）相对于局部坐标系原点的位置。

在图 5.4 中会看到，矩形的左上角与局部坐标系的原点重合，而图 5.5 中，相比图 5.4 而言，矩形的左上角沿着局部坐标系 x 轴的负方向（向左侧）移动了矩形宽度的 0.5 倍的单位，接着沿着局部坐标系 y 轴的负方向（向上侧）移动了矩形高度的 0.5 倍的单位。

如果想让坐标系位于矩形的右下角（根据图 5.4 坐标系位于矩形左上角，图 5.5 坐标系位于矩形的中心的图示，大家可以想象一下局部坐标系位于右下角时的画面），根据上述的规则，很容易得出如下结论：

让矩形的左上角沿着局部坐标系 x 轴的负方向（左侧）移动矩形宽度单位，沿着局部坐标系 y 轴的负方向（上侧）移动矩形高度单位。可以有 9 种排列组合方式，其算法类似。

在实现上述算法前，先将第 4 章中定义的 ETextLayout 枚举重命名一下，让其更具有通用的意义。可以使用 VS Code 中的重命名符号命令来替换名称，具体如图 5.6 所示。

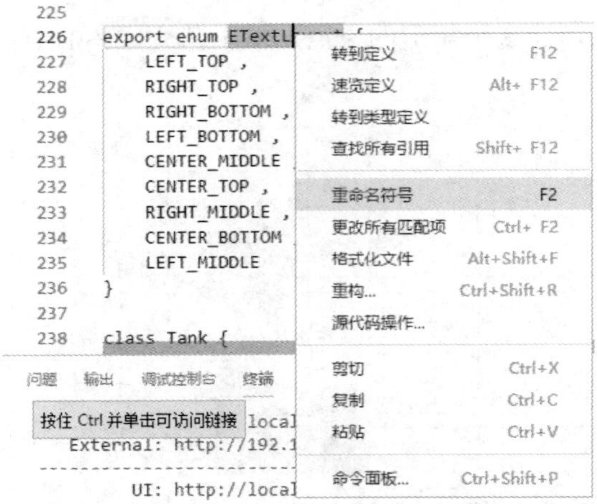

图 5.6　重命名 ETextLayout 为 ELayout

接下来实现一个名为 fillLocalRectWithTitle 的方法。具体代码如下：

```
public fillLocalRectWithTitle ( width : number ,        //要绘制的矩形宽度
        height : number ,                               //要绘制的矩形高度
        title : string = '' ,                           //矩形中显示的字符串
        referencePt : ELayout = ELayout . CENTER_MIDDLE ,
                                                        //坐标系原点位置,默认居中
        layout : ELayout = ELayout . CENTER_MIDDLE ,
```

```
                                        //文字框位置，默认居中绘制文本
        color : string = 'grey' ,       //要绘制矩形的填充颜色
        showCoord : boolean = true
                                        //是否显示局部坐标系，默认为显示局部坐标系
) : void {
    if ( this . context2D !== null ) {
        let x : number = 0 ;
        let y : number = 0 ;
        // 首先根据 referencePt 的值计算原点相对左上角的偏移量
        // Canvas2D 中，左上角是默认的坐标系原点，所有原点变换都是相对左上角的偏移
        switch ( referencePt ) {
            case ELayout . LEFT_TOP :       //Canvas2D 中，默认是左上角为坐标系原点
                x = 0 ;
                y = 0 ;
                break ;
            case ELayout . LEFT_MIDDLE:                 //左中为原点
                x = 0 ;
                y = - height * 0.5 ;
                break ;
            case ELayout . LEFT_BOTTOM :                //左下为原点
                x = 0 ;
                y = - height ;
                break ;
            case ELayout . RIGHT_TOP :                  //右上为原点
                x = - width ;
                y = 0 ;
                break ;
            case ELayout . RIGHT_MIDDLE:                //右中为原点
                x = - width ;
                y = - height * 0.5 ;
                break ;
            case ELayout. RIGHT_BOTTOM :                //右下为原点
                x = - width ;
                y = - height ;
                break ;
            case ELayout . CENTER_TOP :                 //中上为原点
                x = - width * 0.5 ;
                y = 0 ;
                break ;
            case ELayout. CENTER_MIDDLE :               //中中为原点
                x = - width * 0.5 ;
                y = - height * 0.5 ;
                break ;
            case ELayout . CENTER_BOTTOM :              //中下为原点
                x = - width * 0.5 ;
                y = - height ;
                break ;
        }
        // 下面的代码和上一章实现的 fillRectWithTitle 一样
        this . context2D . save ( ) ;
        // 1. 绘制矩形
        this . context2D . fillStyle = color ;
        this . context2D . beginPath ( ) ;
```

```
                this . context2D . rect ( x , y , width , height ) ;
                this . context2D . fill ( ) ;
                // 如果有文字，先根据枚举值计算 x,y 坐标
                if ( title . length !== 0 ) {
                    // 2．绘制文字信息
                    // 在矩形的左上角绘制相关文字信息，使用的是 10px 大小的文字
                    let rect : Rectangle = this . calcLocalTextRectangle ( layout ,
                    title , width , height ) ;
                    // 绘制文本
                    this . fillText ( title , x + rect . origin . x , y + rect . origin .
                    y , 'white' , 'left' , 'top' /*, '10px sans-serif'*/ ) ;
                    // 绘制文本框
                    this . strokeRect ( x + rect . origin . x , y + rect . origin .
                    y , rect .size . width , rect . size . height , 'rgba( 0 , 0 ,
                    0 , 0.5 ) ' ) ;
                    // 绘制文本框左上角坐标（相对父矩形表示）
                    this . fillCircle ( x + rect . origin . x , y + rect . origin .
                    y , 2 ) ;
                }
                // 3．绘制变换的局部坐标系，局部坐标原点总是为 [ 0 , 0 ]
                // 附加一个坐标，x 轴和 y 轴比矩形的 width 和 height 多 20 像素
                // 并且绘制 3 像素的原点
                if ( showCoord ) {
                    this . strokeCoord ( 0 , 0 , width + 20 , height + 20 ) ;
                    this . fillCircle ( 0 , 0 , 3 ) ;
                }
                this . context2D . restore ( ) ;
            }
        }
```

fillLocalRectWithTitle 方法实际上是第 4 章中实现的 fillRectWithTitle 方法的加强版，它们之间的区别是，local 版本增加了坐标的 9 种原点变换，以及取消了 x 和 y 坐标参数，强制让局部坐标系原点位于 [0 , 0] 处，这样可以通过内置的 translate、rotate 和 scale 方法进行局部坐标变换，这样更加方便、强大。后续会有很多比较复杂的例子来演示基于局部坐标系的变换操作。

5.1.6 测试 fillLocalRectWithTitle 方法

下面来测试一下 fillLocalRectWithTitle 方法，将 9 种坐标系变换后的结果全部绘制出来。具体代码如下：

```
// 将 doTransform 中先旋转后平移的代码独立出来，形成 rotateTranslate 方法
public rotateTranslate ( degree : number , layout : ELayout = ELayout .
LEFT_TOP , width : number = 40 , height : number = 20 ) : void {
    if ( this . context2D === null ) {
        return ;
    }
    // 将角度转换为弧度，由此可见，本方法的参数 degree 是以角度而不是弧度表示
    let radians : number = Math2D . toRadian ( degree ) ;
    // 顺时针旋转
```

```
        this . context2D . save ( ) ;
        // 先顺时针旋转 20°
        this . context2D . rotate ( radians ) ;
        // 然后再平移到画布中心
        this . context2D . translate ( this . canvas . width * 0.5 , this . canvas .
        height * 0.5 ) ;
        // 调用新实现的 localRect 方法
        this . fillLocalRectWithTitle ( width , height , '' , layout ) ;
        this . context2D . restore ( ) ;
    }
    // 实现 testFillLocalRectWitTitle 方法,该方法分别在圆的路径上绘制 9 种不同的坐标系
    public testFillLocalRectWithTitle ( ) : void {
        if ( this . context2D !== null ) {
            // 旋转 0°，坐标原点位于左上角（默认）
            this . rotateTranslate ( 0 , ELayout . LEFT_TOP) ;
            // 顺时针旋转，使用 4 种不同的 ELayout 值
            this . rotateTranslate ( 10 , ELayout . LEFT_MIDDLE ) ;
            this . rotateTranslate ( 20 , ELayout . LEFT_BOTTOM ) ;
            this . rotateTranslate ( 30 , ELayout . CENTER_TOP ) ;
            this . rotateTranslate ( 40 , ELayout . CENTER_MIDDLE ) ;
            // 逆时针旋转，使用 4 种不同的 ELayout 值
            this . rotateTranslate ( - 10 , ELayout . CENTER_BOTTOM ) ;
            this . rotateTranslate ( - 20 , ELayout . RIGHT_TOP ) ;
            this . rotateTranslate ( - 30 , ELayout . RIGHT_MIDDLE ) ;
            this . rotateTranslate ( - 40 , ELayout . RIGHT_BOTTOM ) ;
            // 计算半径
            let radius : number = this . distance ( 0 , 0 , this . canvas . width
            * 0.5 , this . canvas . height * 0.5 ) ;
            // 最后绘制一个圆
            this . strokeCircle ( 0 , 0 , radius , 'black' ) ;
        }
    }
```

当运行测试程序后，会得到如图 5.7 所示的结果。

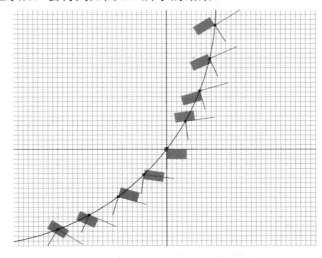

图 5.7　坐标原点变换测试效果

5.1.7 彻底掌控局部坐标系变换

回顾前面实现的 doTransform 方法，当使用 rotateFirst 的不同值（true 和 false）时，得到完全不同的变换效果，由此可以知道：在 Canvas2D 中，translate、rotate 和 scale 不同的调用顺序会产生不同的变换效果。

本节将通过一个更加明确的 Demo 来追踪坐标系变换的过程，理解坐标系变换的三大要点，从而完全掌控坐标系变换的奥秘！

（1）首先定义一个名为 doLocalTransform 的方法，其方法体内定义如下流程：

```
public doLocalTransform ( ) : void {
    if ( this . context2D === null ) {
        return ;
    }
    let width : number = 100 ;              // 在局部坐标系中显示的 rect 的 width
    let height : number = 60 ;              // 在局部坐标系中显示的 rect 的 height
    let coordWidth : number = width * 1.2 ;   // 局部坐标系 x 轴的长度
    let coordHeight : number = height * 1.2 ; // 局部坐标系 y 轴的长度
    let radius : number = 5 ;               // 绘制原点时使用的半径
    this . context2D . save ( ) ;
    /*
    所有局部坐标系变换演示的代码都现在此处
    本节下面的所有绘制代码都写在此处
    */
    this . context2D . restore ( ) ;
}
```

（2）为了在调用上面实现的 fillLocalRectWithTitle 方法时少输入一些函数参数，修改一下 fillLocalRectWithTitle 的默认参数的值，将原点的布局设置从居中改为左上对齐，将最后一个参数 showCoord 默认值改为 false，不显示矩形的局部坐标系。具体代码如下：

```
public fillLocalRectWithTitle (
    width : number ,
    height : number ,
    title : string = '' ,
    referencePt : ELayout = ELayout . LEFT_TOP ,   // 改为参考点位于左上角
    layout : ELayout = ELayout . CENTER_MIDDLE ,
    color : string = 'grey' ,
    showCoord : boolean = false   // 改为不显示依附在当前矩形的局部坐标系
) : void
```

（3）绘制初始化时矩形的状态。具体代码如下：

```
this . strokeCoord ( 0 , 0 , coordWidth , coordHeight ) ;
                                                    // 注意坐标为 [ 0 , 0 ]
this . fillCircle ( 0 , 0 , radius ) ;              // 注意坐标为 [ 0 , 0 ]
this . fillLocalRectWithTitle ( width , height , ' 1、初始状态 ' ) ;
```

当 debug 运行 doLocalTransform 方法后，会得到如图 5.8 所示的效果。

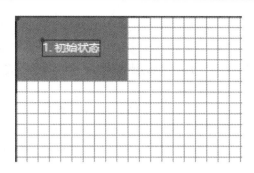

图 5.8　初始化时矩形位于画布左上角

（4）将局部坐标移动到全局坐标系中 x 轴的中心（画布宽度的一半）、y 轴 10 个单位处。具体代码如下：

```
// 将坐标系向右移动到画布的中心，向下移动10个单位，再绘制局部坐标系
this . context2D . translate ( this . canvas . width * 0.5 , 10 ) ;
this . strokeCoord ( 0 , 0 , coordWidth , coordHeight ) ;
                        // 注意坐标为[ 0 , 0 ]，绘制坐标系 x 和 y 轴
this . fillCircle ( 0 , 0 , radius ) ;
                        // 注意坐标为[ 0 , 0 ]， 绘制坐标系原点
```

观察一下上述代码运行后的效果，如图 5.9 所示。

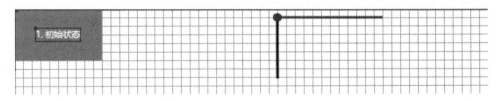

图 5.9　绘制平移后的坐标系

然后在局部坐标系中绘制矩形。具体代码如下：

```
this . fillLocalRectWithTitle ( width , height , ' 2、平移 ') ; //绘制矩形
```

再次运行上述代码后，将会得到如图 5.10 所示的结果。

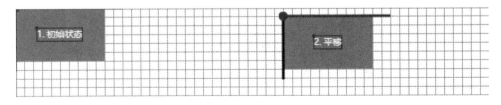

图 5.10　绘制平移后的矩形

通过绘制变换后的局部坐标系（如图 5.9 所示），以及绘制出相对于变换后的局部坐标系所定义的矩形（如图 5.10 所示），可以得到如下两条结论：

- 结论 1：使用 translate、rotate 和 scale 这些方法时，变换的是局部坐标系。
- 结论 2：所有绘制操作都是相对于变换后的局部坐标系所进行的。

如图 5.10 所示，绘制矩形时，是相对于变换后的局部坐标进行的。也就是说，原点在局部坐标中还是[0 , 0]，但是在全局坐标中，矩形的原点变成了[canvas . width * 0.5 , 10]，矩形的左上、右上、右下和左下 4 个顶点坐标相对局部坐标系的原点与 x 和 y 轴之间的方向都是保持不变的。这就是局部坐标变换，在后续操作中会一直验证上述结论。

（5）接下来继续平移操作，将局部坐标变换到全局坐标（整个画布）的中心。具体代码如下：

```
this . context2D . translate ( 0 , this . canvas . height * 0.5 - 10 ) ;
this . strokeCoord ( 0 , 0 , coordWidth , coordHeight ) ;
                              // 注意坐标为[ 0 , 0 ]，绘制坐标系 x 和 y 轴
this . fillCircle ( 0 , 0 , radius ) ;
                              // 注意坐标为[ 0 , 0 ]，绘制坐标系原点
this . fillLocalRectWithTitle ( width , height , ' 3、平移到画布中心 ' ) ;
```

关于上述代码，需要强调的一点是，并不是使用如下代码将局部坐标系平移到画布中心的：

```
this . context2D . translate ( this . canvas . width * 0.5 , this . canvas . height * 0.5 ) ;
```

而是使用下面的代码将局部坐标系平移到画布中心：

```
this . context2D . translate ( 0 , this . canvas . height * 0.5 - 10 ) ;
```

再来总结如下一条极其重要的结论：

- 结论 3：translate、rotate 和 scale 这些局部坐标系变换方法都具有累积性（Accumulation），每次变换操作都是相对上一次结果的叠加。

🔔 **注意**：所谓变换的累积性是指 translate、rotate 和 scale 这些操作都是相对上一次结果的叠加。例如上一次 translate，已经将局部坐标的原点平移到了全局坐标系中 x 轴的中心，因此不再需要移动局部坐标系。而上一次 y 轴只是 10 个单位的平移，现在需要平移到 y 轴的中心，因此需要使用 canva2 . height * 0.5 - 10 这样的算法来计算与上一次 y 轴平移的偏移量。

关键点已经说明白了，运行上述代码，会得到如图 5.11 所示的效果。

（6）然后来看一下旋转，将当前的局部坐标系逆时针旋转-120°后，再继续旋转-130°，并且绘制出两次旋转后的局部坐标系及基于局部坐标系定义的矩形，代码如下：

```
// 将坐标系继续旋转-120°
this . context2D . rotate ( Math2D . toRadian ( - 120 ) ) ;
// 绘制旋转-120°的矩形
this . fillLocalRectWithTitle ( width , height , ' 4、旋转-120度 ' ) ;
this . strokeCoord ( 0 , 0 , coordWidth , coordHeight ) ;
                              // 注意坐标为[ 0 , 0 ]，绘制坐标系 x 和 y 轴
this . fillCircle ( 0 , 0 , radius ) ;
```

```
                              // 注意坐标为[ 0 , 0 ] , 绘制坐标系原点
// 将坐标系在-120°旋转的基础上再旋转-130°, 合计旋转了-250°
this . context2D . rotate ( Math2D . toRadian ( - 130 ) ) ;
this . fillLocalRectWithTitle ( width , height , ' 5 . 旋转-130度 ' ) ;
this . strokeCoord ( 0 , 0 , coordWidth , coordHeight ) ;
                              // 注意坐标为[ 0 , 0 ], 绘制坐标系 x 和 y 轴
this . fillCircle ( 0 , 0 , radius ) ;
                              // 注意坐标为[ 0 , 0 ] , 绘制坐标系原点
```

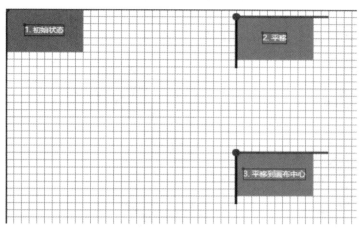

图 5.11 平移到画布中心

上述代码中,连续两次逆时针(负角度)旋转,根据变换的累积性原则,会将位于画布中心的局部坐标系先旋转-120°,再继续旋转-130°,总计旋转-250°。来看一下运行效果,如图 5.12 所示。

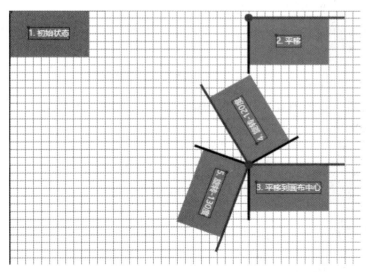

图 5.12 两次旋转后的结果

通过上面的代码及运行结果,验证了上述 3 条重要结论的正确性。

(7)继续沿着局部坐标系 x 轴(红色轴)的正方向平移 100 个单位,沿着局部坐标系 y 轴(蓝色轴)的正方向平移 100 个单位。具体代码如下:

```
// 沿着局部坐标的 x 轴和 y 轴正方向各自平移 100 个单位
this . context2D . translate ( 100 , 100 ) ;
this . fillLocalRectWithTitle ( width , height , ' 6、局部平移 100 个单位 ' ) ;
this . strokeCoord ( 0 , 0 , coordWidth , coordHeight ) ;
                              // 注意坐标为 [ 0 , 0 ],绘制坐标系 x 和 y 轴
```

当运行上述代码后,得到如图 5.13 所示的效果。

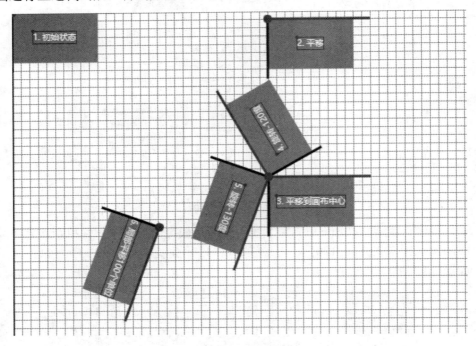

图 5.13　基于局部坐标系的平移操作

(8)最后来看一下使用 scale 方法将局部坐标系 x 轴放大 1.5 倍,y 轴放大 2.0 倍后的显示效果,代码如下:

```
this . context2D .scale ( 1.5 , 2.0 ) ;
                              // 局部坐标系的 x 轴放大 1.5 倍 , y 轴放大 2 倍
this . fillLocalRectWithTitle ( width , height , ' 7、x 轴局部放大 1.5 倍, y 轴局部放大 2 倍 ' ) ;
                              // 同时物体的宽度也会放大 1.5 倍, 高度放大 2 倍
this . strokeCoord ( 0 , 0 , coordWidth , coordHeight ) ;
                              // 注意坐标为 [ 0 , 0 ],绘制坐标系 x 和 y 轴
this . fillCircle ( 0 , 0 , radius ) ;
                              // 注意坐标为 [ 0 , 0 ], 绘制坐标系原点
```

再来看一下如图 5.14 所示的效果。

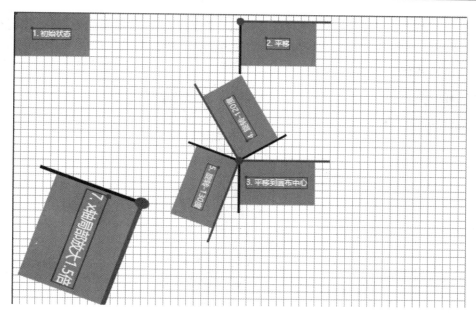

图 5.14　局部坐标系 x 和 y 轴非等比缩放

对比一下图 5.13 和图 5.14，可以很明显地看出 scale 缩放的是局部坐标系，局部坐标系的 x 和 y 轴的线宽都发生了改变，原点也发生明显的形变（因为 x 和 y 轴缩放比例不一致，导致呈现椭圆形），从而导致相对局部坐标系定义的矩形也发生了尺寸变化，符合上面得出的 3 条结论。

（9）现在将上面的缩放代码注释掉，然后调整一下 fillLocalRectWithTitle 方法中的 width 和 height 参数，将其分别放大 1.5 倍和 2.0 倍（此时变换的不是局部坐标系，而是调整物体在局部坐标系中的尺寸，从而导致物体的顶点坐标发生改变）。来看一下效果。具体代码如下：

```
// this . context2D .scale ( 1.5 , 2.0 ) ;
                                //局部坐标系的 x 轴放大 1.5 倍，y 轴放大 2 倍
// this . fillLocalRectWithTitle ( width , height , ' 7、x 轴局部放大 1.5 倍 ' ) ;
                                //同时物体的宽度也会放大 1.5 倍，高度放大 2 倍
// 注释掉上面进行局部坐标系缩放的代码，输入下面的代码
this . fillLocalRectWithTitle ( width * 1.5 , height * 2.0 , ' 8、放大物体尺寸 ' ) ;
                                // 这里是放大物体本身的尺寸，而不是放大局部坐标系的尺寸，一定要注意！！！
this . strokeCoord ( 0 , 0 , coordWidth , coordHeight ) ;
                                // 注意坐标为 [ 0 , 0 ]，绘制坐标系 x 和 y 轴
this . fillCircle ( 0 , 0 , radius ) ;
                                // 注意坐标为 [ 0 , 0 ]，绘制坐标系原点
```

来看一下效果，如图 5.15 所示。

对比一下图 5.14 和图 5.15 会发现，图 5.15 由于没有使用 scale 方法，因此坐标系本身大小没发生任何变化，而文字绘制使用的参考坐标系也因为没有使用 scale 方法，亦没发生尺寸变化。

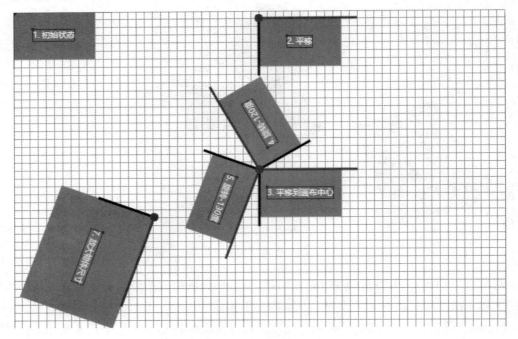

图 5.15　缩放物体尺寸

唯一改变的是矩形本身大小，因为绘制矩形时将其 width 和 height 参数分别扩大了 1.5 倍和 2.0 倍。通过这个例子，更加证明了 translate、rotate 和 scale 变换的是局部坐标系而不是物体本身（物体本身的变换会导致物体在局部坐标系中的顶点坐标发生改变）。

（10）恢复到图 5.14（也就是使用 scale 分别扩大局部坐标系的 x 和 y 轴）所示的源码，但是绘制时，将坐标系的原点位于矩形的左中位置，这时使用 ELayout．LEFT_MIDDLE 来绘制。具体代码如下：

```
this . context2D .scale ( 1.5 , 2.0 ) ;
                                    // 局部坐标系的 x 轴放大 1.5 倍，y 轴放大 2 倍
this . fillLocalRectWithTitle ( width , height , '7、缩放局部坐标系 ', ELayout .
LEFT_MIDDLE) ;              // 使用 LEFT_MIDDEL 来绘制矩形
//this . fillLocalRectWithTitle ( width * 1.5 , height * 2.0 , '7、
放大物体尺寸 ' ) ;
this . strokeCoord ( 0 , 0 , coordWidth , coordHeight ) ;
                                    // 注意坐标为 [ 0 , 0 ]，绘制坐标系 x 和 y 轴
this . fillCircle ( 0 , 0 , radius ) ;
                                    // 注意坐标为 [ 0 , 0 ]，绘制坐标系原点
```

来看一下上述代码的调用结果，如图 5.16 所示。

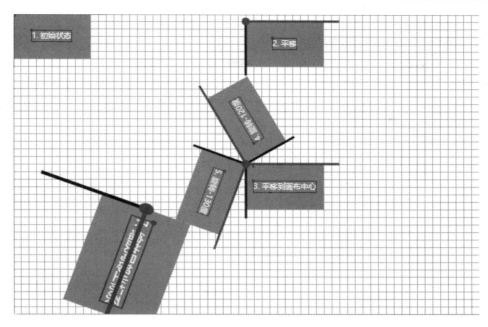

图 5.16　缩放后矩形左中对齐效果

通过（9）和（10）这两步实验会发现：

- 当调用 translate、rotate 和 scale 方法时，操作的是坐标系的变换。
- 使用 fillLocalRectWithTitle 方法绘制图形时，是在变换后的局部坐标系中，相对局部坐标系的原点重新定义顶点坐标数据（请参考 5.1.5 节中实现 fillLocalRectWithTitle 方法中根据 ELayout 的枚举值计算矩形左上角顶点坐标的代码），然后进行 stroke（描边）操作和（或）fill（填充）操作。

大家会看到任何形体（图形），最终总是由顶点坐标集合方式投送到渲染器，然后进行坐标系变换，将顶点变换到实际需要绘制的全局坐标系，然后进行光栅化操作后形成像素片段（图像）最终呈现到屏幕上，这个过程就是从图形转换成图像的过程。

最后再来重新总结本节的 3 个关键结论。

- 结论 1：使用 translate、rotate 和 scale 这些方法时，变换的是局部坐标系。
- 结论 2：所有绘制操作都是相对于变换后的局部坐标系所进行的，可以针对变换后的局部坐标系的原点重新定义顶点坐标，从而达到变换物体（由顶点集合组成的可绘制形体）原点的目的。
- 结论 3：translate、rotate 和 scale 这些局部坐标系变换方法都具有累积性，每次变换操作都是相对上一次结果的叠加。

笔者所掌握的坐标变换关键点就是：如本节所示，将空间变换分为局部坐标系变换和基于局部坐标系变换后的原点变换。

5.1.8 通用的原点变换方法

在前面章节中，大量使用了 fillLocalRectWithTitle 方法来演示坐标系变换和原点变换，其实，仔细观察 fillLocalRectWithTitle 方法中关于 9 个参考点的变换实现源码，会发现有更加通用的解决方案。

如果将 9 个参考点变换使用的 ELayout 枚举换成矩形尺寸的比例关系，会得到更加通用的、相对局部坐标系原点变换的算法。

来看一下该算法的实现过程，为了不干扰以前的 Demo，实现一个新的名为 fillLocalRectWithTitleUV 的方法。具体代码如下：

```
public fillLocalRectWithTitleUV (
    width : number , height : number ,           //矩形尺寸
    title : string = '' ,                         //矩形显示的文字内容
    u : number = 0 , v : number = 0 ,
                                                  //这里使用 u 和 v 参数代替原来的 ELayout 枚举
    layout : ELayout = ELayout . CENTER_MIDDLE ,  //文字框的对齐方式
    color : string = 'grey' ,                     //矩形填充颜色
    showCoord : boolean = true        // 是否显示局部坐标系，默认显示
) : void {
    if ( this . context2D !== null ) {
        // 将原来的 fillLocalRectWithTitle 方法中的 ELayout 9 种处理方式的代码替换
        成如下代码：
        let x : number = - width * u ;
        let y : number = - height * v ;
        // 和 fillLocalRectWithTitle 中的绘制代码一样
        // 因此在此省略
    }
}
```

由此会发现，fillLocalRectWithTitleUV 的实现比 fillLocalRectWithTitle 更加简洁、通用。接下来通过一个 Demo 来演示 fillLocalRectWithTitleUV 的用法，如图 5.17 所示。

接下来我们看一下图 5.17 效果实现的步骤，具体如下所述。

（1）首先来看一下如何将矩形变换到圆的路径上，并且让矩形的朝向正确显示。将这个操作封装成一个名为 translateRotateTranslateDrawRect 的方法，从名称就能知道整个变换的顺序。具体代码如下：

```
// 这个方法名称按照变换顺序取名
// 其形成一个圆的路径，而且绘制物体的朝向和圆路径一致
public translateRotateTranslateDrawRect ( degree : number , u : number = 0 , v : number = 0 , radius = 200 , width : number = 40 , height : number = 20 ) : void {
    if ( this . context2D === null ) {
        return ;
```

```
                    }
                    // 将角度变换为弧度
                    let radians : number = Math2D . toRadian ( degree ) ;
                    // 记录状态
                    this . context2D . save ( ) ;
                        // 将局部坐标系平移到画布的中心
                        this . context2D . translate ( this . canvas . width * 0.5 , this .
                        canvas . height * 0.5 );
                        // 然后再将局部坐标系旋转某个弧度
                        this . context2D . rotate ( radians ) ;
                        // 然后再将位于画布中心旋转后的局部坐标系沿着局部 x 轴的方向平移 250 个单位。
                        this . context2D . translate ( radius , 0 ) ;
                        // 在变换后的局部坐标系中根据 u、v 值绘制矩形,其原点由 u、v 确定
                        this . fillLocalRectWithTitleUV ( width , height , '' , u , v ) ;
                    // 丢弃修改的状态集
                    this . context2D . restore ( ) ;
}
```

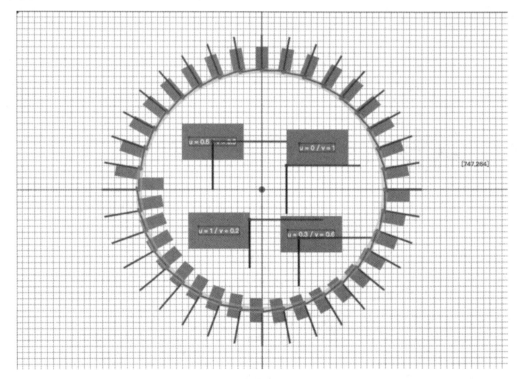

图 5.17　fillLocalRectWithTitleUV Demo 演示效果图

上述变换是一个经典的操作,在此再强调以下几点:
- 通过 translate 将与全局坐标系重合的局部坐标系变换到画布的中心点。
- 以变换后的局部坐标系的原点为转动点,旋转 radians 弧度,形成一个新的旋转后的局部坐标系。

- 将位于画布中心的、旋转后的局部坐标系沿着 x 轴的正方向平移 radius 个单位后形成新的局部坐标系。
- 在最后合成的局部坐标系中绘制矩形，该矩形的原点由 u 和 v 参数确定。

可以将坐标系分为固定的全局坐标和变换的局部坐标系，这样就能更加深刻地了解坐标系变换。

（2）有了 translateRotateTranslateDrawRect 这个关键方法，就可以实现如图 5.17 所示的效果了。具体代码如下：

```
public testFillLocalRectWithTitleUV ( ) : void {

    if ( this . context2D === null ) {
      return ;
    }

    let radius : number = 200 ;                    // 圆路径的半径为 200 个单位
    let steps : number = 18 ;
                        // 将圆分成上下各 18 个等分，-180°～180°，每个等分 10°

    // [ 0 , +180 ]度绘制 u 系数从 0～1，v 系数不变
    // 导致的结果是 x 轴原点一直从左到右变动，y 轴原点一直在上面（top）
    for ( let i = 0 ; i <= steps ; i ++ ) {
      let n : number = i / steps ;
      this . translateRotateTranslateDrawRect ( i * 10 , n , 0 , radius ) ;
    }

    // [ 0 , -180 ]度绘制
    // 导致的结果是 y 轴原点一直从上到下变动，x 轴原点一直在左面（left）
    for ( let i = 0 ; i < steps ; i ++ ) {
        let n : number = i / steps ;
        this . translateRotateTranslateDrawRect ( - i * 10 , 0 , n , radius ) ;
     }
    // 在画布中心的 4 个象限绘制不同 u、v 的矩形，可以看一下 u、v 不同系数产生的不同效果
    this . context2D . save ( ) ;
    this . context2D . translate ( this . canvas . width * 0.5 - radius * 0.4 , this . canvas . height * 0.5 - radius * 0.4 ) ;
    this . fillLocalRectWithTitleUV ( 100 , 60 ,'u = 0.5 / v = 0.5' , 0.5 , 0.5 ) ;
    this . context2D . restore ( ) ;
    this . context2D . save ( ) ;
    this . context2D . translate ( this . canvas . width * 0.5 + radius * 0.2 , this . canvas . height * 0.5 - radius * 0.2 ) ;
    this . fillLocalRectWithTitleUV ( 100 , 60 ,'u = 0 / v = 1' , 0 , 1 ) ;
    this . context2D . restore ( ) ;
    this . context2D . save ( ) ;
    this . context2D . translate ( this . canvas . width * 0.5 + radius * 0.3 , this . canvas . height * 0.5 + radius * 0.4 ) ;
    this . fillLocalRectWithTitleUV ( 100 , 60 ,'u = 0.3 / v = 0.6' , 0.3 , 0.6 ) ;
    this . context2D . restore ( ) ;
    this . context2D . save ( ) ;
    this . context2D . translate ( this . canvas . width * 0.5 - radius *
```

```
            0.1 , this . canvas . height * 0.5 + radius * 0.25 ) ;
        this . fillLocalRectWithTitleUV ( 100 , 60 ,'u = 1 / v = 0.2' , 1 , 0.2 ) ;
        this . context2D . restore ( ) ;
        // 使用 10 个单位线宽，半透明的颜色绘制圆的路径
        this . strokeCircle ( this . canvas . width * 0.5 , this . canvas . height
            * 0.5 , radius , 'rgba( 0 , 255 , 255 , 0.5 )' , 10 ) ;
    }
```

运行上述代码后，会获得图 5.17 所示的效果。从代码及演示效果中可以得到以下几个信息：

- Canvas2D 中的第 1 象限是右下区域，取值范围为[0 , 90]度。
- Canvas2D 中的第 2 象限是左下区域，取值范围为[90 , 180]度。
- Canvas2D 中的第 3 象限是左上区域，取值范围为[-180 , -90]度。
- Canvas2D 中的第 4 象限是右上区域，取值范围为[-90 , 0]度。

而从演示效果可以观察到：

- 0～180°，每 10°绘制一个矩形时，其路径是在下半个圆上（第 1 和第 2 象限上），并且不停地沿着 x 轴正方向从左到右移动原点。
- 而从 0～-180°时，每隔 10°绘制一个矩形时，其路径是在上半个圆（第 4 和第 3 象限上），并且不停地沿着 y 轴正方向从上到下移动原点。
- 4 个象限中又有 4 个显示 u、v 系数的矩形，用于演示不同的 u、v 取值范围。
- 大家可以将 u、v 取正负值，以及超出[0 , 1]范围值时测试一下效果。

关于象限及旋转等相关细节，将在后续章节中详解。

5.1.9　公转（Revolution）与自转（Rotation）

前面章节中比较详细地介绍了局部坐标系变换，以及原点变换的相关知识，本节通过一个公转和自转相关的 Demo 来更加深入地理解坐标系变换，先来看一下效果，如图 5.18 所示。

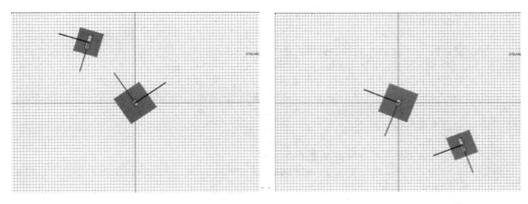

图 5.18　公转与自转

参考图 5.18 所示，该 Demo 的需求描述如下：
- 大一点的正方形位于画布中心进行自转。
- 小一点的正方形自转的同时绕着大一点的正方形以圆形轨道进行公转。

需求就这么简单，这是一个经典的太阳系模拟，只是使用正方形而不是圆形来模拟太阳和其他行星（代码中使用 Moon 来指代），这是因为正方形自转时效果明显，而圆形自转时效果并不明显，下面来看一下代码实现。

（1）首先这是一个动画效果演示，它需要不停地进行角位移的计算，因此需要在 TestApplication 类中声明如下成员变量：

```
private _rotationSunSpeed : number = 50 ;          //太阳自转的角速度，以角度为单位
private _rotationMoonSpeed : number = 100 ;        //月球自转的角速度，以角度为单位
private _revolutionSpeed : number = 60 ;           //月球公转的角速度

private _rotationSun : number = 0 ;                //太阳自转的角位移
private _rotationMoon : number = 0 ;               //月亮自转的角位移
private _revolution : number = 0 ;                 //月亮围绕太阳公转的角位移
```

（2）实现基于时间的更新，因此需要覆写（override）基类的 update 方法。具体代码如下：

```
// override update 函数
public update ( elapsedMsec : number , intervalSec : number ) : void {
    // 角位移公式：v = s * t ;
    this . _rotationMoon += this . _rotationMoonSpeed * intervalSec ;
    this . _rotationSun += this . _rotationSunSpeed * intervalSec ;
    this . _revolution += this . _revolutionSpeed * intervalSec ;
}
```

（3）实现一个名为 rotationAndRevolutionSimulation 的方法，该方法进行太阳自转及月亮的公转和自转变换及绘制操作。具体代码如下：

```
// 公转自转模拟
public rotationAndRevolutionSimulation ( radius : number = 250 ) : void {
    if ( this . context2D === null ) {
        return ;
    }
    // 将自转 rotation 转换为弧度表示
    let rotationMoon : number = Math2D . toRadian ( this._rotationMoon ) ;
    // 将公转 revolution 转换为弧度表示
    let rotationSun : number = Math2D . toRadian ( this . _rotationSun ) ;
    let revolution : number = Math2D . toRadian ( this . _revolution ) ;
    // 记录当前渲染状态
    this . context2D . save ( ) ;
        // 将局部坐标系平移到画布中心
        this . context2D . translate ( this . canvas . width * 0.5 , this . canvas . height * 0.5 );
        this . context2D . save ( ) ;
            // 绘制矩形在画布中心自转
```

```
            this . context2D . rotate ( rotationSun ) ;
                    // 绕局部坐标系原点自转
            this . fillLocalRectWithTitleUV ( 100 , 100 , '自转' , 0.5 , 0.5 );
        this . context2D . restore ( ) ;
        // 公转 + 自转,注意顺序:
        this . context2D . save ( ) ;
            this . context2D . rotate ( revolution ) ;                  // 先公转
            this . context2D . translate ( radius , 0 ) ;
                    //然后沿着当前的 x 轴平移 radius 个单位, radius 半径形成圆路径
            this . context2D . rotate ( rotationMoon ) ;
                    // 一旦平移到圆的路径上,开始绕局部坐标系原点进行自转
            this . fillLocalRectWithTitleUV ( 80 , 80 , '自转+公转' , 0.5 ,
            0.5 ) ;
        this . context2D . restore ( ) ;
        // 恢复上一次记录的渲染状态
        this . context2D . restore ( ) ;
}
```

上述代码中注释足够详细了,但还是需要重点提示的一点是,仔细地体会 translate 和 rotate 的变换顺序,以及 save 和 restore 之间的层次关系:上面的代码层次是先将局部坐标系平移到画布中心,然后太阳在画布中心旋转,月亮的公转半径也是相对画布中心的偏移。

大家可以修改一下_rotationMoonSpeed 和_revolution 这两个成员变量的数值,观察一下自转和公转的效果。

5.1.10　原点变换的另一种方法

在前面的代码中,所有的原点变换都是使用 ELayout 枚举值或 u、v 参数方式来控制的。再来总结一下其特点:

首先进行各种层次的局部坐标系变换(translate、rotate 和 scale 操作变换的局部坐标系)。

然后在绘制代码中,固定变换后的局部坐标系(fillLocalRectWithTitleUV 方法定义要绘制的矩形时,假设局部坐标系是固定不变的局部坐标系),变动的是几何图形的各个顶点相对局部坐标系原点的偏移量,从而形成几何图形原点的变换操作。

如果自己实现一个精灵系统,精灵的各个几何形体是自己定义的,那么可以很容易地使用上面提到的方式来变换各个形体的原点坐标。

但是,如果使用的是现成的第三方库,而且该库已经限定例如矩形原点在左上角,而且由于版权协议等原因,不允许修改源码,在这种情况下,就要使用第二种变换原点的方式了,从而达到绕点旋转或以某点为中心的缩放或平移操作(实际还是使用 translate、rotate 和 scale 这些方法来变换坐标系)。

修改 5.1.9 节中的源码,使用变换坐标系的方式来重新定义要追踪的原点(平移、旋转或缩放时的参考点),先来看一下如图 5.19 所示的效果。

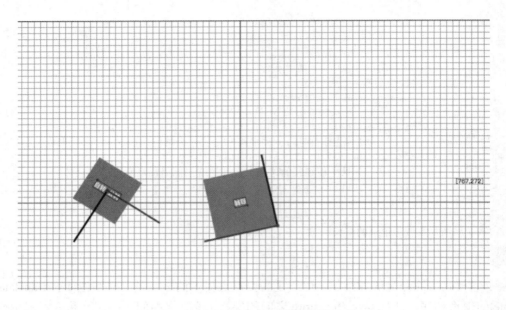

图 5.19　使用 translate 变换旋转参考点

会发现图 5.19 中，虽然旋转时参考点位于矩形（Sun）的中心，并且与全局坐标中的画布中心重合，但是其局部坐标是在矩形的左上角（由于自转关系，在全局坐标中，太阳的局部坐标系是在右下角）。而在图 5.18 中，局部坐标系的原点位于矩形（Sun）的中心。

要实现图 5.19 所示的效果，需要输入如下代码片段：

```
this . context2D . save ( ) ;
    // 绘制矩形在画布中心自转
    this . context2D . rotate ( revolution ) ;
    // 因为要绘制的矩形宽度和高度都为 100 个单位，而原点在矩形左上角
    // 为了将自转的原点位于矩形的中心，同时要让矩形的中心与画布原点重合
    // 因此再次沿着局部坐标系的负 x 轴（左）和负 y 轴（上）各自平移 50 个单位
    // 这样旋转的参考点位于矩形的中心，并且在全局坐标中也和画布原点重合
    this . context2D . translate ( - 50 , - 50 ) ;
    this . fillLocalRectWithTitleUV ( 100 , 100 , '自转' , 0.0 , 0.0 );
this . context2D . restore ( ) ;
```

> 注意：Canvas2D 坐标系中的 y 轴的正方向指向下方，而负方向指向上方，和数学中的描述正好相反。

通过这个简单的例子可以发现，通过 translate、rotate 和 scale 进行局部坐标系的变换，可以完成所有想要做的事情，这也是坐标系变换的精华所在。

本节中使用了几个例子演示了 translate、rotate 和 scale 方法的用法。下一节将实现一个坦克例子，帮助读者进一步掌握 translate、rotate 和 scale 的层次变换用法，以及了解一些常用的三角函数知识。

5.2 坦克 Demo

为了进一步掌握关键的坐标变换操作，在本节中，将实现一个经典的坦克 Demo。先来看一下演示效果，如图 5.20 和图 5.21 所示。

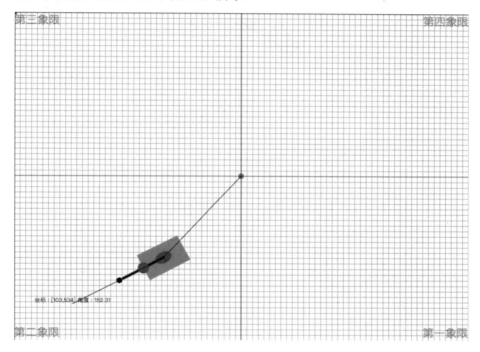

图 5.20　坦克朝着鼠标光标位置朝向正确地移动

根据图 5.20 及图 5.21 所示，来描述一下需求：
- 将整个画布分为四大象限，分别标记出这些象限。
- 坦克本身由四大基本图形组成：直线表示炮管，椭圆表示炮塔，矩形的底座及圆形用于标记坦克的正前方及炮口。
- 当移动鼠标时，整个坦克会自动旋转后朝着鼠标指针方向并不停地移动，到达鼠标指针所在点后停止运行。
- 当按 R 键时，坦克的炮塔顺时针旋转；当按 E 键时，坦克的炮塔逆时针旋转；当按 T 键时，坦克的炮塔重置到初始化状态。
- 在上一节的 drawCoordInfo 方法中增加输出坦克当前的朝向角度相关信息。
- 为了追踪运行路线，绘制出画布中心点到坦克中心点（translate、rotate 和 scale 追踪原点），以及坦克中心点到鼠标指针位置的连线。

这个坦克 Demo，蕴含了几个比较重要的数学操作：

- Canvas2D 中象限及每个象限对应的角度取值范围（在 5.18 节已作说明）。
- 坦克整体朝向，以及移动时涉及 atan2、sin 和 cos 这三个初中学过的三角函数的应用。
- 炮管依赖于底座，但是能够独立控制，此时需要特殊处理，涉及坐标系变换时的层次操作。

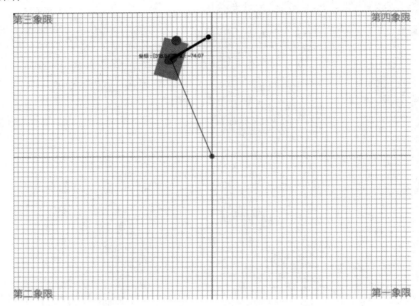

图 5.21　坦克炮塔旋转效果

5.2.1　象限（Quadrant）文字绘制

在 5.1.8 节中提到过，Canvas2D 中的象限是从右下角开始为第 1 象限，然后顺时针依次为第 2、第 3 和第 4 象限，因此可以使用第 4 章实现的 fillText 函数，使用不同的文字布局参数，将其绘制出来，以利于显示参考。具体代码如下：

```
public draw4Quadrant ( ) : void {
  if ( this . context2D === null ) {
    return ;
  }
  this . context2D . save ( ) ;
    this . fillText ("第一象限"    , this . canvas . width , this . canvas .
    height , 'rgba( 0 , 0 , 255 , 0.5 )' , 'right' , 'bottom' , "20px
    sans-serif" ) ;
    this . fillText ("第二象限"    , 0 , this . canvas . height , 'rgba( 0 ,
    0 , 255 , 0.5 )' , 'left' , 'bottom' , "20px sans-serif" ) ;
    this . fillText ("第三象限"    , 0 , 0 , 'rgba( 0 , 0 , 255 , 0.5 )' ,
    'left' , 'top' , "20px sans-serif" ) ;
```

```
        this . fillText ("第四象限"    , this . canvas . width , 0 , 'rgba( 0 ,
            0 , 255 , 0.5 )' , 'right' , 'top' , "20px sans-serif" ) ;
        this . context2D . restore ( ) ;
    }
```

5.2.2 坦克形体的绘制

接下来绘制坦克的形体。坦克本身由四大基本图形组成：直线表示炮管，椭圆表示炮塔，矩形的底座，以及圆形用于标记坦克的正前方。为了让代码更加清晰，将坦克的所有操作、更新及绘制的相关代码都放在 Tank 类中。下面来看一下 Tank 类的成员变量及绘制方法。具体代码如下：

```
class Tank {
    // 坦克的大小尺寸
    public width : number = 80 ;
    public height : number = 50 ;
    // 坦克当前的位置
    // default 情况下为[ 100 , 100 ]
    public x : number = 100 ;
    public y : number = 100 ;
    // 坦克当前的 x 和 y 方向上的缩放系数
    // default 情况下为 1.0
    public scaleX : number = 1.0 ;
    public scaleY : number = 1.0 ;
    // 坦克当前的旋转角度
    public tankRotation : number = 0 ;          //整个坦克的旋转角度，弧度表示
    public turretRotation : number = 0 ;        //炮塔的旋转角度，弧度表示
    // 在 Tank 类中增加一个成员变量，用来标示 Tank 初始化时是否朝着 y 轴正方向
    public initYAxis : boolean = true ;
    public showLine : boolean = false ;
                            //是否显示坦克原点与画布中心点和目标点之间的连线
    public showCoord : boolean = false ;        //是否显示坦克本身的局部坐标系
    public gunLength : number = Math . max ( this . width , this . height ) ;
// 炮管长度，default 情况下，等于坦克的 width 和 height 中最大的一个数值
    public gunMuzzleRadius : number = 5 ;
    public draw ( app : TestApplication ) : void {
        if ( app . context2D === null ) {
            return ;
        }
        // 整个坦克绘制 tank
        app . context2D . save ( ) ;
            // 整个坦克移动和旋转，注意局部变换的经典结合顺序（ trs: translate -> 
            rotate -> scale ）
            app . context2D . translate ( this . x , this . y ) ;
            app . context2D . rotate ( this . tankRotation ) ;
            app . context2D . scale ( this . scaleX , this . scaleY ) ;
            // 绘制坦克的底盘（矩形）
            app . context2D . save ( ) ;
                app . context2D . fillStyle = 'grey' ;
                app . context2D . beginPath ( ) ;
```

```
            app . context2D . rect ( - this . width * 0.5 , - this . height 
            * 0.5 , this . width , this . height ) ; 
            app . context2D . fill ( ) ; 
        app . context2D . restore ( ) ; 
        // 绘制炮塔 turret
        app . context2D . save ( ) ; 
            app . context2D . rotate ( this . turretRotation ) ; 
            // 椭圆炮塔 ellipse 方法
            app . context2D . fillStyle = 'red' ; 
            app . context2D . beginPath ( ) ; 
            app . context2D . ellipse ( 0 , 0 , 15 , 10 , 0 , 0 , Math.PI
            * 2 ) ; 
            app . context2D . fill ( ) ; 
            // 炮管 gun barrel （炮管）
            app . context2D . strokeStyle = 'blue' ; 
            app . context2D . lineWidth = 5 ; 
                                            //炮管需要粗一点,因此为 5 个单位
            app . context2D . lineCap = 'round' ;      // 使用 round 方式
            app . context2D . beginPath ( ) ; 
            app . context2D . moveTo ( 0 , 0 ) ; 
            app . context2D . lineTo ( this . gunLength , 0 ) ; 
            app . context2D . stroke ( ) ; 
            // 炮口，先将局部坐标系从当前的方向，向 x 轴的正方向平移 gunLength
            (数值类型的变量，以像素为单位，表示炮管的长度) 个像素，此时局部坐标系
            在炮管最右侧
            app . context2D . translate ( this . gunLength , 0 ) ; 
            // 然后再从当前的坐标系向 x 轴的正方向平移 gunMuzzleRadius（数值类型
            的变量，以像素为单位，表示炮管的半径) 个像素，这样炮口的外切圆正好和炮
            管相接触
            app . context2D . translate ( this . gunMuzzleRadius , 0 ) ; 
            // 调用自己实现的 fillCircle 方法，内部使用 Canvas2D arc 绘制圆弧方法
            app . fillCircle ( 0 , 0 , 5 , 'black' ) ; 
        app . context2D . restore ( ) ; 
        // 绘制一个圆球，标记坦克正方向，一旦炮管旋转后，可以知道正前方在哪里
        app . context2D . save ( ) ; 
            app . context2D . translate ( this . width * 0.5 , 0 ) ; 
            app . fillCircle ( 0 , 0 , 10 , 'green' ) ; 
        app . context2D . restore ( ) ; 
        // 坐标系是跟随整个坦克的
        if ( this . showCoord ) {
            app . context2D . save ( ) ; 
                app . context2D . lineWidth = 1 ; 
                app . context2D . lineCap = '' ; 
                app . strokeCoord ( 0 , 0 , this . width * 1.2 , this . 
                height * 1.2 ) ; 
            app . context2D . restore ( ) ; 
        }
    app . context2D . restore ( ) ; 
    }
}
```

来调用上面的坦克绘制函数。具体代码如下：

```typescript
// 在 TestApplication 类中增加如下成员变量
private _tank : Tank ;
// 在 TestApplication 类的构造函数中初始化_tank 成员变量, 并且让坦克位于画布中心
this . _tank = new Tank ( ) ;
this . _tank . x = canvas . width * 0.5 ;
this . _tank . y = canvas . height * 0.5 ;
// 在 TestApplication 类中增加 drawTank 方法
public drawTank ( ) : void {
    this . _tank . draw ( this ) ;
}
// 在 TestApplication 类中覆写(override) render 方法, 当调用 TestApplication
类的 start 方法后就会进入动画的 update 和 render 循环
public render ( ) : void {
    // 由于 canvas . getContext 方法返回的 CanvasRenderingContext2D 可能会是 null
    // 因此 VS Code 会强制要求 null 值检查, 否则报错
    if ( this . context2D !== null ) {
        let centX : number
        // 每次重绘都先清屏
        this . context2D . clearRect ( 0 , 0 , this . canvas .width ,this .
        canvas .height ) ;
        // 调用第 4 章实现的背景网格绘制方法
        this . strokeGrid ( ) ;
        // 绘制中心原点和 x 与 y 轴
        this . drawCanvasCoordCenter ( ) ;
        this . draw4Quadrant ( ) ;
        this . drawTank ( ) ;
    }
}
```

当按 F5 键以调试模式运行上述代码后, 就会得到如图 5.22 所示的效果。

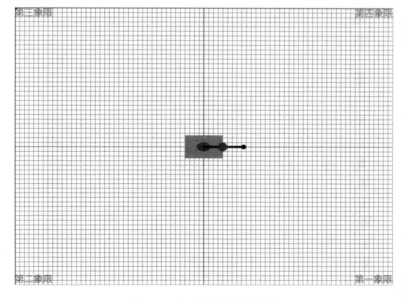

图 5.22　坦克初始化效果图

5.2.3　坦克及炮塔的旋转

从需求描述及 5.2.2 节中坦克绘制的实现代码，可以了解到坦克作为一个整体，可以控制整个坦克的位置、缩放和方向。而炮管依赖坦克本身的变换，同时也允许独自旋转，这是很经典的整体与局部的层次变换关系，可以通过 CanvasRenderingContext2D 渲染上下文对象的 save 和 restore 来实现这种整体与局部的层次变换关系，关键代码如下：

```
// 整个坦克绘制 tank
// 整个坦克作为变换操作的第一层
app . context2D . save ( ) ;
    // 整个坦克移动和旋转，注意局部变换的经典结合顺序（ trs: translate -> rotate -> scale ）
    app . context2D . translate ( this . x , this . y ) ;
    app . context2D . rotate ( this . tankRotation ) ;
    app . context2D . scale ( this . scaleX , this . scaleY ) ;
    // 上面的代码控制了整个坦克的平移、旋转和缩放操作
    // 炮塔作为第二层，受第一层的 save 所影响
    app . context2D . save ( ) ;
    // 很重要的一点，炮塔作为整个坦克的一部分，是在上一级变换（trs）后的累积操作
        app . context2D . rotate ( this . turretRotation ) ;
        // 炮塔由 3 部分组成：椭圆底座+炮管+炮口
        // 绘制底座
        // 绘制炮管
        // 绘制炮口
    app . context2D . restore ( ) ;
app . context2D . restore ( ) ;
```

那么来做个实验，设置整个坦克的位置在画布中心，旋转 30°，然后扩大两倍，同时炮塔旋转-30°。具体代码如下：

```
this . _tank = new Tank ( ) ;
// 让坦克位于画布中心
this . _tank . x = canvas . width * 0.5 ;
this . _tank . y = canvas . height * 0.5 ;
// 让坦克按比例整体扩大两倍
this . _tank . scaleX = 2 ;
this . _tank . scaleY = 2 ;
// 分别旋转坦克和炮管，将角度转换为弧度表示
this . _tank . tankRotation = Math2D . toRadian ( 30 ) ;
this . _tank . turretRotation = Math2D . toRadian ( - 30 ) ;
```

来看一下上述代码运行后的结果，如图 5.23 所示，坦克位于画布中心，与图 5.22 相比，坦克整体扩大了两倍，同时整个坦克顺时针旋转了 30°，而炮塔作为整个坦克的一部分，也会同时旋转 30°。但是后面又设置炮塔继续独自逆时针旋转 30°，这样一正一负方向旋转，炮塔应为 0°，于是和 x 轴重合并指向 x 轴的正方向，图 5.23 所示的结果验证了其正确性。

这是一个比较经典的、正确理解层次变换的 Demo，建议读者亲自动手调试并理解变换的细节和原理。

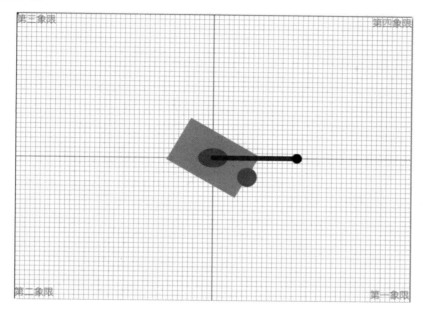

图 5.23　坦克与炮管的旋转

5.2.4　计算坦克的朝向

根据坦克 Demo 的需求描述，当移动鼠标时，整个坦克会自动旋转后朝着鼠标指针方向，具体效果可参考图 5.24 所示。这是一个很经典的朝向计算应用，可以使用 TypeScript 或 JavaScript 内置的 Math 类的静态方法 atan2 来实现上述需求。

需要在 Tank 类中增加要朝向的某个点（例如鼠标指针的位置）的成员变量，代码如下：

```
public targetX : number = 0 ;
public targetY : number = 0 ;
```

继续实现一个名为_lookAt 的私有函数，该方法使用 atan2 函数计算出坦克中心点的位置与鼠标指针之间的方位角，并将该方位角赋值给 tankRotation 成员变量。具体代码如下：

```
private _lookAt ( ) : void {
    // 将鼠标点的 x 和 y 变换为相对坦克坐标系原点的表示值
    let diffX : number = this . targetX - this . x ;
    let diffY : number = this . targetY - this . y ;
    // 通过 atan2 方法，计算出方位角，以弧度表示
    let radian = Math . atan2 ( diffY , diffX ) ;
```

```
        // 设置坦克将要朝向的方向
        this . tankRotation = radian ;
    }
```

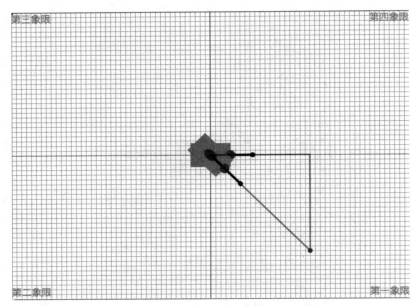

图 5.24　使用 atan2 方法计算方位角

在 Tank 类中增加一个处理鼠标指针移动的方法。具体代码如下：

```
public onMouseMove ( evt : CanvasMouseEvent ) : void {
    // 每次鼠标指针移动时，记录当前鼠标指针在 Canvas2D 画布中的位置
    this . targetX = evt . canvasPosition . x ;
    this . targetY = evt . canvasPosition . y ;
    // 并且根据鼠标指针的位置与坦克本身的位置，使用 atan2 方法计算出方位夹角
    this . _lookAt ( ) ;
}
```

只要在 TestApplication 类的构造函数中设置 isSupportMouseMove 为 true，并且覆写（override）基类鼠标移动的事件处理方法，代码如下：

```
// 覆写（override）基类的 dispatchMouseMove 方法
protected dispatchMouseMove ( evt : CanvasMouseEvent ) : void {
    // 必须要设置 this . isSupportMouseMove = true 才能处理 moveMove 事件
    this . _mouseX = evt . canvasPosition . x ;
    this . _mouseY = evt . canvasPosition . y ;
    this . _tank . onMouseMove ( evt ) ;
}
```

为了更好地了解当前鼠标相对坦克坐标系的位置信息及旋转角度，在 TestApplication 类的 render 函数中绘制出相关信息。具体代码如下：

```
public render ( ) : void {
    // 由于 canvas . getContext 方法返回的 CanvasRenderingContext2D 可能会是 null
```

```
        // 因此 VS Code 会强制要求进行 null 值检查，否则报错
        if ( this . context2D !== null ) {
            let centX : number
            // 每次重绘都先清屏
            this . context2D . clearRect ( 0 , 0 , this . canvas . width ,this .
            canvas .height ) ;
            // 调用第 4 章实现的背景网格绘制方法
            this . strokeGrid ( ) ;
            // 绘制中心原点和 x、y 轴
            this . drawCanvasCoordCenter ( ) ;
            this . draw4Quadrant ( ) ;
            this . drawTank ( ) ;
            // 坐标信息总是在最后绘制
            // 1. 显示鼠标当前位置（相对坦克坐标系的表示，而不是全局表示！！）
            // 2. 显示当前坦克方位角度，使用 Number . toFix ( 2 )方法，将浮点数保留两
            位小数
            this . drawCoordInfo (
                '坐标 : [' + ( this . _mouseX - this . _tank . x ) . toFixed
                ( 2 ) + ',' + ( this . _mouseY - this . _tank . y ) . toFixed
                ( 2 ) + "] 角度 : " + Math2D . toDegree (this . _tank . tank
                Rotation ) . toFixed ( 2 ) ,
                this . _mouseX ,
                this . _mouseY
            ) ;
        }
    }
```

一切准备就绪，当运行上述代码后，会看到坦克一直位于画布的中心，随着鼠标指针的移动，不停地跟着鼠标指针进行转动，并且实时地显示当前鼠标指针的坐标位置信息（相对 tank 坐标系的偏移）及当前的旋转角度。

可以不停地移动鼠标（或者移动到四个象限的分界线上）来观察鼠标指针坐标信息及坦克的旋转角度，从而了解 atan2 内部实现的一些关键信息（以当前 Demo 为例）：

（1）Math 类的 atan2 静态方法签名如下：

```
/**
* Returns the angle (in radians) from the X axis to a point . （相对 x 轴的
正方向的有符号弧度）
* @param y A numeric expression representing the cartesian y-coordinate .
* @param x A numeric expression representing the cartesian x-coordinate .
*/
public atan2 ( y : number , x : number ) : number ;
```

以坦克 Demo 为例，atan2 的参数[x , y]的取值为[this . _mouseX - this . _tank . x , this . _mouseY - this . _tank . y]，是鼠标指针坐标点相对坦克坐标系原点的偏移表示，而不是相对整个画布左上角的偏移，这一点一定要注意！

（2）当鼠标指针坐标点在象限内时，atan2 的返回值如下：

- 在第 1 象限中时，点[x , y]的值[+ , +]，返回夹角的弧度为 (0 , Math . PI / 2)，角度表示则为(0 , 90)。

- 在第 2 象限中时，点[x,y]的值[-,+]，返回夹角的弧度为（Math.PI/2，Math.PI），角度表示则为(90,180)。
- 在第 3 象限中时，点[x,y]的值[-,-]，返回夹角的弧度为 (-Math.PI,-Math.PI/2)，角度表示则为(-180,-90)。
- 在第 4 象限中，点[x,y]的值[+,-]，返回夹角的弧度为 (-Math.PI/2,0)，角度表示则为(-90,0)。

（3）当鼠标指针坐标点在象限的分界线上或原点上时，atan2 的返回值如下：
- 如果 y 为 0 且 x 为正，即[+,0]时，则返回夹角的弧度为 0，角度表示则为 0°。
- 如果 y 为 0 且 x 为负，即[-,0]时，则返回夹角的弧度为 Math.PI，角度表示则为 180°。
- 如果 x 为 0 且 y 为正，即[0,+]时，则返回夹角的弧度为 Math.PI/2，角度表示则为 90°。
- 如果 x 为 0 且 y 为负，即[0,-]时，则返回夹角的弧度为-Math.PI/2，角度表示则为-90°。
- 当鼠标指针坐标点位于坦克局部坐标系的原点[0,0]时，则返回夹角的弧度为 0，角度表示则为 0°。

（4）atan2 与 atan 方法的区别如下：

当将_lookAt 方法中的 Math.atan2 (diffY , diffX) 这句代码替换为 Math.atan (diffY / diffX)后，可以发现运行程序后，在第一象限[0，Math.PI/2]和第四象限[0,-Math.PI/2]中表现正确，而在第二象限和第三象限中表现不正确。由此可知，atan 返回的弧度值处于[-Math.PI/2，Math.PI/2]之间，也就是[-90，+90]°。

5.2.5 坦克朝着目标移动

当使用 atan2 方法计算出坦克本身和鼠标指针坐标点之间的夹角弧度，并且将该夹角弧度值赋值给坦克的 tankRotation 成员变量后，意味着坦克已经发生了旋转，接下来，需要做的是沿着旋转后的方向一直运行，当到达鼠标指针坐标点时停止运行，如图 5.25 所示。

可以使用 sin 和 cos 函数来计算斜向运动时的 x 和 y 分量，来看一下如下代码：

```
// Tank 类中增加如下成员变量，用于表示线性移动时的速率
public linearSpeed : number = 100.0 ;
private _moveTowardTo ( intervalSec : number ) : void {
    // 将鼠标点的 x 和 y 变换到相对坦克坐标系原点的表示
    let diffX : number = this . targetX - this . x ;
    let diffY : number = this . targetY - this . y ;
    // linearSpeed 的单位是：像素 / 秒
    let currSpeed : number = this . linearSpeed * intervalSec ;
                                // 根据时间差计算出当前的运行速度
```

```
        // 关键点 1：判断坦克是否要停止运动
        // 如果整个要运行的距离大于当前的速度，说明还没到达目的地，可以继续刷新坦克的位置
        if ( ( diffX * diffX + diffY * diffY ) > currSpeed * currSpeed ) {
            // 关键点 2：使用 sin 和 cos 函数计算斜向运行时 x、y 分量
            this . x = this . x + Math . cos ( this . tankRotation  ) * currSpeed ;
            this . y = this . y + Math . sin (  this . tankRotation ) * currSpeed ;
        }
    }
    // 在 Tank 类中实现 update 公开方法，TestApplication 类的 update 覆写方法调用本方法
    // 就可以不停地更新坦克的位置
    public update ( intervalSec : number ) : void {
        this . _moveTowardTo (intervalSec ) ;
    }
```

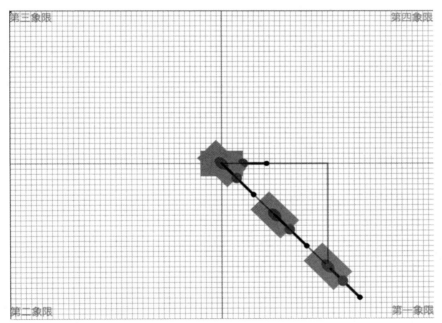

图 5.25　使用 sin 和 cos 函数朝着目标移动

- 当运行上面的代码后，会看到随着鼠标指针的移动，坦克会正确地移动。

5.2.6　使用键盘控制炮塔的旋转

可以在 Tank 类中增加一个名为 onKeyPress 的公开方法，用于实现键盘事件的处理，来看一下如下实现代码：

```
//Tank 增加一个成员变量
//turretRotateSpeed 用于控制炮塔的旋转速度，初始化时旋转速度设置为每秒旋转 2°角度表示
//由于三角函数中要使用弧度为单位，因此需要调用 Math2D.toRadian( 2 ) 方法，将角度转
```

换为弧度表示
```
public turretRotateSpeed : number = Math2D . toRadian ( 2 ) ;
public onKeyPress ( evt : CanvasKeyBoardEvent ) : void {
    if ( evt . key === 'r' ) {
        this . turretRotation += this . turretRotateSpeed ;
    } else if ( evt . key === 't' ) {
        this . turretRotation = 0 ;
    } else if ( evt . key === 'e' ) {
        this . turretRotation -= this . turretRotateSpeed ;
    }
}
// TestApplication 类中覆盖（override）基类的 dispatchKeyPress
protected dispatchKeyPress ( evt : CanvasKeyBoardEvent ) : void {
    this . _tank . onKeyPress ( evt ) ;
}
```

当运行代码后，按 R 键是顺时针旋转炮塔，按 E 键是逆时针旋转炮塔，而按 T 键可将炮塔恢复到初始化状态。

最后来绘制坦克的原点与画布的中心连线，以及坦克原点与鼠标指针坐标点的连线，可以在 Tank 类的 draw 函数的最后添加如下代码：

```
// 整个坦克绘制 tank
app . context2D . save ( ) ;
    ......................
    ......................
app . context2D . restore ( ) ;
app . context2D . save ( ) ;
    // 绘制坦克原点到画布中心的连线
    app . strokeLine ( this . x , this . y , app . canvas . width * 0.5 ,
    app . canvas . height * 0.5 ) ;
    // 绘制坦克原点到目标点（鼠标指针位置）的连线
    app . strokeLine ( this . x , this . y , this . targetX , this .targetY ) ;
app . context2D . restore ( ) ;
```

上述代码中使用的坐标都是相对画布左上角的偏移表示（全局坐标系），因此这段代码的 save 和 restore 对与整个 Tank 的 save 和 restore 对处于同一层次。

至此实现了全部的需求，当运行坦克 Demo 的代码后，就会得到如图 5.20 所示的结果。

5.2.7 初始朝向的重要性

参考图 5.22，当坦克始化时，朝向为 x 轴正方向（此时 tankRotation 为 0 弧度），上述代码运行的结果非常正确。

现在来做个实验，如图 5.26 所示，让坦克初始化时朝向为 y 轴正方向（此时 tankRotation 仍为 0 弧度），看看会是怎样的效果？

第 5 章 Canvas2D 坐标系变换

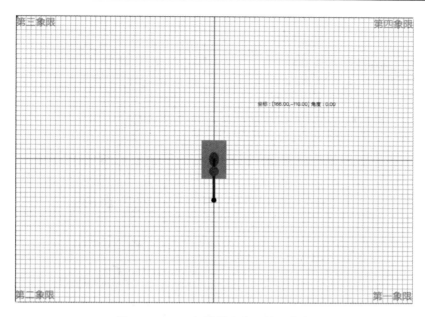

图 5.26 Tank 初始朝向为 y 轴正方向

为了演示上述效果，扩展一下 Tank 的 draw 函数，让其支持初始化时朝着 y 轴正方向显示。具体代码如下：

```
// 在 Tank 类中增加一个成员变量，用来标示 Tank 初始化时是否朝着 y 轴正方向
public initYAxis : boolean = true ;
public draw ( app : TestApplication ) : void {
    if ( app . context2D === null ) {
        return ;
    }
    // 整个坦克绘制 tank
    app . context2D . save ( ) ;
        // 整个坦克移动和旋转，注意局部变换的经典结合顺序（ trs: translate -> rotate -> scale ）
        app . context2D . translate ( this . x , this . y ) ;
        app . context2D . rotate ( this . initYAxis ? this . tankRotation - Math . PI * 0.5 : this . tankRotation ) ;
        app . context2D . scale ( this . scaleX , this . scaleY ) ;
        // 绘制坦克的底盘（矩形）
        app . context2D . save ( ) ;
            app . context2D . fillStyle = 'grey' ;
            app . context2D . beginPath ( ) ;
            if ( this . initYAxis ) {
                // 交换 width 和 height，这样就不需要修改 TestApplication 中的示例代码了
                app . context2D . rect ( - this . height * 0.5 , - this . width * 0.5 , this . height , this . width ) ;
            } else {
                app . context2D . rect ( - this . width * 0.5 , - this . height * 0.5 , this . width , this . height ) ;
```

```
        }
        app.context2D.fill();
app.context2D.restore();
// 绘制炮塔 turret
app.context2D.save();
    app.context2D.rotate(this.turretRotation);
    // 椭圆炮塔 ellipse 方法
    app.context2D.fillStyle = 'red';
    app.context2D.beginPath();
    if(this.initYAxis){
        // 当朝着 y 轴正方向时,椭圆的 radiuX < radiuY
        app.context2D.ellipse(0,0,10,15,0,0,Math.PI * 2);
    } else {
        // 当朝着 x 轴正方向时,椭圆的 radiuX > radiuY
        app.context2D.ellipse(0,0,15,10,0,0,Math.PI * 2);
    }
    app.context2D.fill();
    // 炮管 gun barrel (炮管)
    app.context2D.strokeStyle = 'blue';
    app.context2D.lineWidth = 5; // 炮管需要粗一点,因此 5 个单位
    app.context2D.lineCap = 'round'; // 使用 round 方式
    app.context2D.beginPath();
    app.context2D.moveTo(0, 0);
    if(this.initYAxis){
        // 当朝着 y 轴正方向时,炮管是沿着 y 轴正方向绘制的
        app.context2D.lineTo(0, this.gunLength);
    } else {
        // 当朝着 x 轴正方向时,炮管是沿着 x 轴正方向绘制的
        app.context2D.lineTo(this.gunLength, 0);
    }
    app.context2D.stroke();
    // 炮口,先将局部坐标系从当前的方向,向 x 轴的正方向平移 gunLength(数值
    类型的变量,以像素为单位,表示炮管的长度)个像素,此时局部坐标系在炮管最右侧
    if(this.initYAxis){
        // 当朝着 y 轴正方向时,炮口是沿着 y 轴正方向绘制的
        app.context2D.translate(0, this.gunLength);
        app.context2D.translate(0, this.gunMuzzleRadius);
    } else {
        // 当朝着 x 轴正方向时,炮口是沿着 x 轴正方向绘制的
        app.context2D.translate(this.gunLength, 0);
        app.context2D.translate(this.gunMuzzleRadius, 0);
    }
    // 调用自己实现的 fillCircle 方法,内部使用 Canvas2D arc 绘制圆弧方法
    app.fillCircle(0, 0, 5, 'black');
app.context2D.restore();
// 绘制一个圆球,标记坦克正方向,一旦炮管旋转后,可以知道正前方在哪里
app.context2D.save();
    if(this.initYAxis){
        // 当朝着 y 轴正方向时,标记坦克前方的圆球是沿着 y 轴正方向绘制的
        app.context2D.translate(0, this.height * 0.5);
    } else {
        // 当朝着 x 轴正方向时,标记坦克前方的圆球是沿着 x 轴正方向绘制的
```

```
                app . context2D . translate ( this . width * 0.5 , 0 ) ;
            }
            app . fillCircle ( 0 , 0 , 10 , 'green' ) ;
        app . context2D . restore ( ) ;
        // 坐标系是跟随整个坦克的
        if ( this . showCoord ) {
            app . context2D . save ( ) ;
                app . context2D . lineWidth = 1 ;
                app . context2D . lineCap = '' ;
                app . strokeCoord ( 0 , 0 , this . width * 1.2 , this . height
                    * 1.2 ) ;
            app . context2D . restore ( ) ;
        }
    app . context2D . restore ( ) ;
    if ( this . showLine === false ) {
        return ;
    }
    app . context2D . save ( ) ;
        app . strokeLine ( this . x , this . y , app . canvas . width * 0.5 ,
            app . canvas . height * 0.5 ) ;
        app . strokeLine ( this . x , this . y , this . targetX , this .targetY ) ;
    app . context2D . restore ( ) ;
}
```

运行上述代码后，会发现初始化时，如图 5.26 所示，坦克位于画布中心，并朝着 y 轴正方向，但是当移动鼠标后，坦克的朝向和鼠标指针的移动方向并不相符。这是因为 Tank 类的_lookAt 方法及_moveTowardTo 方法是针对坦克初始化朝着 x 轴正方向时编写的。

5.2.8 朝向正确的运行

本节来修正一下上面碰到的问题，有两种解决方案：

第一种解决方案：修改坦克类的_lookAt 方法及_moveTowardTo 方法。具体代码如下：

```
private _lookAt ( ) : void {
    let diffX : number = this . targetX - this . x ;
    let diffY : number = this . targetY - this . y ;
    let radian = Math . atan2 ( diffY , diffX ) ;
    if ( this . initYAxis ) {
        radian -= Math . PI / 2 ;
                        // 坦克朝着 y 轴正方向和朝着 x 轴正方向之间相差-90 度
    }
    this . tankRotation = radian ;
}
// 下面的代码演示了两种修正方式
// 第一种：修改 sin 和 cos 代码
private _moveTowardTo ( intervalSec : number ) : void {
    let diffX : number = this . targetX - this . x ;
    let diffY : number = this . targetY - this . y ;
    // linearSpeed 的单位是：像素 / 秒
```

```
    // 如果整个要运行的距离大于当前的速度，说明还没到达目的地，可以继续刷新坦克的位置
    let currSpeed : number = this . linearSpeed * intervalSec ;
                                             // 根据时间差计算出当前的运行速度
    // 如果整个要运行的距离大于当前的速度，说明还没到达目的地，可以继续刷新坦克的位置
    if ( ( diffX * diffX + diffY * diffY ) > currSpeed * currSpeed ) {
        if ( this . initYAxis ) {
            this . x = this . x - Math . sin ( this . tankRotation  ) * 
            currSpeed ;
            this . y = this . y + Math . cos ( this . tankRotation ) * currSpeed ;
        } else {
            this . x = this . x + Math . cos ( this . tankRotation  ) * currSpeed ;
            this . y = this . y + Math . sin ( this . tankRotation ) * currSpeed ;
        }
    }
}
// 第二种：不修改 sin 和 cos，而是增加 90°旋转
private _moveTowardTo ( intervalSec : number ) : void {
    let diffX : number = this . targetX - this . x ;
    let diffY : number = this . targetY - this . y ;
    // linearSpeed 的单位是：像素/秒
    // 如果整个要运行的距离大于当前的速度，说明还没到达目的地，可以继续刷新坦克的位置
    let currSpeed : number = this . linearSpeed * intervalSec ;
                                             // 根据时间差计算出当前的运行速度
    if ( ( diffX * diffX + diffY * diffY ) > currSpeed * currSpeed ) {
        let rot : number = this . tankRotation ;
        if ( this . initYAxis ) {
            rot += Math . PI / 2 ;
        }
        this . x = this . x + Math . cos ( rot ) * currSpeed ;
        this . y = this . y + Math . sin ( rot ) * currSpeed ;
    }
}
```

当运行上述代码后，坦克就能够正确地运行了。

第二种解决方案：在 draw 函数中通过坐标系变换的方式从而避免修改 _lookAt 和 moveTowardTo 方法。具体代码如下：

```
// 在 Tank 类的 draw 函数中找到如下代码
// 整个坦克移动和旋转，注意局部变换的经典结合顺序 ( trs: translate -> rotate -> scale )
app . context2D . translate ( this . x , this . y ) ;
// 将原来的 app . context2D . rotate ( this . tankRotation ) ; 修改为如下代码：
app . context2D . rotate ( this . initYAxis ? this . tankRotation - Math . PI * 0.5 : this . tankRotation ) ;
app . context2D . scale ( this . scaleX , this . scaleY ) ;
```

当运行上述代码后，坦克就能以正确的朝向运行了。下面来了解一下变换的过程。

坦克通过 draw 函数定义在局部坐标系中，此时坦克的朝向初始化时是对着坦克局部坐标系 y 轴的正方向。

在绘制时，先将坦克的局部坐标系变换到画布的[x , y]坐标处。

_lookAt 方法和_moveFowardTo 方法是针对坦克初始化朝着 x 轴正方向这种情况编写的代码，但是目前实际情况是坦克初始化时朝着 y 轴正方向，因此从 y 轴正方向变换到 x 轴正方向，需要逆时针（负）旋转 90°（弧度为 Math.PI/2），所以首先需要调用 app.context2D.rotate (-Math.PI * 0.5)，这样坦克的局部坐标系就符合_lookAt 方法和_moveFowardTo 方法处理的要求了。

然后需要加上当前坦克本身的旋转角度，因此需要再调用 app.context2D.rotate (tankRotation)。我们知道，两次连续的旋转，可以直接将角度相加，因此可以调用 app.context2D.rotate (tankRotation - Math.PI * 0.5)。

最后是处理缩放，使用 app.context2D.scale(this.scale X, this, scale Y)来缩放整个坐标系，例如我们设置 this.scale X 和 this.scale Y 的值分别为 2.0 和 1.5，会发现程序没有任何问题。

必须要使用上述的变换顺序才能获得正确的结果，如果修改变换顺序，会是完全不同的运行效果。

由此可见，通过局部坐标系的变换，以最小的代码修改量就能解决朝向问题，因此局部坐标系的变换应该掌握！

5.2.9　坦克朝着目标移动效果的生成代码

来看一下图 5.25 所示效果的生成代码，具体如下：

```
// 在 TestApplication 类中增加三角形绘制代码
public drawTriangle ( x0 : number , y0 : number , x1 : number , y1 : number ,
x2 : number , y2 : number , stroke : boolean = true ) : void {
   if ( this . context2D === null ) {
      return ;
   }
   this . context2D . save ( ) ;
      this . context2D . lineWidth = 3 ;
      this . context2D . strokeStyle = 'rgba( 0 , 0 , 0 , 0.5 )';
      this . context2D . beginPath ( ) ;
      this . context2D . moveTo ( x0 , y0 ) ;
      this . context2D . lineTo ( x1 , y1 ) ;
      this . context2D . lineTo ( x2 , y2 ) ;
      this . context2D . closePath ( ) ;

      if ( stroke ) {
         this . context2D . stroke ( ) ;
      } else {
         this . context2D . fill ( ) ;
      }

      this . fillCircle ( x2 , y2 , 5 ) ;
   this . context2D . restore ( ) ;
```

```
}
// 调用 TestApplication 类来绘制图 5.25 所示的效果
//获取 canvas 元素
let canvas : HTMLCanvasElement = document . getElementById ( 'canvas' ) as
HTMLCanvasElement ;
let app : TestApplication = new TestApplication ( canvas ) ;
let ptX : number = 600 ;
let ptY : number = 500 ;
app . strokeGrid ( ) ; // 绘制背景网格
app . drawCanvasCoordCenter ( ) ; //绘制中心坐标系和原点
app . draw4Quadrant ( ) ; // 绘制四个象限文字
// 在画布中心绘制坦克
app . drawTank ( ) ; // 绘制坦克
// 绘制旋转后并且位于画布中心的坦克
app . _tank . tankRotation = Math . atan2 ( ptX - app . canvas . width *
0.5 , ptY - app . canvas . height * 0.5 ) ;
app . drawTank ( ) ;

//计算出点[ ptX , ptY ] 与 坦克原点之间的距离（也就是三角形斜边的长度）
let len : number = app . distance ( ptX , ptY , app . canvas . width * 0.5 ,
app . canvas . height * 0.5 ) ;
// 计算出斜边一半时的坐标，然后在该坐标处绘制坦克
app . _tank . x = app . _tank . x + Math . cos ( app . _tank . tankRotation )
* len * 0.5;
app . _tank . y = app . _tank . y + Math . sin (  app . _tank . tankRotation )
* len * 0.5 ;
app . drawTank ( ) ;
// 接下来要继续将坦克绘制到斜边的末尾，上面的代码已经将坦克的坐标更新到了斜边一半的位置
app . _tank . x = app . _tank . x + Math . cos ( app . _tank . tankRotation )
* len * 0.5;
app . _tank . y = app . _tank . y + Math . sin ( app . _tank . tankRotation )
* len * 0.5 ;
app . drawTank ( ) ;
// 绘制平面直角三角形
app . drawTriangle ( app . canvas . width * 0.5 , app . canvas . height *
0.5 , ptX , app . canvas . height * 0.5 , ptX , ptY ) ;
```

5.3 本章总结

本章介绍了多个 Demo，目的是让大家了解 Canvas2D 中局部坐标系变换的相关知识。下面再来总结一下变换的几个要点：

- 使用 translate、rotate 和 scale 这些方法时，变换的是依附在物体本体上的局部坐标系。
- 所有绘制操作都是相对于变换后的物体的局部坐标系进行的。也就是说，在 draw 函数中定义物体时，其顶点坐标都是相对于局部坐标系原点的偏移。这样的好处是，

只要定义一次后，通过 translate、rotate 和 scale 等操作，便可以将物体变换到任意位置，朝着任意角度，以及任意大小，这是一个非常棒的特性，可分离渲染数据源（不变性）及对渲染数据源的操作（可变性）。
- translate、rotate 和 scale 这些局部坐标系变换方法都具有累积性，每次变换操作都是相对上一次结果的叠加，可以通过精心设计 save 和 restore 的层次来修改这种累积性。

第 6 章 向量数学及基本形体的点选

上一章中，了解到图形需要经过一系列坐标系变换后才能生成图像并显示在屏幕上，而整个坐标系的变换在数学上的表示则涉及向量和矩阵的相关知识。在本章中，重点关注向量的相关概念，以及一些基本运算，并且通过一个向量投影的 Demo 来了解向量的这些基本运算是如何被应用的。

在完成向量投影 Demo 后，会发现稍微延伸一下，就能获得点与线段的碰撞检测算法，而在点与线段的碰撞检测算法的内部又会引用到点与圆的碰撞检测算法。那么就沿着这个思路，将点与矩形、点与椭圆、点与三角形，以及点与凸多边形的碰撞检测算法都了解一下。

6.1 向量数学

在图形或游戏开发中将会广泛地使用向量（Vector）。可以用向量来表示物体在局部坐标系中定义的顶点位置，也可以表示空间中表面的方向，或者用来定义物理中力、力矩、速度及加速度等，向量数学在图形学和物理学中有着广泛的应用。

在本节中，将要了解向量的相关概念，以及一些基本运算。

6.1.1 向量的概念

向量的定义：向量是具有方向（Direction）和大小（Length / Magnitude）的空间变量。

如图 6.1 所示，在四个象限中分别绘制了四个不同方向和不同大小（长度）的向量，这四个向量的方向由箭头指明，大小分别是 100、150、200 和 250。

从几何的角度看，可以将向量想象成具有方向的直线段。

- 圆点标记出向量的尾部（Tail）。
- 箭头标记出向量的头部（Head）。
- 从尾部到头部的连线指明了向量的方向（Direction）。
- 从尾部到头部之间的距离表示向量的大小（Length / Magnitude）。

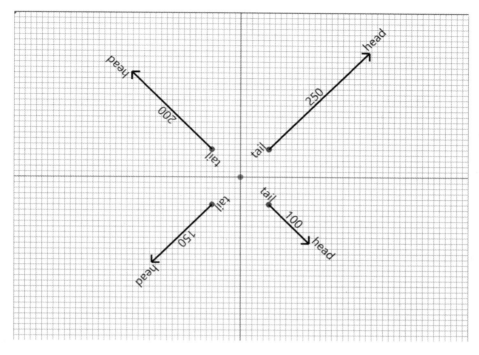

图 6.1 向量

在二维空间中，可以用 $a = [x, y]$ 来表示一个向量，习惯上用粗斜体的小写字母来区分向量和标量（Scalar）。例如 $a = [x, y]$ 中，粗体小写字母 a 表示一个向量，该向量由两个标量 x 和 y 组成。

所谓标量，是指只有大小，没有方向的数值变量，TypeScript 中的 number 类型就是标量，在本书中，标量使用非粗体小写字母表示。

在程序中，为了支持向量的相关操作，专门定义了一个名为 vec2 的类表示二维向量。具体代码如下：

```
export class vec2 {
    // 使用 float32Array 强类型数组，不需要进行引用类型到值类型，以及值类型到引用类
    型的转换，效率比较高
    public values : Float32Array ;
    // 构造函数
    public constructor ( x : number = 0 , y : number = 0 ) {
        this . values = new Float32Array ( [ x , y ] ) ;
    }

    // 静态的 create 方法
    public static create ( x : number = 0 , y : number = 0 ) : vec2 {
        return new vec2( x , y ) ;
    }
    // 复制当前的向量到 result
    public static copy(src : vec2 , result : vec2 | null = null ) : vec2 {
```

```
        if ( result === null ) result = new vec2 ( ) ;
        result . values [ 0 ] = src . values [ 0 ] ;
        result .values [ 1 ] = src . values [ 1 ] ;
        return result ;
    }
    // 为了 debug 输出, override toString 方法
    // 当调用例如 console . log 方法时, 会自动调用如下定义的 toString 方法
    public toString ( ) : string {
        return " [ " + this . values [ 0 ] + " , " + this . values [ 1 ] +
          " ] " ;
    }
    // 方便地 x 和 y 读写操作
    public get x ( ) : number { return this . values [ 0 ] ; }
    public set x ( x : number ) { this . values [ 0 ] = x ; }
    public get y ( ) : number { return this . values [ 1 ] ; }
    public set y ( y : number ) { this . values [ 1 ] = y ; }
    // 为了重用向量, 有时需要重置向量的 x , y 值
    public reset ( x : number = 0 , y : number = 0 ) : vec2 {
        this . values [ 0 ] = x ;
        this . values [ 1 ] = y ;
        return this ;
    }
    // 为了避免浮点数误差, 使用 EPSILON 进行容差处理, 默认情况下为 0.00001
    public equals ( vector : vec2 ) : boolean {
        if ( Math . abs ( this . values[ 0 ] - vector . values [ 0 ] ) > EPSILON )
            return false ;
        if ( Math . abs( this . values[ 1 ] - vector.values[ 1 ] ) > EPSILON )
            return false ;
        return true ;
    }
}
```

上述代码定义了 vec2 类的成员变量,以及最基本的一些常用方法和属性。

6.1.2 向量的大小与方向

已知一个向量 $a = [x, y]$,那么该向量的大小可以使用 $\|a\|$ 来表示,其值为一个标量,可以由 $\|a\| = \sqrt{x^2 + y^2}$ 这个公式计算得出。

可以看到,计算向量的大小实际上就是勾股定理的应用。

下面来实现计算向量大小的代码,具体如下:

```
// 返回没有开根号的向量大小
public get squaredLength ( ) : number {
    let x = this . values [ 0 ] ;
    let y = this . values [ 1 ] ;
    return ( x * x + y * y ) ;
}
// 返回真正的向量大小
public get length ( ) : number {
```

```
        return Math . sqrt ( this . squaredLength ) ;
}
```

接下来看一下如何计算向量的方向。具体代码如下：

```
// 调用本方法后会在内部修改当前向量的 x 和 y 值，修改后的向量大小为 1.0（单位向量或叫
方向向量），并返回未修改前向量的大小
public normalize ( ) : number {
    // 计算出向量的大小
    let len : number = this . length ;
    // 对 0 向量的判断与处理
    if ( Math2D . isEquals ( len , 0 ) ) {
        // alert( "长度为 0，并非方向向量!!!" ) ;
        console . log ( " the length = 0 ") ;
        this . values [ 0 ] = 0 ;
        this . values [ 1 ] = 0 ;
        return 0 ;
    }
    // 如果已经是单位向量，直接返回 1.0
    if ( Math2D . isEquals ( len , 1 ) ) {
        console . log ( " the length = 1 ") ;
        return 1.0 ;
    }

    // 否则计算出单位向量（也就是方向）
    this . values [ 0 ] /= len ;
    this . values [ 1 ] /= len ;

    // 同时返回向量的大小
    return len ;
}
```

通过调用 vec2 的 normalize 方法，会获得一个大小为 1.0 的单位向量（零向量除外），该向量表示方向。例如，在全局坐标系中：
- x 轴的正方向向量为 [1 , 0]。
- x 轴的负方向向量为 [- 1 , 0]。
- y 轴的正方向向量为 [0 , 1]。
- y 轴的负方向向量为 [0 , - 1]。

由于这几个特殊向量比较重要，将其实现为公开访问级别的静态成员变量。具体代码如下：

```
public static xAxis = new vec2 ( 1 , 0 ) ;
public static yAxis = new vec2 ( 0 , 1 ) ;
public static nXAxis = new vec2 ( - 1 , 0 ) ;
public static nYAxis = new vec2 ( 0 , - 1 ) ;
```

如果想仅获得某个 vec2 的大小，可以调用 vec2 的 length 实例方法。

如果想获得某个 vec2 的大小并将该 vec2 改成单位方向向量，可以使用 vec2 的 normalize 实列方法。

6.1.3 向量的加减法及几何含义

到目前为止，一直在操作的是单个向量的两个分量 x 和 y。接下来看一下两个向量之间的运算。首先看一下两个向量之间的加法。具体代码如下：

```
// 公开静态方法
public static sum ( left : vec2 , right : vec2 , result : vec2 | null = null ) : vec2 {
    // 如果输出参数 result 为 null，则分配内存给 result 变量
    if ( result === null ) result = new vec2 ( ) ;
    // x 和 y 分量分别相加，结果仍旧是一个向量
    result . values [ 0 ] = left . values [ 0 ] + right . values [ 0 ] ;
    result . values [ 1 ] = left . values [ 1 ] + right . values [ 1 ] ;
    // 返回相加后的向量 result
    return result ;
}
// vec2 类的公开实例方法：加
public add ( right : vec2 ) : vec2 {
    // this + right = this
    // 会修改 this 的 x 和 y 分量
    // 不需要重新分配内存空间，效率相对较高
    vec2 . sum ( this , right , this ) ;
    return this ;
}
```

由上面的代码可知，向量的加法非常简单，仅仅是两个向量的分量各自相加，形成一个新的向量。

来看一下向量加法的几何含义，如图 6.2 所示。

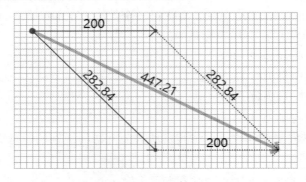

图 6.2 向量加法

在图 6.2 中，向量 *a*（实线且大小为 200）与向量 *b*（实线且大小为 282.84）相加的几何解释就是：

平移向量，使向量 *b* 的尾部（向量 *b* 的圆点部分）连接向量 *a* 的头部（向量 *a* 箭头部分），接着从向量 *a* 的尾部向向量 *b* 的头部画一个向量（大小为 446.21），该向量就是向

量 $a + b$ 形成的新的向量,这就是向量加法的三角形法则。而在图 6.2 所示的向量加法几何特性图示中,使用的是向量加法的平行四边形法则绘制的。

向量加法有两个非常重要的运算规则:
- 向量加法交换律,即 $a + b = b + a$,从图 6.2 中可以看到,可以平移向量 a,也可以平移向量 b,结果相同。
- 向量加法结合律,即 $(a + b) + c = a + (b + c)$。

向量加法很有用。举个例子,在力学中,如果有两个力(向量)同时施加在某个物体上,那么该物体受到的合力就是这两个力(向量)相加。

对向量加法来说,最后需要了解的是,既然可以平移向量,说明向量的位置是无关紧要的,就向量而言,大小和方向才是关键。如果两个向量,大小相同,方向相同,则这两个向量相等(可以使用 vec2 类的 equals 实例方法来判断两个向量是否相等,内部会做浮点数容差处理)。

接下来看一下向量减法的源码,如下:

```
// 公开静态方法:差
public static difference ( end : vec2 , start : vec2 , result : vec2 | null = null ) : vec2 {
    // 如果输出参数 result 为 null,则分配内存给 result 变量
    if ( result === null ) result = new vec2 ( ) ;
    // x 和 y 分量分别相减,结果仍旧是一个向量
    result . values [ 0 ] = end . values [ 0 ] - start . values [ 0 ] ;
    result . values [ 1 ] = end . values [ 1 ] - start . values [ 1 ] ;
    return result ;
}
// vec2 类的实例方法:减
public substract ( another : vec2 ) : vec2 {
    // this - right = this
    // 会修改 this 的 x 和 y 分量
    // 不需要重新分配内存空间,效率相对较高
    vec2 . difference ( this , another , this ) ;
    return this ;
}
```

由上面的代码可以知,向量的减法也是非常简单的,仅仅是两个向量的分量各自相减,形成一个新的向量。

来看一下向量减法的几何含义,如图 6.3 所示,可以固定任意一个向量,平移另外一个向量,让两个向量的尾部重合,此时如果是:
- 向量 a(实线且大小为 200)减去向量 b(实线且大小为 282.84),则从向量 b(实线且大小为 282.84)的头部向着向量 a(实线且大小为 200)的头部画一个新的向量(粗线且大小为 200),如图 6.3 左图所示。
- 向量 b(实线且大小为 282.84)减去向量 a(实线且大小为 200),则从向量 a(实线且大小为 200)的头部向着向量 b(实线且大小为 282.84)的头部画一个新的向量(粗线且大小为 200),如图 6.3 右图所示。

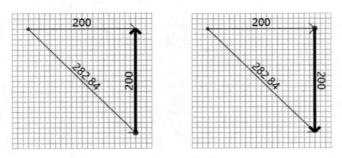

图 6.3 向量减法

会发现,向量的减法是具有方向性的,并不满足交换律。

6.1.4 负向量及几何含义

要计算一个向量的负向量,只需要将该向量的每个分量取反,下面来看一下实现代码:

```
// 会修改 this 向量的两个分量,返回的是修改后的向量:this 指针
public negative () : vec2 {
    this . values [ 0 ] = - this . values [ 0 ] ;
    this . values [ 1 ] = - this . values [ 1 ] ;
    return this ;
}
```

向量变负的几何含义:得到一个和原向量方向相反,大小相同的向量,如图 6.4 所示。

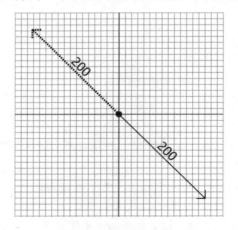

图 6.4 向量及负向量

6.1.5 向量与标量乘法及几何含义

向量与标量不能相加,但是它们能相乘。当一个标量和一个向量相乘时,将得到一个

新的向量，该向量与原向量平行，但长度不同或方向相反。

如图 6.5 所示，画布中心左侧的向量的方向都是相同的（平行），向量的大小则以每 20 个单位递增。而画布中心右侧的向量，方向相反（仍旧平行），其大小也是以每 20 个单位递增。

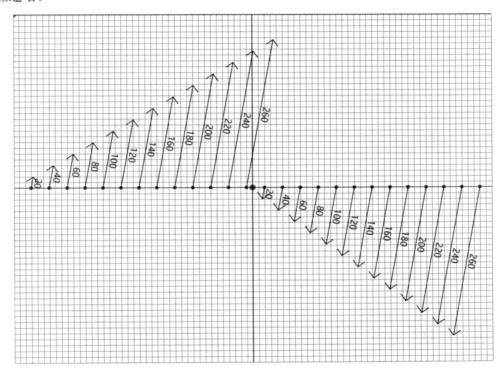

图 6.5　向量与标量乘法

来看一下向量与标量相乘的代码，如下：

```
public static scale (direction : vec2 , scalar : number , result : vec2 | null = null ) : vec2 {
    if ( result === null ) result = new vec2 ( ) ;
    result . values [ 0 ] = direction . values [ 0 ] * scalar ;
    result . values [ 1 ] = direction . values [ 1 ] * scalar ;
    return result ;
}
```

向量与标量相乘操作非常重要，来了解一些细节：

- 向量与标量相乘的本质是缩放向量，因此实现的静态方法名为 scale。现在大家应该知道标量的英文为什么叫 Scalar 了，因为标量（Scalar）用来缩放（Scale）向量（Vector）。
- 当用正数表示的标量来缩放向量时，改变的是原向量的大小，方向与原向量不变。如图 6.5 所示，画布中心左侧的所有向量的方向不变，标量取值分别为：[20 ，

40，…，260]。
- 当用负数表示的标量来缩放向量时，改变的是原向量的大小，并且方向与原向量正好相反。如图 6.5 所示，画布中心右侧的所有向量的方向与原向量相反，其标量取值分别为：[－20，－40，…，－260]。
- 负向量可以被认为是标量与向量相乘的特殊情况，乘以标量-1，即 $(-1)a = -a$。
- 向量和标量相乘，符合交换律：$sa = as$。
- 标量 0 与向量相乘，得到零向量：$0a = [0, 0]$。
- 向量和标量相乘，符合分配律 1：$s(a + b) = sa + sb$。
- 向量和标量相乘，符合分配律 2：$(s1 + s2)a = s1a + s2a$。
- 当缩放向量时，其方向最好使用单位向量（normalize 后的向量），但不是必须的。

接下来再来看 scaleAdd 静态方法。具体代码如下：

```
public static scaleAdd ( start : vec2 , direction : vec2 , scalar : number ,
result : vec2 | null = null ) : vec2 {
    if ( result === null ) result = new vec2() ;
    vec2 . scale ( direction , scalar , result ) ;
                    // result 中存储的是缩放后的向量
    return vec2 . sum ( start , result , result ) ;
                    // start + result = result,然后将 result 返回给调用者
}
```

公开静态方法 scaleAdd 是一个非常重要的操作，其公式为：result = start + direction × scalar，作用是将一个点（start），沿着 direction 给定的方向，移动 scalar 个单位。

6.1.6 向量标量相乘取代三角函数 sin 和 cos 的应用

回顾一下 5.2.5 节（坦克朝着目标移动）中的需求描述：

当坦克旋转到某个角度后，然后要做的是沿着旋转后的方向一直运行，当到达鼠标点时停止运行。

当时是使用了三角函数中的 sin 和 cos 方法来实现上述需求的。具体代码如下：

```
// 涉及朝向运动的位置变量
public x : number = 100 ;
public y : number = 100 ;
public targetX : number = 0 ;
public targetY : number = 0 ;
private _moveTowardTo ( intervalSec : number ) : void {
    // 将鼠标点的 x 和 y 变换到相对坦克坐标系原点的表示
    let diffX : number = this . targetX - this . x ;
    let diffY : number = this . targetY - this . y ;
    // linearSpeed 的单位是：像素 / 秒
    let currSpeed : number = this . linearSpeed * intervalSec ;
                    // 根据时间差计算出当前的运行速度
    // 关键点 1：判断坦克是否要停止运动
    // 如果整个要运行的距离大于当前的速度，说明还没到达目的地，可以继续刷新坦克的位置
```

```
        if ( ( diffX * diffX + diffY * diffY ) > currSpeed * currSpeed ) {
            // 关键点 2：使用 sin / cos 函数计算斜向运行时 x / y 分量
            this . x = this . x + Math . cos ( this . tankRotation  ) * currSpeed ;
            this . y = this . y + Math . sin ( this . tankRotation ) * currSpeed ;
        }
    }
```

一直以来，三角函数是比较耗时的操作，为了优化这些操作，很多时候可以使用诸如 sin、cos 和 tan 查找表等技术来加速三角函数的计算。

现在有了向量和标量相乘操作，可以替换上述代码中的三角函数，从而提高代码的运行效率。那么来看一下具体做法，代码如下：

```
// 涉及朝向运动的位置变量改成向量表示
public pos : vec2 = new vec2 ( 100 , 100 ) ;
public target : vec2 = new vec2 ( ) ;
private _moveTowardTo ( intervalSec : number ) : void {
    // 首先计算坦克当前的位置到鼠标点之间的向量
    let dir : vec2 = vec2 . difference ( this . target , this . pos ) ;
    dir . normalize ( ) ;              // 将该向量 normalize 成单位方向向量
    // 调用 vec2 . scaleAdd 方法，表示将当前的坦克位置沿着单位方向移动 this . linearSpeed * intervalSec 个单位
    this . pos = vec2 . scaleAdd ( this . pos , dir , this . linearSpeed * intervalSec ) ;
}
```

然后运行坦克 Demo，会发现程序的运行正确无误。在向量标量相乘版本的朝向运动代码中，没有任何耗时的三角函数操作，全部都是标量（一个二维向量由两个标量组成）的加法、减法，以及乘法操作。

6.1.7 向量的点乘及几何含义

向量能与标量相乘，向量也能和向量相乘。两个向量相乘被称为点乘（也常称为点积或内积）。

在书写的过程中，可以使用 *a*·*b* 来表示点乘。在向量与标量相乘时，可以省略乘号，但是在表示两个向量的点乘时，是不能省略·符号（点乘符号）。

在 vec2 类中，使用如下代码来表示向量的点乘：

```
// 公开的静态函数：点积
public static dotProduct ( left : vec2 , right : vec2 ) : number {
    return left . values [ 0 ] * right . values[ 0 ] + left . values[ 1 ] * right . values [ 1 ] ;
}
// 公开的实例函数：内积
public innerProduct ( right : vec2 ) : number {
    // 调用静态方法
    return vec2 . dotProduct ( this , right ) ;
}
```

从上述代码可以看到，两个向量的点乘就是对应分量的乘积的和，其返回的结果是一个标量（number 类型）。

向量的点乘具有如下代数性质：
- 符合交换律，即 $a \cdot b = b \cdot a$。
- 符合加法分配律，即 $a \cdot (b+c) = a \cdot b + a \cdot c$。
- 标量乘法结合律，即 $(aa) \cdot b = a(ab) = a(a \cdot b)$。
- Schwarz 不等式，即 $a \cdot b <= \|a\| \|b\|$。
- 三角不等式，即 $\|a+b\| <= \|a\| + \|b\|$。
- 最后一个性质，即 $\|a\| = a \cdot a$。

接下来看一下向量点乘的几何含义，如图 6.6 所示。

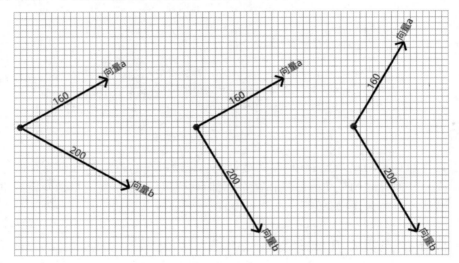

图 6.6 向量的点乘

两个向量 a 和 b，夹角为 θ，根据余弦定律：
$$\|a-b\|^2 = \|a\|^2 + \|b\|^2 - 2\|a\|\|b\|\cos\theta 。$$
将上述表达式的左侧 $\|a-b\|^2$ 展开，写成 $\|a\|^2 + \|b\|^2 - 2(a \cdot b)$，则可以得到：
$$\|a\|^2 + \|b\|^2 - 2(a \cdot b) = \|a\|^2 + \|b\|^2 - 2\|a\|\|b\|\cos\theta 。$$
从而就可以导出如下公式：
$$a \cdot b = \|a\|\|b\|\cos\theta 。$$
根据上面的公式，可以写成如下形式：
$$\cos\theta = a \cdot b / (\|a\|\|b\|) \quad \text{其中}\ (\|a\|\|b\|)\ \text{的值总是正数}。$$
由此得到如下重要的信息：
- 当 $a \cdot b > 0$ 时，向量 a 和 b 的夹角 θ 为锐角（因为 $0 <= \theta < Math.PI / 2$ 的余弦值大于 0），可以认为两个向量方向基本相同，如图 6.6 左侧图示。

- 当 ***a*** · ***b*** = 0 时，向量 ***a*** 和 ***b*** 的夹角 θ 为直角（因为 Math.PI / 2 的余弦值等于 0），可以认为两个向量相互垂直，如图 6.6 中间所示的图示。
- 当 ***a*** · ***b*** < 0 时，向量 ***a*** 和 ***b*** 的夹角 θ 为钝角（因为 Math.PI / 2 < θ <= Math.PI]的余弦值小于 0），可以认为两个向量方向基本相反，如图 6.6 右侧图示。
- 如果向量 ***a***、***b*** 中任意一个向量为零向量，则 ***a*** · ***b*** 的结果也为 0，这意味着零向量和任意其他向量都是相互垂直的。

6.1.8 向量的夹角及朝向计算

上一节中，得到了这个公式：cosθ = ***a*** · ***b*** / (‖ ***a*** ‖ ‖ ***b*** ‖)，那么能够很容易计算出向量 ***a*** 与向量 ***b*** 之间的夹角。具体代码如下：

```
public static getAngle ( a : vec2 , b : vec2 , isRadian : boolean = false ) : number {
    let dot : number = vec2 . dotProduct ( a , b ) ;
    et radian : number = Math . acos ( dot / ( a . length * b . length ) ) ;
    if ( isRadian === false ) {
        radian = Math2D . toDegree ( radian ) ;
    }
    return radian ;
}
```

上述代码使用了上面提到的公式，并且用 Math.acos 这个方法计算两个向量之间的夹角，其返回值根据 isRadian 参数的取值来判断是返回弧度还是返回角度。

接下来看一下朝向（Orietation）计算，为了与夹角区分，使用朝向来表示物体的方向。具体代码如下：

```
public static getOrientation (from : vec2 , to : vec2 , isRadian : boolean = false ) : number {
    let diff : vec2 = vec2 . difference ( to , from ) ;
    let radian = Math . atan2 ( diff . y , diff . x ) ;
    if ( isRadian === false ) {
        radian = Math2D . toDegree ( radian ) ;
    }
    return radian ;
}
```

会看到，这段代码就是 5.2.4 节中 _lookAt 函数中使用的代码。

关于计算夹角和朝向之间的区别，在下一节的 Demo 中进行相关说明。

6.2 向量投影 Demo

在上一节中，了解了向量的长度与方向、加法与减法、标量乘法、点积，以及夹角和朝向计算。这些操作基本涵盖了向量的常用运算方法，本节将实现一个向量投影的 Demo，用到了上面提到的所有操作,具体效果如图 6.7、图 6.8 和图 6.9 所示。

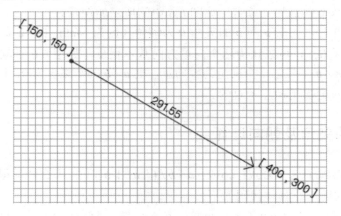

图 6.7 鼠标指针位于向量区域外面

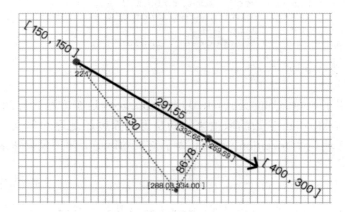

图 6.8 鼠标指针位于向量下方

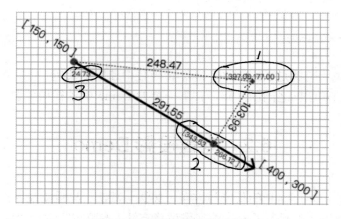

图 6.9 鼠标指针位于向量上方

6.2.1 Demo 的需求描述

当鼠标指针的位置在向量的区域范围之外（如图 6.7 所示），则正常显示该向量。但当鼠标指针的位置移动到向量区域范围内，则会加粗显示该向量，并且会：

（1）标记出鼠标指针位置（圆圈表示）、坐标信息，以及线段起点到鼠标指针处的向量。

（2）标记出鼠标指针位置在向量上的投影点（圆圈），坐标信息，以及从投影点到鼠标指针位置之间的向量。

（3）标记出鼠标指针位置与线段起点之间以角度表示的夹角。

（4）所有的坐标信息是相对全局坐标系原点（左上角）的偏移表示。

6.2.2 绘制向量

为了形象化显示向量，实现 drawVec 方法，用来绘制一个向量。具体代码如下：

```
// 沿着局部坐标系 x 轴的正方向，绘制长度为 len 的向量
// 参数 len：要绘制的向量的长度，例如 291.55
// 参数 arrowLen：要绘制的向量的箭头长度
// 参数 beginText / endText：表示向量尾部和头部的信息，例如 [ 150 , 150 ]和 [ 400 , 300 ]
// 参数 lineWidth：用来加粗显示向量
// 参数 isLineDash：是否以虚线方式显示向量
// 参数 showInfo：是否显示向量的长度
// 参数 alpha：是否以半透明方式显示向量
public drawVec (len : number , arrowLen : number =10 ,beginText : string = '' , endText : string = '' , lineWidth : number = 1 , isLineDash : boolean = false , showInfo : boolean = true , alpha : boolean = false ) : void {
    if ( this . context2D === null ) {
        return ;
    }
    // 当绘制向量的负向量时，len 是负数
    // 此时如果不做如下处理，会导致向量的箭头绘制错误
    if ( len < 0 ) {
        arrowLen = - arrowLen ;
    }
    this . context2D . save ( ) ;
    // 设置线宽
    this . context2D . lineWidth = lineWidth ;
    // 设置是否虚线绘制
    if ( isLineDash ) {
        this . context2D . setLineDash ( [ 2 , 2 ] ) ;
    }

    // 绘制向量的起点圆圈，如果加粗显示，那么向量的起点也要加大
    if ( lineWidth > 1 ) {
```

```typescript
            this . fillCircle ( 0 , 0 , 5 ) ;
        } else {
            this . fillCircle ( 0 , 0 , 3 ) ;
        }

        // 绘制向量和箭头
        this . context2D . save ( ) ;
            // 设置是否半透明显示向量
            if ( alpha === true ) {
                this . context2D . strokeStyle = 'rgba( 0 , 0 , 0 , 0.3 )' ;
            }

            // 绘制长度为 len 的线段表示向量
            this . strokeLine ( 0 , 0 , len , 0 ) ;

            // 绘制箭头的上半部分
            this . context2D . save ( ) ;
                this . strokeLine ( len , 0 , len - arrowLen , arrowLen ) ;
            this . context2D . restore ( ) ;
            // 绘制箭头的下半部分
            this . context2D . save ( ) ;
                this . strokeLine ( len , 0 , len - arrowLen , - arrowLen ) ;
            this . context2D . restore ( ) ;
        this . context2D . restore ( ) ;
        // 绘制线段的起点，终点信息
        let font : FontType = "15px sans-serif" ;
        if ( beginText != undefined && beginText . length != 0 ) {
            if ( len > 0 ) {
                this . fillText ( beginText , 0 , 0 , 'black' , 'right' ,
                'bottom' , font ) ;
            } else {
                this . fillText ( beginText , 0 , 0 , 'black' , 'left' ,
                'bottom' , font ) ;
            }
        }
        len = parseFloat ( len . toFixed ( 2 ) ) ;
        if ( beginText != undefined && endText . length != 0 ) {
            if ( len > 0 ) {
                this . fillText ( endText , len , 0 , 'black' , 'left' ,
                'bottom' , font ) ;
            } else {
                this . fillText ( endText , len , 0 , 'black' , 'right' ,
                'bottom' , font ) ;
            }
        }
        // 绘制向量的长度信息
        if ( showInfo === true ) {
            this . fillText ( Math . abs ( len ) .toString ( ) , len * 0.5 ,
            0 , 'black' , 'center' , 'bottom' , font ) ;
        }
    this . context2D . restore ( ) ;
}
```

接下来看一个更加方便、常用的绘制向量的方法 drawVecFromLine，该方法从一条以起点和终点表示的有向线段计算出向量，并调用上面的 drawVec 方法将该向量绘制出来。具体代码如下：

```
// 一个更常用的绘制向量的方法
// 从两个点计算出一个向量，然后调用 drawVec 绘制该向量
// 返回值：当前向量与 x 正方向的夹角，以弧度表示
public drawVecFromLine ( start : vec2 , end : vec2 , arrowLen : number =
10 , beginText : string = '' , endText : string = '' , lineWidth : number = 1 ,
isLineDash : boolean = false , showInfo : boolean = false , alpha : boolean
= false ) : number{
    // 获取从 start-end 形成的向量与 x 轴正方向[0 , 1]之间以弧度表示的夹角
    let angle : number = vec2 . getOrientation ( start , end , true ) ;
    if ( this . context2D !== null ) {
        // 计算出向量之间的差，注意方向
        let diff : vec2 = vec2 . difference ( end , start ) ;
        // 计算出向量的大小
        let len : number = diff . length ;
        this . context2D . save ( ) ;
        // 局部坐标系原点变换到 start
        this . context2D . translate ( start . x , start . y ) ;
        // 局部坐标系旋转 angle 弧度
        this . context2D . rotate ( angle ) ;
        // 调用 drawVec 方法
        this . drawVec ( len , arrowLen , beginText , endText , lineWidth ,
        isLineDash , showInfo , alpha) ;
        this . context2D . restore ( ) ;
    }
    return angle ;
}
```

6.2.3 向量投影算法

图 6.7 和图 6.8 所示的效果是经典的向量投影算法。简单地说，就是将一个点（鼠标位置表示）投影到由起点和终点所形成的向量上。该算法比较常用，将其作为 Math2D 辅助类的静态公开方法。具体代码如下：

```
// 将一个点 pt 投影到 start 和 end 形成的线段上
// 返回值：true 表示 pt 在线段起点和终点之间，此时 closePoint 输出参数返回线段上的
投影点坐标
//        false 表示在线段起点或终点之外，此时 closePoint 输出参数返回线段的起点或终点
// 本方法使用了向量的 difference、normalize、dotProduct、scaleAdd（scale 和
sum）方法
public static projectPointOnLineSegment ( pt : vec2 , start : vec2 , end :
vec2 , closePoint : vec2 ) : boolean {
    // 向量的 create 方法
    let v0 : vec2 = vec2 . create ( ) ;
    let v1 : vec2 = vec2 . create ( ) ;
    let d : number = 0 ;
    // 向量减法，形成方向向量
```

```typescript
        vec2 . difference ( pt , start , v0 ) ;
                          // 线段的起点到某个点（例如鼠标位置点）的方向向量
        vec2 . difference ( end , start , v1 ) ;                    //获取线段
        // 使用向量的 normalize 方法，原向量变成单位向量，并返回原向量的长度
        // 需要注意的是，normalize 起点到终点线段形成的向量
        // 要投影到哪个向量，就要将这个向量 normalize 成单位向量
        d = v1 . normalize ( ) ;
        // 将 v0 投影在 v1 上，获取投影长度 t
        let t : number = vec2 . dotProduct ( v0 , v1 ) ;
        // 如果投影长度 t < 0，说明鼠标位置点在线段的起点范围之外
        // 处理的方式是：
        // closePt 输出线段起点并且返回 false
        if ( t < 0 ) {
            closePoint . x = start . x ;
            closePoint . y = start . y ;
            // console . log ( t . toString ( ) + " < 0 " ) ;
            return false ;
        } else if ( t > d ) // 如果投影长度 > 线段的长度，说明鼠标位置点超过线段终点范围
        {
            // closePt 输出线段起点并且返回 false
            closePoint . x = end . x ;
            closePoint . y = end . y ;
            // console . log ( t . toString ( ) + " > " + d . toString ( ) ) ;
            return false ;
        } else {
            // 说明鼠标位置点位于线段起点和终点之间
            // 使用 scaleAdd 方法计算出相对全局坐标（左上角）的坐标偏移信息
            // 只有此时才返回 true
            vec2 . scaleAdd ( start , v1 , t , closePoint ) ;
            // console . log ( " 0 <= " + t . toString ( ) + " <= " + d . toString
 ( ) ) ;
            return true ;
        }
    }
```

6.2.4　投影效果演示代码

有了上面绘制向量的方法，以及计算投影点的方法，就可以完成整个 Demo 了。来看一下具体的实现效果，首先在 TestApplication 中声明如下一些成员变量：

```typescript
public lineStart : vec2 = vec2 . create ( 150 , 150 ) ;         //线段起点
public lineEnd : vec2 = vec2 . create ( 400 , 300 ) ;           //线段终点
public closePt : vec2 = vec2 . create ( ) ;  //输出参数，预先内存分配，可以重用
private _hitted : boolean = false ;        //鼠标位置是否在线段的起点和终点范围内
```

其次，需要实时地知道当前的鼠标位置，因此可以覆写（override）TestApplication 类中的 dispatchMouseMove 方法。具体代码如下：

```typescript
protected dispatchMouseMove ( evt : CanvasMouseEvent ) : void {
    // 必须设置 this . isSupportMouseMove = true 才能处理 moveMove 事件
    this . _mouseX = evt . canvasPosition . x ;
    this . _mouseY = evt . canvasPosition . y ;
    // 调用 projectPointOnLineSegment 方法
    // 将结果保存在_hitted 和 closePt 这两个成员变量中
    this . _hitted = Math2D . projectPointOnLineSegment (
        vec2 . create (evt . canvasPosition .x ,evt .canvasPosition .y ),
        this . lineStart , this . lineEnd , this . closePt ) ;
}
```

然后根据_hitted 的状态及 closePt 的内容，按照 6.2.1 节的需求描述，来绘制投影效果，所有代码都在 drawMouseLineProjection 方法中实现，具体如下：

```typescript
public drawMouseLineProjection ( ) : void {
    if ( this . context2D != null ) {
        // 鼠标位置在线段范围外的绘制效果
        if ( this . _hitted === false ) {
            this . drawVecFromLine ( this . lineStart , this . lineEnd , 10 ,
                this . lineStart . toString ( ) , this . lineEnd . toString ( ) ,
                1 , false , true ) ;
        } else { // 鼠标位置在线段范围内
            let angle : number = 0 ;
            let mousePt : vec2 = vec2 . create ( this . _mouseX , this .
            _mouseY ) ;
            this . context2D . save ( ) ;
                // 绘制向量
                angle = this . drawVecFromLine ( this . lineStart , this .
                lineEnd , 10 , this . lineStart . toString ( ) , this . lineEnd .
                toString ( ) , 3 , false , true ) ;
                // 绘制投影点
                this .fillCircle (this . closePt .x ,this .closePt .y , 5 ) ;
                // 绘制线段起点到鼠标点向量
                this . drawVecFromLine ( this . lineStart , mousePt , 10 , '' ,
                '' , 1 , true , true , false ) ;
                // 绘制鼠标点到投影点的线段
                this . drawVecFromLine ( mousePt , this . closePt , 10 , '' ,
                '' , 1 , true , true , false ) ;
            this . context2D . restore ( ) ;
            // 绘制投影点的坐标信息（相对左上角的表示）
            this . context2D . save ( ) ;
                this . context2D . translate ( this . closePt . x , this .
                closePt . y ) ;
                this . context2D . rotate ( angle ) ;
                this . drawCoordInfo ( '[' + ( this . closePt . x ) . toFixed
                ( 2 ) +' , ' + ( this . closePt . y ) . toFixed ( 2 ) +" ]",
                0 , 0 , "center" , "top" ) ;
            this . context2D . restore ( ) ;
            // 计算出线段与鼠标之间的夹角，以弧度表示
            angle = vec2 . getAngle ( vec2 . difference ( this . lineEnd ,
            this . lineStart ) , vec2 . difference ( mousePt , this .
            lineStart ) , false ) ;
            // 绘制出夹角信息
```

```
                this . drawCoordInfo ( angle . toFixed ( 2 ) , this . lineStart .
                x + 10, this . lineStart . y + 10 , "center" , "top" ) ;
            }
        }
    }
```

只要在 TestApplication 类中的 render 方法中调用 drawMouseLineProjection 方法，就能获得如图 6.7、图 6.8 和图 6.9 所示的效果。

6.2.5　向量 getAngle 和 getOrientation 方法的区别

在 drawMouseLineProjection 方法中使用了如下代码：

```
angle = this . drawVecFromLine ( this . lineStart , this . lineEnd , 10 ,
this . lineStart . toString ( ) , this . lineEnd . toString ( ) , 3 , false ,
true ) ;
```

其中，drawVecFromLine 方法中调用了 vec2 . getOrientation 方法获得 angle，并将 anlge 从 drawVecFromLine 方法中返回给调用方。

此外还使用了如下代码

```
angle = vec2 . getAngle ( vec2 . difference ( this . lineEnd , this .
lineStart ) , vec2 . difference ( mousePt , this . lineStart ) , false ) ;
```

为了更好地了解 getOrientation 和 getAngle 方法之间的区别，参考如图 6.10 所示的效果，可以知道：

- getOrientation 方法内部使用的是 Math 类的 atan2 方法，atan2 方法返回的总是与 x 轴正方向向量之间的夹角，其取值范围为 [- Math . PI , Math . PI]之间。
- getAngle 方法内部使用 Math 类的 acos 方法，acos 方法返回的是两个向量之间的夹角，其返回值的取值范围为 [0 , Math . PI]。
- 可以将 getOrientation 看作绝对方向的表示，能唯一地确定物体的方向。而 getAngle 返回的是相对方向，由于其取值范围，无法确定角度的旋转方向（是逆时针旋转还是顺时针旋转）。

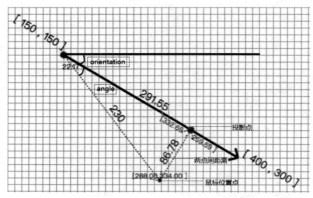

图 6.10　angle 和 orientation 图示

6.3 点与基本几何形体的碰撞检测算法

我们将上节介绍的向量投影 Demo 稍微延伸，就会得到一个很重要的基础算法，即点与线段的碰撞检测。除了线段之外，本节中还将实现点与圆、矩形、椭圆、三角形，以及多边形的碰撞检测算法。

6.3.1 点与线段及圆的碰撞检测

点与线段的碰撞检测是一个比较基础和重要的算法，该算法可以由两部分组成：
- 用 Math2D . projectPointOnLineSegment 方法确定一个点是否在一条线段的表示区间，并且返该点的投影点，前面实现了该方法。
- 再在 Math2D 类中实现静态的 isPointInCircle 方法，该方法判断一个点是否和一个圆发生碰撞。该算法原理的图示可以参考图 6.10，结合下面的 isPointInCircle 方法实现。

来看一下点与线段碰撞检测算法的具体代码，如下：

```
// 圆由圆心坐标和半径表示
public static isPointInCircle ( pt : vec2 , center : vec2 , radius : number ) : boolean {
    let diff : vec2 = vec2 . difference ( pt , center ) ;
    let len2 : number = diff . squaredLength ;
    // 如果一个点在圆的半径范围之内，说明发生了碰撞
    // 避免使用 Math . sqrt 方法
    if ( len2 <= radius * radius ) {
        return true ;
    }
    return false ;
}
public static isPointOnLineSegment ( pt : vec2 , start : vec2 , end : vec2 , radius : number = 2 ) : boolean {
    let closePt : vec2 = vec2 . create ( ) ;
    if ( Math2D . projectPointOnLineSegment ( pt , start , end , closePt ) === false ) {
        return false ;
    }
    //需要进行点与圆的碰撞检测
    return Math2D . isPointInCircle ( pt , closePt , radius ) ;
}
```

来测试一下，在 TestApplication 中实现如下代码：

```
public drawMouseLineHitTest ( ) : void {
    if ( this . context2D != null ) {
```

```typescript
            // 鼠标位置在线段范围外的绘制效果
            if ( this . _hitted === false ) {
                this . drawVecFromLine ( this . lineStart , this . lineEnd , 10 ,
                this . lineStart . toString ( ) , this . lineEnd . toString ( ) ,
                1 , false , true ) ;
            } else { // 鼠标位置在线段范围内
                let mousePt : vec2 = vec2 . create ( this . _mouseX , this .
                _mouseY ) ;

                this . context2D . save ( ) ;
                // 绘制向量
                this . drawVecFromLine ( this . lineStart , this . lineEnd ,
                10 , this . lineStart . toString ( ) , this . lineEnd . toString
                ( ) , 3 , false , true ) ;
                // 绘制投影点
                this . fillCircle (this . closePt .x ,this . closePt .y ,5 ) ;
                this . context2D . restore ( ) ;
            }
        }
    }

    protected dispatchMouseMove ( evt : CanvasMouseEvent ) : void {
        // 必须要设置 this . isSupportMouseMove = true 才能处理 moveMove 事件
        this . _mouseX = evt . canvasPosition . x ;
        this . _mouseY = evt . canvasPosition . y ;

        this . _hitted = Math2D . isPointOnLineSegment (vec2 . create ( evt .
        canvasPosition . x , evt . canvasPosition . y ) , this . lineStart , this .
        lineEnd ) ;
    }

    public render ( ) : void {
        // 由于 canvas . getContext 方法返回的 CanvasRenderingContext2D 可能会是 null
        // 因此 VS Code 会强制要求 null 值检查,否则报错
        if ( this . context2D !== null ) {
            let centX : number
            // 每次重绘,都先清屏
            this . context2D . clearRect ( 0 , 0 , this . canvas .width ,this .
            canvas .height ) ;
            // 调用第 4 章实现的背景网格绘制方法
            this . strokeGrid ( ) ;
            this . drawMouseLineHitTest ( )
            this . drawCoordInfo (
                '[' + (this . _mouseX ) . toFixed ( 2 ) +','+ (this . _mouseY ) .
                toFixed ( 2 ) + " ]" ,
                this . _mouseX ,
                this . _mouseY
            ) ;
        }
    }
```

运行上述程序后,当鼠标移动到线段上就会被选中,并加粗显示,否则就正常显示。

6.3.2 点与矩形及椭圆的碰撞检测

关于点与矩形的碰撞检测算法非常简单。具体代码如下:
```
public static isPointInRect ( ptX : number , ptY : number , x : number ,
y : number , w : number , h : number ) : boolean {
    // 一个点在矩形的上下左右范围之内,则发生碰撞
    if ( ptX >= x && ptX <= x + w && ptY >= y && ptY <= y + h ) {
        return true ;
    }
    return false ;
}
```

接下来看一下点与椭圆的碰撞检测。假设椭圆的中心点定义的坐标值为[$centerX$, $centerY$],并且半径分别为[$radiusX$, $radiusY$],在这种情况下,一个点 $P(pX,pY)$ 如果在椭圆的内部,那么要满足如下公式:

$$\frac{(pX - centerX)^2}{radiusX^2} + \frac{(pY - centerY)^2}{radiusY^2} \leqslant 1.0$$

有了上述公式,就可以在 Math2D 类中实现 isPointInEllipse 静态方法。具体代码如下:
```
public static isPointInEllipse ( ptX : number , ptY : number , centerX :
number ,centerY : number ,radiusX : number ,radiusY : number ) :boolean {
    let diffX = ptX - centerX ;
    let diffY = ptY - centerY ;
    let n : number = ( diffX * diffX ) / ( radiusX * radiusX ) + ( diffY *
diffY ) / ( radiusY * radiusY ) ;
    return ( n <= 1.0 ) ;
}
```

6.3.3 点与三角形的碰撞检测

如图 6.11 所示为点与三角形的关系,从图中可以知道,点 **P** 与[v0 , v1 , v2]形成的三角形之间的关系有两种,要么点 **P** 在三角形内部,要么点 **P** 在三角形的外部。

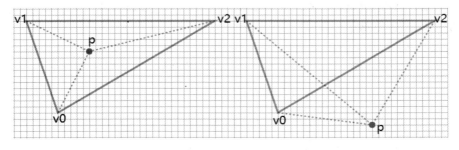

图 6.11 点与三角形的关系图

来观察一下它们之间的区别,你会发现:
- 如果点 ***P*** 在[v0 , v1 , v2]形成的三角形内部,那么三角形的三个顶点和 ***P*** 形成的三个子三角形[v0 , v1 , *p*]、[v1 , v2 , *p*]和[v2 , v0 , *p*]的顶点顺序都是按照顺时针排列的。
- 如果点 ***P*** 在[v0 , v1 , v2]形成的三角形外部,那么三角形的三个顶点和 ***P*** 形成的三个子三角形[v0 , v1 , *p*]、[v1 , v2 , *p*]和[v2 , v0 , *p*]的顶点顺序总有一个不是按照顺时针顺序排列的,例如图 6.11 右侧图像中,[v2 , v0 , *p*]形成的子三角形是非顺时针(逆时针)排列的。
- 更加通用的算法描述是,如果点 ***P*** 与[v0 , v1 , v2]形成的三个子三角形的顶点排列顺序一致,那么该点 ***P*** 肯定在[v0 , v1 , v2]形成的三角形的内部,否则该点 ***P*** 就在三角形的外部。

得到了上述算法后,接下来可以使用向量叉乘(或叉积)的运算来判断三角形顶点顺序。由于向量叉乘比较特别,只能使用 3D 向量,现在假设有两个 3D 向量 $a=[x_0,y_0,z_0]$ 和 $b=[x_1,y_1,z_1]$,那么

$$a \otimes b = \begin{bmatrix} x_0 \\ y_0 \\ z_0 \end{bmatrix} \otimes \begin{bmatrix} x_1 \\ y_1 \\ z_1 \end{bmatrix} = \begin{bmatrix} y_0 z_1 - z_0 y_1 \\ z_0 x_1 - x0 z_1 \\ x_0 y_1 - y_0 x_1 \end{bmatrix}$$

为了将 2D 向量 vec2 以 3D 向量的形式来表示,可以将 3D 向量的 z 分量设置为 0,例如 $a=[x_0,y_0,0]$,$b=[x_1,y_1,0]$,套用上述叉积公式,会得到:

$$a \otimes b = \begin{bmatrix} x_0 \\ y_0 \\ 0 \end{bmatrix} \otimes \begin{bmatrix} x_1 \\ y_1 \\ 0 \end{bmatrix} = \begin{bmatrix} y_0 0 - 0 y_1 \\ 0 x_1 - x_0 0 \\ x_0 y_1 - y_0 x_1 \end{bmatrix} = \begin{bmatrix} 0 \\ 0 \\ x_0 y_1 - y_0 x_1 \end{bmatrix}$$

可以看到,对于 2D 向量的叉积来说,其 *x* 和 *y* 分量总是为 0,但是对于点与三角形碰撞检测算法来说,叉积后的 *z* 分量 $x_0 y_1 - y_0 x_1$ 才是最关键的,先把 *z* 分量的计算作为 vec2 的一个静态方法。具体代码如下:

```
// 公开的静态函数:叉积,返回标量
public static crossProduct ( left : vec2 , right : vec2 ) : number {
    return left . x * right . y - left .y * right .x ;
}
```

接下来,在 Math2D 中实现一个名为 sign 的静态方法,该方法用来计算三角形两条边向量的叉积。具体代码如下:

```
// 计算三角形两条边向量的叉积
public static sign ( v0 : vec2 , v1 : vec2 , v2 : vec2 ) : number {
    // e1 = v2 -> v0 边向量
    let e1 : vec2 = vec2 . difference ( v0 , v2 ) ;
    // e2 = v2 -> v1 边向量
    let e2 : vec2 = vec2 . difference ( v1 , v2 ) ;
    // 获取 e1 cross e2 的值
    return vec2 . crossProduct ( e1 , e2 ) ;
}
```

有了上面的 sign 方法，就能实现点与三角形的碰撞检测算法。具体代码如下：
```
public static isPointInTriangle ( pt : vec2 , v0 : vec2 , v1 : vec2 , v2 : vec2 ) {
    // 计算三角形三个顶点与点 pt 形成的三个子三角形的边向量的叉积
    let b1 : boolean = Math2D . sign ( v0 , v1 , pt ) < 0.0 ;
    let b2 : boolean = Math2D . sign ( v1 , v2 , pt ) < 0.0 ;
    let b3 : boolean = Math2D . sign ( v2 , v0 , pt ) < 0.0 ;
    // 三角形三条边的方向都一致，说明点在三角形内部
    // 否则就在三角形外部
    return (( b1 === b2 ) && ( b2 === b3 ) ) ;
}
```

通过上面的代码，结合上图 6.11 会发现，实际上并不关心子三角形两条边向量叉积的数值大小，更关心的是叉积的正负性，只要三个子三角形的两条边向量的叉积的正负性都一致的话，表示三个子三角形的顶点排列顺序一致，那么该点 *P* 肯定在[v0 , v1 , v2]形成的三角形的内部，否则肯定在外部。

由于本书主要涉及 2D 相关知识，因此对向量叉积的性质不做进一步介绍，若有兴趣，可以自行查阅相关资料。

6.3.4　点与任意凸多边形的碰撞检测

如图 6.12 所示，可以将任意的凸多边形很方便地分解成三角形，然后依次调用上一节实现的点与三角形碰撞检测算法，这样就能获得点与任意凸多边形碰撞检测的算法。

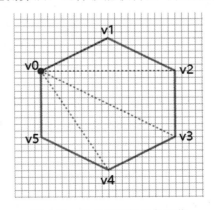

图 6.12　凸多边形的三角形化

具体代码如下：
```
 // 凸多边形由点的集合表示
public static isPointInPolygon ( pt : vec2 , points : vec2 [ ] ) : boolean
{
    // 三角形是最小的凸多边形
    if ( points . length < 3 ) {
        return false ;
```

```
        }
        // 以point[0]为共享点,遍历多变形点集,构建三角形,调用isPointInTriangle
        方法
        // 一旦点与某个三角形发生碰撞,就返回true
        for ( let i : number = 2 ; i < points . length ; i ++ ) {
            if ( Math2D . isPointInTriangle ( pt , points [ 0 ] , points [ i
 - 1 ] , points [ i ] ) ) {
                return true ;
            }
        }
        // 没有和多边形中的任何三角形发生碰撞,返回false
        return false ;
    }
```

从以上代码可以看到,具有 n 个顶点的多边形可以化分解成 $n-2$ 个三角形。接下来再看一下判断一个多边形是不是凸多边形的算法。如图 6.13 所示,先来了解凸多边形和凹多边形的区别。

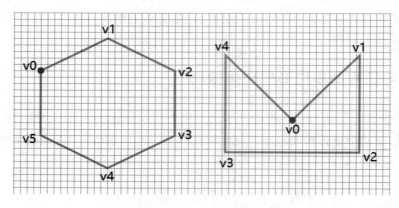

图 6.13 凸多边形和凹多边形

参考图 6.13 会发现,左侧的凸多边形(六边形)顶点形成的 6 个子三角形分别是[v0 , v1 , v2]、[v1 , v2 , v3]、[v2 , v3 , v4]、[v3 , v4 , v5]、[v4 , v5 , v0]和[v5 , v0 , v1],这 6 个子三角形顶点的顺序都是顺时针排列的。

而观察上图右侧的凹多边形,由 5 个顶点组成,形成的 5 个子三角形分别是[v0 , v1 , v2]、[v1 , v2 , v3]、[v2 , v3 , v4]、[v3 , v4 , v0]和[v4 , v0 , v1],会发现最后一个三角形[v4 , v0 , v1]的顶点顺序是逆时针排列,而其他的三角形都是顺时针排列。

如何判断一个三角形的顶点顺序,已经在上一节中了解过了,那么直接来看一下判断凸多边形的算法。具体代码如下:

```
    public static isConvex ( points : vec2 [ ] ) : boolean {
        // 第一个三角形的顶点顺序
        let sign : boolean = Math2D . sign ( points [ 0 ] , points [ 1 ] , points
 [ 2 ] ) < 0 ;
        // 从第二个三角形开始遍历
        let j : number , k : number ;
```

```
        for ( let i : number = 1 ; i < points . length ; i ++ ) {
            j = (i + 1) % points . length ;
            k = (i + 2) % points . length ;
            // 如果当前的三角形的顶点方向和第一个三角形的顶点方向不一致，说明是凹多边形
            if ( sign !== Math2D . sign ( points [ i ] , points [ j ] , points [ k ] ) < 0) {
                return false ;
            }
        }
        // 凸多边形
        return true ;
    }
```

6.4　附录：图示代码

本章中的所有图示都是通过程序绘制得到的，其绘制过程也是深入理解向量相关概念的过程，因此特增加本节，将所有的图示代码罗列在此处。

6.4.1　图 6.1 向量概念图示源码

绘制四个方向不同，大小不同的向量。具体代码如下：

```
// 向量概念
let center : vec2 = vec2 . create ( app . canvas . width * 0.5 , app . canvas . height * 0.5 ) ;
let offset : vec2 = vec2 .create ( 50 , 50 ) ;
if ( app . context2D ) {
    app . context2D . save ( ) ;
        app . context2D . translate ( center . x + offset . x , center . y + offset . y ) ;
        app . context2D . rotate ( Math2D . toRadian ( 45 ) )
        app . drawVec ( 100 , 10 , "tail" , "head" , 3 , false ) ;
    app . context2D . restore ( ) ;
    app . context2D . save ( ) ;
        app . context2D . translate ( center .x - offset . x , center . y + offset . y) ;
        app . context2D . rotate ( Math2D . toRadian ( 135) )
        app . drawVec ( 150 , 10 , "tail" , "head" , 3 ) ;
    app . context2D . restore ( ) ;
    app . context2D . save ( ) ;
        app . context2D . translate ( center .x - offset . x , center . y - offset . y ) ;
        app . context2D . rotate ( Math2D . toRadian ( -135) )
        app . drawVec ( 200 , 10 , "tail" , "head" , 3 ) ;
    app . context2D . restore ( ) ;
    app . context2D . save ( ) ;
        app . context2D . translate ( center . x + offset . x , center . y - offset . y ) ;
        app . context2D . rotate ( Math2D . toRadian ( - 45) )
```

```
        app . drawVec ( 250 , 10 , "tail" , "head" , 3 ) ;
    app . context2D . restore ( ) ;
}
```

6.4.2 图 6.2 和图 6.3 向量加减法图示源码

向量加法和简法的图示代码，当 isAdd 设为 true 时，绘制的是向量加法的几何性质，否则，为向量减法的几何性质。具体代码如下：

```
// 向量加减法
let isAdd : boolean = true;
if ( app . context2D ) {
    let start : vec2 = vec2 .create ( 200 , 100 ) ;
    let end0 : vec2 = vec2 . create ( 400 , 100 ) ;
    let end1 : vec2 = vec2 . create ( 400 , 300 ) ;
    // 绘制同一个原点的两个向量
    app . drawVecFromLine ( start , end0 , 10 , undefined , undefined , 1 , false , true ) ;
    app . drawVecFromLine ( start , end1 , 10 , undefined , undefined , 1 , false , true) ;
    if ( isAdd === true ) {
        // 加法
        // 首尾相连法加法，将 start->end1 向量与 start-end0 向量尾部相连
        app . drawVecFromLine ( end0 , vec2 . sum ( end0 , vec2 . difference ( end1 , start ) ) , 10 , undefined , undefined , 1 , true , true ) ;
        app . drawVecFromLine ( end1 , vec2 . sum ( end1 , vec2 . difference ( end0 , start ) ) , 10 , undefined , undefined , 1 , true , true ) ;
        let sum : vec2 = vec2 . sum ( end0 , end1 ) ; //
        sum = vec2 . difference ( sum , start ) ; // 相对 start 的偏移向量
        app . drawVecFromLine ( start , sum , 10 , undefined , undefined , 5 , false , true , true ) ;
    } else {
        // 减法
        app . drawVecFromLine ( end0 , end1 , 10 , undefined , undefined , 5 , false , true , false ) ;
    }
}
```

6.4.3 图 6.4 负向量图示源码

绘制两个向量，它们之间的关系是大小相等，方向相反，用来演示向量和负向量的几何性质。具体代码如下：

```
// 负向量，大小相等，方向相反
if ( app . context2D ) {
    //在[ 500 , 400 ]绘制长度为 200 的向量
    app . context2D . save ( ) ;
        app . context2D . translate ( 400 , 300 ) ;
        app . context2D . rotate ( Math2D . toRadian ( 45 ) )
        app . drawVec ( 200 ) ;
```

```
        app . context2D . restore ( ) ;
//在[ 500 , 400 ]绘制长度为 200 的向量
app . context2D . save ( ) ;
    app . context2D . translate ( 400 , 300 ) ;
    app . context2D . rotate ( Math2D . toRadian ( 45 ) )
    app . drawVec ( - 200 , 10 , '' , '' , 3 , true ) ;
app . context2D . restore ( ) ;
}
```

6.4.4 图 6.5 向量与标量相乘图示源码

下面的代码用来绘制向量与标量相乘的几何效果:

```
//向量的缩放(向量与标量乘法)
if ( app . context2D ) {
    let curr : number = 0 ;
    for ( let i : number = 1 ; i < 14 ; i ++ ) {
        app . context2D . save ( ) ;
            app . context2D . translate ( curr = i * 30 , 300 ) ;
            app . context2D . rotate ( Math2D . toRadian ( 100 ) ) ;
            app . drawVec ( - 20 * i , 10 ) ;
        app . context2D . restore ( ) ;
    }
    for ( let i : number = 1 ; i < 14 ; i ++ ) {
        app . context2D . save ( ) ;
            app . context2D . translate ( curr + i * 30 , 300 ) ;
            app . context2D . rotate ( Math2D . toRadian ( 100 ) )
            app . drawVec ( 20 * i , 10 ) ;
        app . context2D . restore ( ) ;
    }
}
```

6.4.5 图 6.6 向量的点乘图示源码

向量的点乘运算的几何性质图示源码。具体代码如下:

```
// 向量点乘
if ( app . context2D ) {
    app . context2D . save ( ) ;
        app . context2D . translate ( 20 , 300 ) ;
        app . context2D . save ( ) ;
            app . context2D . rotate ( Math2D . toRadian ( 30 ) ) ;
            app . drawVec ( 200 , 10 , undefined , "向量b" , 3 , false ) ;
        app . context2D . restore ( ) ;
        app . context2D . save ( ) ;
            app . context2D . rotate ( Math2D . toRadian ( - 30 ) ) ;
            app . drawVec ( 160 , 10 , undefined , "向量a" , 3 , false ) ;
        app . context2D . restore ( ) ;
    app . context2D . restore ( ) ;
    app . context2D . save ( ) ;
        app . context2D . translate ( 300 , 300 ) ;
```

```
        app . context2D . save ( ) ;
            app . context2D . rotate ( Math2D . toRadian ( - 30 ) ) ;
            app . drawVec ( 160 , 10 , undefined , "向量a" , 3 , false ) ;
        app . context2D . restore ( ) ;
        app . context2D . save ( ) ;
            app . context2D . rotate ( Math2D . toRadian ( 60 ) ) ;
            app . drawVec ( 200 , 10 , undefined , "向量b" , 3 , false ) ;
         app . context2D . restore ( ) ;
    app . context2D . restore ( ) ;
    app . context2D . save ( ) ;
        app . context2D . translate ( 550 , 300 ) ;
        app . context2D . save ( ) ;
            app . context2D . rotate ( Math2D . toRadian ( - 60 ) ) ;
            app . drawVec ( 160 , 10 , undefined , "向量a" , 3 , false ) ;
        app . context2D . restore ( ) ;
        app . context2D . save ( ) ;
            app . context2D . rotate ( Math2D . toRadian ( 60 ) ) ;
            app . drawVec ( 200 , 10 , undefined , "向量b" , 3 , false ) ;
        app . context2D . restore ( ) ;
    app . context2D . restore ( ) ;
}
```

6.4.6　图 6.11 点与三角形的关系图示源码

下面的代码可以显示一个点在三角形内部还是外部的几何效果，用来理解三角形顶点顺序的重要性。具体代码如下：

```
public drawTriangle ( x0 : number , y0 : number , x1 : number , y1 : number ,
x2 : number , y2 : number , ptX : number = x2 , ptY : number = y2 ,
drawSubTriangle : boolean = false , stroke : boolean = true ) : void {
    if ( this . context2D === null ) {
        return ;
    }
    this . context2D . save ( ) ;
       this . context2D . lineWidth = 3 ;
       this . context2D . strokeStyle = 'rgba( 0 , 0 , 0 , 0.5 )';
       this . context2D . beginPath ( ) ;
       this . context2D . moveTo ( x0 , y0 ) ;
       this . context2D . lineTo ( x1 , y1 ) ;
       this . context2D . lineTo ( x2 , y2 ) ;
       this . context2D . closePath ( ) ;
       if ( stroke ) {
          this . context2D . stroke ( ) ;
       } else {
          this . context2D . fill ( ) ;
       }
       if ( drawSubTriangle === true ) {
          this . context2D . lineWidth = 2 ;
          this . context2D . setLineDash ( [ 3 , 3 ] ) ;
          this . strokeLine ( x0 , y0 , ptX , ptY) ;
          this . strokeLine ( x1 , y1 , ptX , ptY ) ;
          this . strokeLine ( x2 , y2 , ptX , ptY ) ;
```

```
        this . fillCircle ( ptX , ptY , 5 ) ;
    this . context2D . restore ( ) ;
}
```

6.4.7　图 6.12 和图 6.13　凹凸多边形图示源码

下面用 drawPolygon 方法绘制一个多边形的同时，还可以进行三角扇形化显示。具体代码如下：

```
public drawPolygon ( points : vec2 [ ] , ptX : number , ptY : number , drawSubTriangle : boolean = false ) : void {
    if ( this . context2D === null ) return ;
    this . context2D . save ( ) ;
        this . context2D . strokeStyle = 'rgba( 0 , 0 , 0 , 0.5 )';
        this . context2D . lineWidth = 3 ;
        this . context2D . translate ( ptX , ptY ) ;
        // 绘制多边形
        this . context2D . beginPath ( ) ;
        this . context2D . moveTo ( points [ 0 ] . x , points [ 0 ] . y ) ;
        for ( let i = 1 ; i < points . length ; i ++ ) {
            this . context2D . lineTo ( points [ i ] . x , points [ i ] . y ) ;
        }
        this . context2D . closePath ( ) ;
        this . context2D . stroke ( ) ;
        // 绘制虚线，形成子三角形
        if ( drawSubTriangle === true ) {
            this . context2D . lineWidth = 2 ;
            this . context2D . setLineDash ( [ 3 , 3 ] ) ;
            for ( let i : number = 1 ; i < points . length - 1 ; i ++ ) {
                this . strokeLine ( points [ 0 ] . x , points [ 0 ] . y , points [ i ] . x , points [ i ] . y ) ;
            }
        }
        this . fillCircle ( points [ 0 ] . x , points [ 0 ] . y , 5 , 'red') ;
    this . context2D . restore ( ) ;
}
//图 6.12 和图 6.13 凹凸多边形图示源码
if ( app . context2D ) {
    app . context2D . save ( ) ;
    app . context2D . translate ( - 250 , 0 ) ;
    // 绘制凸多边形并进行三角扇形化显示
    app . drawPolygon ( [ vec2 . create (-100 , -50 ) ,
        vec2 . create ( 0 , - 100 ) ,
        vec2 . create ( 100 , -50 ) ,
        vec2 . create ( 100 , 50 ) ,
        vec2 . create ( 0 , 100 ) ,
        vec2 . create ( - 100 , 50 ) ] , 400 , 300 , true ) ;
    app . context2D . restore ( ) ;
    // 绘制凹多边形，不进行扇形化显示
    app . context2D . save ( ) ;
    app . context2D . translate ( 425 , 325 ) ;
```

```
    app . drawPolygon ( [
       vec2 . create ( 0 , 0 ) ,
       vec2 . create ( 100 , - 100 ) ,
       vec2 . create ( 100 , 50 ) ,
       vec2 . create ( -100 , 50 ) ,
       vec2 . create ( - 100 , - 100 )
    ] , 0 , 0 ) ;
    app . context2D . restore ( ) ;
}
```

6.5 本章总结

本章主要涉及的是向量，以及点与基本几何形体碰撞检测方面的相关知识点。对于游戏或图形编程来说，这是必须要具备的基本数学知识。

6.1 节主要讲解二维向量的相关知识点。向量是具有大小和方向的空间变量。而以前常用的数值都是标量，其仅有大小，而没有方向。在向量一节中，知道了如何计算向量的大小、向量的方向、向量的加减法、负向量、向量的缩放，以及向量的点乘。同时更加详细地解释了向量的上述这些操作相对应的几何含义，这是本节的关键点。只有深刻地理解向量的几何含义，才能灵活地应用向量来解决问题。本节提供了一个例子来演示向量的作用，该例子通过向量的有向缩放特性，完全替换耗时的 sin 和 cos 三角函数，这是一个很棒的操作。

6.2 节是向量投影的应用 Demo，通过编程的方式，动态地演示了向量投影的几何效果。向量投影用到了 6.1 中向量相关的所有操作。

6.3 节主要关注点与基本几何形体之间的碰撞检测算法，涉及点与线段、点与圆、点与矩形、点与椭圆、点与三角形，以及点与凸多边形之间的碰撞检测，并且讲解了多边形的三角形化，以及如何判断凸多边形的算法。

由于本章中的所有图示都是通过程序绘制得到，在绘制的过程中涉及向量的很多操作，因此将这些绘制代码作为本章的附加内容罗列出来。

以上就是本章的主要内容。

第 7 章　矩阵数学及贝塞尔曲线

第 5 章花了很长的篇幅讲解 Canvas2D 中坐标系变换的相关知识,用到了例如 translate、scale 和 rotate 这些坐标系变换方法。实际上,在 Canvas2D 内部,这些变换方法都是由矩阵实现的,而且 Canvas2D 还具有一个矩阵堆栈来维持矩阵变换的层次性。本章中将会详细地介绍矩阵,以及矩阵堆栈的相关知识。在了解完矩阵的相关知识后,将会介绍有趣的贝塞尔曲线的相关内容。

7.1　矩 阵 数 学

矩阵(Matrix)是 m 个行(Row)和 n 个列(Column)构成的数组。矩阵用大写的粗斜体英文字母表示,例如 $\boldsymbol{M}$。

m 行 n 列的矩阵被称为 $m \times n$ 矩阵,当 $m=n$ 时的矩阵称为方矩阵(Square Matrix)。

在游戏或计算机图形学中,2×2 矩阵(二维旋转矩阵),3×3 矩阵(可以表示二维仿射变换矩阵或三维旋转矩阵)和 4×4 矩阵(可以表示三维仿射变换矩阵或投影矩阵)是最常见的。由于 4×4 矩阵主要用于三维环境中,因此主要关注 3×3 仿射变换矩阵。

7.1.1　矩阵乘法

Canvas2D 中二维 3×3 仿射变换矩阵及其矩阵乘法具有如下表现形式:

$$\boldsymbol{C} = \boldsymbol{A} \times \boldsymbol{B} = \begin{bmatrix} a_0 & a_2 & a_4 \\ a_1 & a_3 & a_5 \\ 0 & 0 & 1 \end{bmatrix} \times \begin{bmatrix} b_0 & b_2 & b_4 \\ b_1 & b_3 & b_5 \\ 0 & 0 & 1 \end{bmatrix}$$

上述两个 3×3 仿射变换矩阵相乘的结果如下:

$$\boldsymbol{C} = \begin{bmatrix} a_0 b_2 + a_2 b_1 & a_0 b_2 + a_2 b_3 & a_0 b_4 + a_2 b_5 + a_4 \\ a_1 a_0 + a_3 b_1 & a_1 a_2 + a_3 b_3 & a_1 a_4 + a_3 b_5 + a_5 \\ 0 & 0 & 1 \end{bmatrix}$$

下面来实现一下矩阵类。具体代码如下:

```
export class mat2d {
    public values : Float32Array ;        // 使用强类型数组来表示矩阵的各个元素
    public constructor ( a : number = 1 , b : number = 0 , c : number = 0 ,
```

```
    d : number = 1 , x : number = 0 , y : number = 0 ) {
        this . values = new Float32Array ( [ a , b , c , d , x , y ] ) ;
    }
    public static create ( a : number = 1 , b : number = 0 , c : number =
0 , d : number = 1 , x : number = 0 , y : number = 0 ) : mat2d {
        return new mat2d ( a , b , c , d , x , y ) ;
    }
}
```

通过上述源码可以看到,实现的矩阵并没有使用 3×3 数组来表示,而是使用了六个浮点数组成的强类型数组,因为 3×3 仿射变换矩阵的最后一行总是为[0,0,1],因此不需要多浪费三个浮点数来存储这三个常量。

接下来实现矩阵最关键的一个操作,即矩阵乘法。具体代码如下:

```
public static multiply ( left : mat2d , right : mat2d , result : mat2d |
null = null ) : mat2d {
    if ( result === null ) result = new mat2d() ;
    let a0 : number = left . values [ 0 ] ;
    let a1 : number = left . values [ 1 ] ;
    let a2 : number = left . values [ 2 ] ;
    let a3 : number = left . values [ 3 ] ;
    let a4 : number = left . values [ 4 ] ;
    let a5 : number = left . values [ 5 ] ;

    let b0 : number = right . values [ 0 ] ;
    let b1 : number = right . values [ 1 ] ;
    let b2 : number = right . values [ 2 ] ;
    let b3 : number = right . values [ 3 ] ;
    let b4 : number = right . values [ 4 ] ;
    let b5 : number = right . values [ 5 ] ;

    // 参考上面矩阵乘法的结果
    result . values [ 0 ] = a0 * b0 + a2 * b1 ;
    result . values [ 1 ] = a1 * b0 + a3 * b1 ;
    result . values [ 2 ] = a0 * b2 + a2 * b3 ;
    result . values [ 3 ] = a1 * b2 + a3 * b3 ;
    result . values [ 4 ] = a0 * b4 + a2 * b5 + a4 ;
    result . values [ 5 ] = a1 * b4 + a3 * b5 + a5 ;

    return result ;
}
```

来看一下矩阵乘法的一些性质:
- 矩阵乘法不符合交换律,即 $A \times B \neq B \times A$,因此在进行矩阵乘法运算时必须要一直注意乘法的顺序。
- 矩阵乘法符合分配律,即 $A \times (B + C) = A \times B + A \times C$。
- 矩阵乘法符合结合律,即 $A \times (B \times C) = (A \times B) \times C$。
- 矩阵的乘法是矩阵最重要的运算,在后续章节会经常用到矩阵的乘法。

矩阵还有加法、减法、矩阵与标量相乘运算等,由于这些运算在图形或游戏中基本不会用到,因此忽略这些操作。

7.1.2 单位矩阵

在标量乘法中，任何数乘以 1，不会改变该标量的值。同样地，在矩阵中也存在着这样一种矩阵，称为单位矩阵，一般使用大写粗体 I 来标记单位矩阵。任何矩阵和单位矩阵 I 相乘，矩阵保持不变，其公式如下：

$$M \times I = M$$

单位矩阵 I 除了对角线上的元素为 1 外，其他元素都为 0，如下

$$I = \begin{bmatrix} 1 & 0 & 0 \\ 0 & 1 & 0 \\ 0 & 0 & 1 \end{bmatrix}$$

来看一下单位矩阵的实现代码，具体如下：

```
public identity () : void {
    this . values [ 0 ] = 1.0 ;
    this . values [ 1 ] = 0.0 ;
    this . values [ 2 ] = 0.0 ;
    this . values [ 3 ] = 1.0 ;
    this . values [ 4 ] = 0.0 ;
    this . values [ 5 ] = 0.0 ;
}
```

可以看到，identity 方法是 mat2d 类的实例方法，当调用该方法后，会将矩阵重置为单位矩阵。

还要注意的一点是，在 mat2d 构造函数中，默认情况下，初始化构造的也是单位矩阵。

7.1.3 矩阵求逆

对于矩阵的求逆，需要知道如下几点：
- 只有方矩阵才能求逆，本书中使用的都是方矩阵。
- 即使是方矩阵，也不是都能求逆。
- 方矩阵 M 的逆矩阵可以被记作 M^{-1}，也是一个方矩阵。
- $I = M \times M^{-1} = M^{-1} \times M$，矩阵 M 和它的逆矩阵 M^{-1} 相乘的结果是一个单位矩阵，并且符合矩阵乘法的交换律（如上所示，一般矩阵相乘不符合乘法交换律）。
- 方矩阵 M 的逆的逆等于原矩阵 M，即：$(M^{-1})^{-1} = M$。
- 单位矩阵 I 的逆矩阵是它本身，即：$I^{-1} = I$。
- 矩阵乘积的逆等于每个矩阵的逆的相反顺序的乘积，例如，$(A \times B \times C)^{-1} = C^{-1} \times B^{-1} \times A^{-1}$，可以扩展到 n 个矩阵的情况下。
- 常用的方矩阵求逆算法有伴随矩阵算法和高斯消元算法，本书中使用伴随矩阵方式求矩阵的逆。

由于逆矩阵的算法比较复杂，此处就不再展开。下面来看一下代码是如何实现的。

```
// 计算矩阵的行列式
public static determinant ( mat : mat2d ) : number {
    return mat . values [ 0 ] * mat . values [ 3 ] - mat . values [ 2 ] *
    mat . values [ 1 ] ;
}
// 求矩阵 src 的逆矩阵，将结算后的逆矩阵从 result 参数中输出
// 如果有逆矩阵，返回 true；否则返回 false
// 下面的代码中使用：伴随矩阵 / 行列式 的方式来求矩阵的逆
public static invert ( src : mat2d , result : mat2d ) : boolean {
    // 1．获取要求逆的矩阵的行列式
    let det : number = mat2d . determinant ( src ) ;
    // 2．如果行列式为 0，则无法求逆，直接返回 false
    if ( Math2D . isEquals (det , 0 ) ) {
       return false ;
    }
    // 3．使用：伴随矩阵 / 行列式 的算法来求矩阵的逆
    // 由于计算机中除法效率较低，先进行一次除法，求行列式的倒数
    // 后面代码就可以直接乘以行列式的倒数，这样避免了多次除法操作
    det = 1.0 / det ;
    // 4．下面的代码中， * det 之前的代码都是求标准伴随矩阵的源码
    //    最后乘以行列式的倒数，获得每个元素的正确数值
    result . values [ 0 ] = src . values [ 3 ] * det ;
    result . values [ 1 ] = - src . values [ 1 ] * det ;
    result . values [ 2 ] = - src . values [ 2 ] * det ;
    result . values [ 3 ] = src . values [ 0 ] * det ;
    result . values [ 4 ] = ( src . values [ 2 ] * src . values [ 5 ] - src .
    values [ 3 ] * src . values [ 4 ] ) * det ;
    result . values [ 5 ] = ( src . values [ 1 ] * src . values [ 4 ] - src .
    values [ 0 ] * src . values [ 5 ] ) * det ;
    // 如果矩阵求逆成功，返回 true
    return true ;
}
```

7.1.4　用矩阵变换向量

如果要让矩阵作用在向量上，必须要让矩阵和向量相乘才能起作用。如果要让矩阵和向量相乘，则必须要将向量写成矩阵的形式。向量写成矩阵的形式，有两种选择：

- 行向量形式：$[x \ y]$。
- 列向量形式：$\begin{bmatrix} x \\ y \end{bmatrix}$。

本书中使用列向量的表现形式，矩阵 M 与列向量 v 相乘的结果仍旧是个列向量，具体过程如下：

$$v' = M \times v = \begin{bmatrix} m_0 & m_2 & m_4 \\ m_1 & m_3 & m_5 \\ 0 & 0 & 1 \end{bmatrix} \times \begin{bmatrix} x \\ y \\ 1 \end{bmatrix} = \begin{bmatrix} m_0 x + m_2 y + m_4 \\ m_1 x + m_3 y + m_5 \\ 1 \end{bmatrix}$$

下面来实现一个名为 transform 的矩阵和向量相乘的方法,将其作为一个静态公开方法定义在 Math2D 类中。具体代码如下:

```
public static transform (mat : mat2d ,pt : vec2, result : vec2 | null = null ) : vec2 {
    if ( result === null ) result = vec2 . create ( ) ;
    result . values [ 0 ] = mat . values [ 0 ] * pt . values [ 0 ] + mat . values[ 2 ] * pt . values [ 1 ] + mat . values [ 4 ] ;
    result . values [ 1 ] = mat . values [ 1 ] * pt . values [ 0 ] + mat . values[ 3 ] * pt . values[ 1 ] + mat . values[ 5 ] ;
    return result ;
}
```

7.1.5 平移矩阵及其逆矩阵

在第 5 章中,一直在使用 Canvas2D 中的 translate、scale 和 rotate 方法。实际上,Canvas2D 中这些方法都是在操作矩阵。下面首先来看一下 Canvas2D 中的 translate 平移方法的矩阵实现方式,以及该平移矩阵作用于某个向量上的操作,具体效果如下:

$$v' = T \times v = \begin{bmatrix} 1 & 0 & tx \\ 0 & 1 & ty \\ 0 & 0 & 1 \end{bmatrix} \times \begin{bmatrix} x \\ y \\ 1 \end{bmatrix} = \begin{bmatrix} x + tx \\ y + ty \\ 1 \end{bmatrix}$$

在上述的公式中,T 表示平移矩阵,v 表示要变换的向量,而 v' 表示变换后的向量。我们先来实现平移矩阵,在 mat2d 类中,输入如下代码:

```
public static makeTranslation (tx : number ,ty : number ,result : mat2d | null = null ) : mat2d {
    if ( result === null ) result = new mat2d ( ) ;
    result . values [ 0 ] = 1 ;
    result . values [ 1 ] = 0 ;
    result . values [ 2 ] = 0 ;
    result . values [ 3 ] = 1 ;
    // 会看到平移矩阵只需要设置第 4 个和第 5 个元素的值
    result . values [ 4 ] = tx ;
    result . values [ 5 ] = ty ;
    return result ;
}
```

接下来测试一下,让一个向量 v = [100 , 200] 平移 [200 , 100] 个单位后的结果。具体代码如下:

```
let T : mat2d = mat2d . makeTranslation ( 200 , 100 ) ;
                    // 生成一个 tx = 100 , ty = 200 的平移矩阵
let v : vec2 = vec2 . create ( 100 , 200 ) ;
```

```
                                    // 生成一个 x = 100, y = 200 的向量
// 调用 7.3.4 节实现的 transform 方法, 该方法将矩阵的效果作用于某个向量
let tv = Math2D . transform ( T , v ) ;
                                    // tv = transformedV, 也就是变换后的向量
// 输出: T v = v' = tv = [ 300 , 300 ]
alert ( tv ) ;
```

下面看一下对平移矩阵 T 求逆矩阵 T^{-1} 及将该逆矩阵 T^{-1} 作用于向量 v' 上的相关操作及结果。在上面的代码中,已知平移矩阵 $T = \begin{bmatrix} 1 & 0 & 200 \\ 0 & 1 & 100 \\ 0 & 0 & 1 \end{bmatrix}$,并且得到被平移后的向量 $v' = \begin{bmatrix} 300 \\ 300 \\ 1 \end{bmatrix}$,现在需求是求出平移前的向量 v 的值,可以在上面的代码中继续输入如下的代码:

```
let invertT : mat2d = mat2d . create ( ) ;   //分配要计算的逆矩阵的内存
mat2d . invert ( T , invertT ) ;             //求 T 的逆矩阵 invertT
v = Math2D . transform ( invertT , tv ) ;    //Math2D.transform(T,v)的逆变换
alert ( v ) ;                                //此时得到的 v 应该是[100,200]
```

通过上述两段代码,来看一下其矩阵和向量变换的流程,结果如下:

$$v' = T \times v = \begin{bmatrix} 1 & 0 & 200 \\ 0 & 1 & 100 \\ 0 & 0 & 1 \end{bmatrix} \times \begin{bmatrix} 100 \\ 200 \\ 1 \end{bmatrix} = \begin{bmatrix} 100+200 \\ 200+100 \\ 1 \end{bmatrix} = \begin{bmatrix} 300 \\ 300 \\ 1 \end{bmatrix}$$

其逆变换:

$$v = T^{-1} \times v' = \begin{bmatrix} 1 & 0 & -200 \\ 0 & 1 & -100 \\ 0 & 0 & 1 \end{bmatrix} \times \begin{bmatrix} 300 \\ 300 \\ 1 \end{bmatrix} = \begin{bmatrix} 300-200 \\ 300-100 \\ 1 \end{bmatrix} = \begin{bmatrix} 100 \\ 200 \\ 1 \end{bmatrix}$$

可以发现,平移矩阵的逆矩阵就是:

$$T^{-1} = \begin{bmatrix} 1 & 0 & tx \\ 0 & 1 & ty \\ 0 & 0 & 1 \end{bmatrix}^{-1} = \begin{bmatrix} 1 & 0 & -tx \\ 0 & 1 & -ty \\ 0 & 0 & 1 \end{bmatrix}$$

因此,可以使用 mat2d 类的公开静态方法 invert 求平移矩阵的逆矩阵,也可以使用如下方式获取逆矩阵:

```
let invertT : mat2d = mat2d . makeTranslation ( - 200 , - 100 ) ;
                                                    // 参数取反
```

7.1.6 缩放矩阵及其逆矩阵

下面看一下缩放矩阵,其表达方式如下:

$$v' = T \times v = \begin{bmatrix} s_x & 0 & 0 \\ 0 & s_y & 0 \\ 0 & 0 & 1 \end{bmatrix} \times \begin{bmatrix} x \\ y \\ 1 \end{bmatrix} = \begin{bmatrix} s_x x \\ s_y y \\ 1 \end{bmatrix}$$

根据上述公式来实现一个缩放矩阵。具体代码如下：

```
public static makeScale ( sx : number ,sy : number ,result :mat2d | null = null ) : mat2d {
    if ( Math2D . isEquals ( sx , 0 ) || Math2D . isEquals ( sy , 0 ) ) {
        alert ( " x 轴或 y 轴缩放系数为 0 " ) ;
        throw new Error ( " x 轴或 y 轴缩放系数为 0 " ) ;
    }
    if ( result === null ) result = new mat2d() ;
    result . values [ 0 ] = sx ;
    result . values [ 1 ] = 0 ;
    result . values [ 2 ] = 0 ;
    result . values [ 3 ] = sy ;
    result . values [ 4 ] = 0 ;
    result . values [ 5 ] = 0 ;
    return result ;
}
```

下面来测试缩放矩阵，让一个向量 v = [100 , 200] 在 x 轴上放大 3 倍，在 y 轴上缩小 0.5 倍后，得到变换后的向量 v'（代码中用 tv 表示 v'）。具体代码如下：

```
let S : mat2d = mat2d . makeScale ( 3 , 0.5 ) ;         //构造 scale 矩阵
let v : vec2 = vec2 .create ( 100 , 200 ) ;             //创建[100 , 200]向量
// 将向量 v 变换到 S 空间中
let tv : vec2 = Math2D . transform ( S , v ) ;
                                                        //tv = transformedV,也就是变换后的向量
alert ( tv ) ;                                          //此时 tv = [300 , 100]
```

接下来继续测试逆矩阵，将 v' = [300 , 100] 变换回原来的向量 v。具体代码如下：

```
let invertS : mat2d = mat2d . create ( ) ; // // 分配要计算的逆矩阵的内存
mat2d . invert ( S , invertS ) ;      // 求 T 的逆矩阵 invertT
v = Math2D . transform ( invertS , tv ) ;
                                      //Math2D . transform (S , v)的逆变换
alert ( v ) ;                         //此时得到的 v 应该是[100 , 200]
```

通过上述测试代码，可以发现缩放矩阵的逆矩阵就是：

$$S^{-1} = \begin{bmatrix} s_x & 0 & 0 \\ 0 & s_y & 0 \\ 0 & 0 & 1 \end{bmatrix}^{-1} = \begin{bmatrix} 1.0/s_x x & 0 & 0 \\ 0 & 1.0/s_y & 0 \\ 0 & 0 & 1 \end{bmatrix}$$

不使用通用的 invert 方法求逆，而是构造上述的逆矩阵来验证是否正确。具体代码如下：

```
// 不使用通用的 invert 方法求 S = mat2d . makeScale ( 3 , 0.5 )的逆矩阵
// 而是使用[ 1 / sx , 1 / sy ]方式获取逆矩阵
let invertS : mat2d = mat2d . makeScale ( 1.0 / 3 , 1.0 / 0.5 ) ;
```

```
v = Math2D . transform ( invertS , tv ) ;
                                       // Math2D . transform (S , v)的逆变换
alert ( v ) ;                          //此时得到的 v 的确是[100 , 200]
```

7.1.7 旋转矩阵及其逆矩阵

相对于平移矩阵和缩放矩阵，旋转矩阵比较复杂，下面来看一下，在直角平面坐标系下，将某个点 P 绕原点 O 旋转 a 弧度的变换矩阵是如何推导出来的，参考图 7.1 所示的效果。

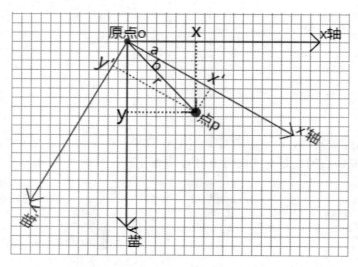

图 7.1 旋转矩阵推导示意图

根据图 7.1 来看一下已知条件：
- 以 **O** 为原点，有两个坐标系：XY 坐标系和 $X'Y'$ 坐标系，这两个坐标系之间的关系是 XY 坐标系顺时针旋转 a 弧度后变成 $X'Y'$ 坐标系，a 弧度是作为已知条件。
- 已知点 **P** 在 $X'Y'$ 坐标系中的坐标为$[x' , y']$。
- 已知 $r = \sqrt{x'^2 + y'^2}$。

现在问题是：求点 P 在 XY 坐标系中的坐标$[x , y]$是多少？
来看一下上面问题的求解过程：
∵ x' 和 y' 已知，且 $c = a + b$
∴ $\sin (b) = \sin (c - a) =$ 对边 / 斜边 = Opposite / Hypotenuse = y' / r
∵ 根据三角函数中 sin 的差角公式：$\sin (c - a) = \sin (c) \cos (a) - \cos (c) \sin (a)$
∴ $\sin (c) \cos (a) - \cos (c) \sin (a) = y' / r$
∴ $y' = r (\sin (c) \cos (a) - \cos (c) \sin (a))$
∴ $y' = r \sin (c) \cos (a) - r \cos (c) \sin (a)$

∵ cos (b) = cos (c - a) =临边 / 斜边= Adjacent / Hypotenuse = x' / r
∵ 根据三角函数中 cos 的和角公式：cos (c - a) = cos (c) cos (a) + sin (c) sin (a)
∴ cos (c) cos (a) + sin (c) sin (a) = x' / r
∴ x' = r (cos (c) cos (a) + sin (c) sin (a))
∴ x' = r cos (c) cos (a) + r sin (c) sin (a)

∵ sin (c) = y / r 且 cos (c) = x / r
∴ y = r sin (c)且 x = r cos (c)
∴ x' = x cos (a) + y sin (a) = cos (a) x + sin (a) y
∴ y' = y cos (a) - x sin (a) = - sin (a) x + cos (a) y

∴ 写成矩阵形式：$\begin{bmatrix} x' \\ y' \end{bmatrix} = \begin{bmatrix} \cos(a) & \sin(a) \\ -\sin(a) & \cos(a) \end{bmatrix} \times \begin{bmatrix} x \\ y \end{bmatrix}$

∵ x'、y'和弧度 a 已知，要求的是 x 和 y
∴ $\begin{bmatrix} x \\ y \end{bmatrix} = \begin{bmatrix} \cos(a) & \sin(a) \\ -\sin(a) & \cos(a) \end{bmatrix}^{-1} \times \begin{bmatrix} x' \\ y' \end{bmatrix}$

一直以来都是使用 3×3 的仿射变换矩阵来表示二维变换，因此改写一下，具体如下：

$$\begin{bmatrix} x \\ y \\ 1 \end{bmatrix} = \begin{bmatrix} \cos(a) & -\sin(a) & 0 \\ \sin(a) & \cos(a) & 0 \\ 0 & 0 & 1 \end{bmatrix} \times \begin{bmatrix} x' \\ y' \\ 1 \end{bmatrix}$$

至此，已经推导出旋转矩阵的结果。现在来了解一下上述旋转矩阵求逆的 3 种方式及优缺点。

1. 通用的invert方法求矩阵的逆

优点：通用省心，适合所有情况。
缺点：计算量稍微大一些。

2. 角度取负求旋转矩阵的逆

$$\begin{bmatrix} \cos(a) & \sin(a) \\ -\sin(a) & \cos(a) \end{bmatrix}^{-1} = \begin{bmatrix} \cos(-a) & \sin(-a) \\ -\sin(-a) & \cos(a) \end{bmatrix} = \begin{bmatrix} \cos(a) & -\sin(a) \\ \sin(a) & \cos(a) \end{bmatrix}$$

优点：计算量相对较小。
缺点：必须知道弧度（后面会看到不通过输入的弧度来构造旋转矩阵的算法）才能构造逆矩阵，并且只能用于旋转矩阵（不能包含缩放及平移等效果）。

3. 转置（Transpose）矩阵方式求旋转矩阵的逆

$$\begin{bmatrix} \cos(a) & \sin(a) \\ -\sin(a) & \cos(a) \end{bmatrix}^{-1} = \begin{bmatrix} \cos(a) & -\sin(a) \\ \sin(a) & \cos(a) \end{bmatrix}$$

如上所示，我们会发现转置矩阵就是行列交换，即等号左侧矩阵的第一行变为等号右侧矩阵的第一列，等号左侧矩阵的第二行变为等号右侧矩阵的第二列。

优点：速度最快，并且不需要知道弧度。

缺点：只能用于旋转矩阵（不能包含缩放及平移等效果）。

接下来看一下旋转矩阵及其逆矩阵的实现。具体代码如下：

```
public static makeRotation ( radians : number , result : mat2d | null = null ) : mat2d {
    if ( result === null ) result = new mat2d ( ) ;
    let s : number = Math . sin ( radians ) , c : number = Math . cos ( radians ) ;
    result . values [ 0 ] = c ;
    result . values [ 1 ] = s ;
    result . values [ 2 ] = -s ;
    result . values [ 3 ] = c ;
    result . values [ 4 ] = 0 ;
    result . values [ 5 ] = 0 ;
    return result ;
}
// 会修改 this 指向的数据
public onlyRotationMatrixInvert ( ) : mat2d {
    let s : number = this . values [ 1 ] ;
    // 矩阵的第 1 个元素和第 2 个元素值交换
    this . values [ 1 ] = this . values [ 2 ] ;
    this . values [ 2 ] = s ;
    return this ;
}
```

看一下测试代码，使用了三种求旋转矩阵的逆矩阵方式，具体如下：

```
// 创建一个顺时针旋转 30 度的旋转矩阵
let R : mat2d = mat2d . makeRotation ( Math2D . toRadian ( 30 ) ) ;
// 创建一个变换前的向量 v
let v : vec2 = vec2 . create ( 100 , 200 ) ;
// 将向量 v 变换为 tv
let tv : vec2 = Math2D . transform ( R , v ) ;
// 输出 tv 的数值, tv = [ -13.3975 , 223.205 ]
alert ( tv ) ;

// 第 1 种方式：使用相反的旋转角度构造旋转矩阵的逆矩阵
let invertR : mat2d = mat2d . makeRotation ( Math2D . toRadian ( - 30 ) ) ;
// 将 tv = [ -13.3975 , 223.205 ]变换回原来的数值
v = Math2D . transform ( invertR , tv ) ;
// 此时 v 肯定是[ 100 , 200 ]
alert ( v ) ;

// 第 2 种方式：使用通用的 invert 方法构造矩阵的逆矩阵
```

```
let invertR2 : mat2d = mat2d . create ( ) ;
mat2d . invert ( R , invertR2 ) ;
// 将 tv = [ -13.3975 , 223.205 ]变换回原来的数值
v = Math2D . transform ( invertR2 , tv ) ;
// 此时 v 肯定是[ 100 , 200 ]
alert ( v ) ;

// 第 3 种方式：转置方式构造旋转矩阵的逆矩阵
// 将旋转矩阵 R 复制到 invertR2
mat2d . copy ( R , invertR2 ) ;
// 在 invertR2 上调用 onlyRotationMatrixInvert 方法
// 该方法使用转置方式求旋转矩阵的逆，会修改矩阵本身
invertR2 . onlyRotationMatrixInvert ( ) ;
// 将 tv = [ -13.3975 , 223.205 ]变换回原来的数值
v = Math2D . transform ( invertR2 , tv ) ;
// 此时 v 肯定是[ 100 , 200 ]
alert ( v ) ;
```

7.1.8　从两个单位向量构建旋转矩阵

在上一节中，推导出旋转矩阵为 $\begin{bmatrix} \cos(a) & -\sin(a) \\ \sin(a) & \cos(a) \end{bmatrix}$，而在向量点乘一节中，得到了 $\cos(a) = \boldsymbol{a} \cdot \boldsymbol{b} / (\|\boldsymbol{a}\| \|\boldsymbol{b}\|)$ 这个公式，其中 $\boldsymbol{a}$ 和 $\boldsymbol{b}$ 是两个向量。再进一步，如果向量 $\boldsymbol{a}$ 和 $\boldsymbol{b}$ 都是单位向量，那么前面的公式可以进一步简化为 $\cos(a) = \boldsymbol{a} \cdot \boldsymbol{b}$。

如果能将 $\sin(a)$ 也表示成如 $\cos(a)$ 类似方式，就能完全避免使用昂贵的三角函数计算，全部使用向量 $\boldsymbol{a}$ 和 $\boldsymbol{b}$ 获得旋转矩阵。

幸运的是，有类似的公式 $\sin(a) = \|\boldsymbol{a} \otimes \boldsymbol{b}\| / \|\boldsymbol{a}\| \|\boldsymbol{b}\|$，同样的，如果向量 $\boldsymbol{a}$ 和 $\boldsymbol{b}$ 都是单位向量，则可以简化为 $\sin(a) = \|\boldsymbol{a} \otimes \boldsymbol{b}\|$。

这里需要说明一下的是 $\otimes$ 符号，该符号表示向量 $\boldsymbol{a}$ 叉乘（Cross）向量 $\boldsymbol{b}$，由于叉乘属于 3D 数学范畴，因此本书不详细讨论，如各位读者有兴趣，可以自行查阅相关资料。

来重新捋一下，对于单位向量 $\boldsymbol{a}$ 和 $\boldsymbol{b}$，可以得到如下两个重要的公式：

- $\sin(a) = \|\boldsymbol{a} \otimes \boldsymbol{b}\| = a.x * b.y - b.x * a.y$
- $\cos(a) = \boldsymbol{a} \bullet \boldsymbol{b} = a.x * b.x + a.y * b.y$

那么就可以从两个单位向量 $\boldsymbol{a}$ 和 $\boldsymbol{b}$ 得到一个旋转矩阵：

$$\begin{bmatrix} \cos(a) & -\sin(a) \\ \sin(a) & \cos(a) \end{bmatrix} = \begin{bmatrix} a.xb.x + a.yb.y & -(a.xb.y - b.xa.y) \\ a.xb.y - b.xa.y & a.xb.x + a.yb.y \end{bmatrix}$$

接下来看一下实现代码，首先在 vec2 类中增加 sinAngle 和 cosAngle 这两个方法。

```
public static sinAngle ( a : vec2 , b : vec2 , norm : boolean = false ) :
number {
    if ( norm === true ) {
        a . normalize ( ) ;
        b . normalize ( ) ;
    }
```

```
        return ( a . x * b . y - b . x * a . y ) ; // 参考上面 sin（a）公式
    }
    public static cosAngle ( a : vec2 , b : vec2 , norm : boolean = false ) : number {
        if ( norm === true ) {
            a . normalize ( ) ;
            b . normalize ( ) ;
        }
        return vec2 . dotProduct ( a , b ) ; // 参考上面 cos（a）公式
    }
```

在 mat2d 类中，从两个单位向量构造旋转矩阵的静态公开方法 makeRotationFromVectors 代码如下：

```
// norm 参数用来指明是否要 normlize 两个向量
public static makeRotationFromVectors(v1:vec2 , v2 :vec2 ,norm :boolean = false ,result : mat2d | null = null ) : mat2d {
    if ( result === null ) result = new mat2d ( ) ;
    result . values [ 0 ] = vec2 . cosAngle ( v1 , v2 , norm ) ;
    result . values [ 1 ] = vec2 . sinAngle ( v1 , v2 , norm ) ;
    result . values [ 2 ] = - vec2 . sinAngle ( v1 , v2 , norm ) ;
    result . values [ 3 ] = vec2 . cosAngle ( v1 , v2 , norm ) ;
    result . values [ 4 ] = 0 ;
    result . values [ 5 ] = 0 ;
    return result ;
}
```

7.1.9　使用 makeRotationFromVectors 方法取代 atan2 的应用

在 6.1.6 节中，使用向量与标量相乘的方式，将第 5 章坦克 Demo 中的 sin 和 cos 这些三角函数都替换了，那么在本节中，更进一步，将计算坦克朝向的 atan2 这个三角函数也进行替换。为了将坦克 Demo 替换为向量及矩阵版本，需要重新创建一个名为 TankWithMatrix 的类，然后把修改的一些代码列出来。具体代码如下：

```
class TankWithMatrix {
    // 坦克当前的位置
    // default 情况下为 [ 100 , 100 ]
    public pos : vec2 = new vec2 ( 100 , 100 ) ;
    // 坦克当前的 x 和 y 方向上的缩放系数
    // default 情况下为 1.0
    public scale : vec2 = new vec2 ( 1 , 1 ) ;
    // 坦克当前的旋转角度，使用旋转矩阵表示
    public tankRotation : mat2d = new mat2d ( ) ;
    public target : vec2 = new vec2 ( ) ;
    public initYAxis : boolean = false ;
    // 其他成员变量与原来 Tank 类一样，不再列出
}
```

可以看到，原来 Tank 类使用标量表示的位置、缩放及 target 现在都用向量表示，而原先的 tankRotation 也是标量，表示旋转的弧度，现在则改成 mat2d 表示的旋转矩阵。

接下来要修改一下_lookA 方法，该方法原来代码使用 atan2 计算坦克的朝向，现在的代码修改成如下所示：

```
private _lookAt ( ) : void {
   // 坦克与鼠标位置形成的方向向量
   let v : vec2 = vec2 . difference ( this . target , this . pos ) ;
   v . normalize ( ) ;
   // 构造从 v 向 x 轴的旋转矩阵
   this . tankRotation = mat2d . makeRotationFromVectors ( v , vec2 . xAxis ) ;
}
```

关于旋转后的运动相关代码，则和 6.1.6 节中源码一致，此处省略。

现在要解决一个问题：如何将自己实现的 mat2d 矩阵传递给 CanvasRenderingContext2D 上下文渲染对象？

5.1 节中提到过，在 Canvas2D 中，CanvasRenderingContext2D 渲染上下文对象使用如下五个方法来进行局部坐标系的变换。

```
translate ( x : number , y : number ) : void ;          // 局部坐标系的平移操作
rotate ( angle : number ) : void ;                     // 局部坐标系的旋转操作
scale ( x : number , y : number ) : void ;             // 局部坐标系的缩放操作
transform ( m11 : number , m12 : number , m21 : number , m22 : number , dx :
number , dy : number ) : void ;                        // 矩阵相乘操作
setTransform ( m11 : number , m12 : number , m21 : number , m22 : number ,
dx : number , dy : number ) : void ;                   // 设置变换矩阵操作
```

在第 5 章中仅仅使用了 translate、scale 和 rotate 方法，另外的 transform 和 setTransform 两个方法并没有使用到。现在将使用到 transform 方法，在 TestApplication 类中封装 transform 方法。具体代码如下：

```
public transform ( mat : mat2d ) : void {
   if ( this . context2D === null ) {
      return ;
   }
   this . context2D . transform (
      mat . values [ 0 ] ,
      mat . values [ 1 ] ,
      mat . values [ 2 ] ,
      mat . values [ 3 ] ,
      mat . values [ 4 ] ,
      mat . values [ 5 ] ) ;
}
```

上述 transform 方法中，将自己实现的 mat2d 矩阵中的各个元素赋值给 CanvasRenderingContext2D 对象所持有的当前矩阵，这样就解决了上面提到的问题，即如何将我们自己实现的 mat2d 矩阵传递给 canvasRederingContext2D 上下文渲染对象。

关于 CanvasRenderingContext2D 中 transform 和 setTransform 的区别，会在后续的矩阵堆栈相关内容中进行解答。

那么接下来要修改的是坦克 Demo 中的 draw 方法，要修改的代码具体如下：

```
// 整个坦克移动和旋转，注意局部变换的经典结合顺序（ trs: translate -> rotate -> scale ）
app . context2D . translate ( this . pos . x , this . pos . y ) ;
// 原本是 app . context2D . translate ( 弧度 )，现在改成如下调用方式
app . transform ( this . tankRotation ) ;
                                       // tankRotation 目前不是弧度表示，而是旋转矩阵
app . context2D . scale ( this . scale . x , this . scale . y ) ;
```

运行上述代码后，发现朝向并不正确。这种情况下，第一个反应就是旋转是有正负方向的，尝试一下求逆操作。

在上面的 _lookAt 方法中求得 tankRotation 旋转矩阵后使用通用的 invert 方法获得逆矩阵，调用具体代码如下：

```
mat2d . invert ( this . tankRotation , this . tankRotation ) ;
```

或者调用 onlyRotationMatrixInvert 方法，代码如下：

```
this . tankRotation . onlyRotationMatrixInvert ( ) ;
```

由于是使用矩阵存储旋转相关信息，而不是使用弧度或角度方式，因此没法快速地使用负角方式构造旋转矩阵的逆矩阵。

现在运行应用程序，会发现坦克能够朝向正确地运行。

通过矩阵求逆的方式，解决了物体朝向问题。实际上，还可以使用如下代码来获得同样的效果：

```
private _lookAt ( ) : void {
    // 坦克与鼠标位置形成的方向向量
    let v : vec2 = vec2 . difference ( this . target , this . pos ) ;
    v . normalize ( ) ;
    // 构造从 v 向 x 轴的旋转矩阵
    // this . tankRotation = mat2d . makeRotationFromVectors ( v , vec2 . xAxis ) ;
    // 构造从 x 轴向 v 向量的旋转矩阵，而不是上面的 v 向量向 x 轴方向的旋转矩阵
    this . tankRotation = mat2d . makeRotationFromVectors (vec2 .xAxis , v );
}
```

运行上述修改后的代码，会发现物体的朝向也是正确无误的。

最后来看一下当 initYAxis 为 true 时，坦克初始化朝向为 y 轴正方向时的处理，代码如下：

```
private _lookAt ( ) : void {
    // 坦克与鼠标位置形成的方向向量
    let v : vec2 = vec2 . difference ( this . target , this . pos ) ;
    v . normalize ( ) ;
    if ( this . initYAxis === true ) {
        // 构造从 y 轴向 v 向量的旋转矩阵，而不是上面的 v 向量向 y 轴方向的旋转矩阵
        this . tankRotation = mat2d.makeRotationFromVectors (vec2.yAxis,v);
    } else {
```

```
            // 构造从 x 轴向 v 向量的旋转矩阵,而不是上面的 v 向量向 x 轴方向的旋转矩阵
            this.tankRotation = mat2d.makeRotationFromVectors(vec2.xAxis,v);
        }
    }
```

7.1.10 仿射变换

前面一直提到使用的是仿射变换（Affine Transformation）这个词，现在是时候来了解一下仿射变换的相关内容，如图 7.2 所示。

$$\left[\begin{array}{cc|c} a_0 & a_2 & a_4 \\ a_1 & a_3 & a_5 \\ \hline 0 & 0 & 1 \end{array}\right]$$

图 7.2 二维仿射变换矩阵

如图 7.2 所示是一直使用的二维仿射变换矩阵的一般形式，这个矩阵中的每个元素都有特殊的含义：

- $\begin{bmatrix} a0 & a2 \\ a1 & a3 \end{bmatrix}$ 部分可以对图形进行缩放和旋转等线性变换。
- $\begin{bmatrix} a4 \\ a5 \end{bmatrix}$ 部分可以对图形进行平移变换。
- 左下的 $\begin{bmatrix} 0 & 0 \end{bmatrix}$ 部分实际可以对图形进行投影变换，但是在二维中，不太使用投影变换，因此将其都设置为 0，表示取消投影变换。
- 最后右下角的 1 部分表示对图形的整体缩放变换，一般总是将其设置为 1。

将包含平移的变换称为仿射变换，线性变换（缩放和旋转等）是仿射变换的特殊形式，即平移部分为 0。

本书是以实战为主，因此关于仿射变换和与之相关的齐次坐标系（Homogeneous coordinate system）等内容，请读者自行查阅相关资料，此处不做细节描述。

我们只要知道，使用了仿射变换及齐次坐标系后，就能使用矩阵乘法来统一操作图形的缩放、旋转和平移等变换。

7.1.11 矩阵堆栈

在前面章节中已经获得了平移矩阵、缩放矩阵和旋转矩阵，接下来实现一个矩阵堆栈类，用来模拟 Canvas2D 中的 translate、scale 和 rotate 操作，这样就能够清晰地了解 Canvas2D 中矩阵变换实现的底层细节。具体代码如下：

```
export class MatrixStack {
    // 持有一个矩阵堆栈
    private _mats : mat2d [ ] ;
    // 构造函数
    public constructor ( ) {
        // 初始化矩阵堆栈后 push 一个单位矩阵
        this . _mats = [ ] ;
        this . _mats . push ( new mat2d ( ) ) ;
    }
```

```typescript
// 获取栈顶的矩阵（也就是当前操作矩阵）
// 矩阵堆栈操作的都是当前堆栈顶部的矩阵
public get matrix ( ) : mat2d {
    if ( this . _mats . length === 0 ) {
        alert ( " 矩阵堆栈为空 " ) ;
        throw new Error ( " 矩阵堆栈为空 " ) ;
    }
    return this . _mats [ this . _mats . length - 1 ] ;
}
// 复制栈顶的矩阵，将其 push 到堆栈中成为当前操作矩阵
public pushMatrix ( ) : void {
    let mat : mat2d = mat2d . copy ( this . matrix ) ;
    this . _mats . push ( mat ) ;
}
// 删除栈顶的矩阵
public popMatrix ( ) : void {
    if ( this . _mats . length === 0 ) {
        alert ( " 矩阵堆栈为空 " ) ;
        return ;
    }
    this . _mats . pop ( ) ;
}
// 将堆栈顶部的矩阵设置为单位矩阵
public loadIdentity ( ) : void {
    this . matrix . identity ( ) ;
}
// 将参数 mat 矩阵替换堆栈顶部的矩阵
public loadMatrix ( mat : mat2d ) : void {
    mat2d . copy ( mat , this . matrix ) ;
}
// 将栈顶（当前矩阵）矩阵与参数矩阵相乘
// 其作用是更新栈顶元素，累积变换效果
// 是一个关键操作
public multMatrix ( mat : mat2d ) : void {
    mat2d . multiply ( this . matrix , mat , this . matrix ) ;
}
public translate ( x : number = 0 , y : number = 0 ) : void {
    let mat : mat2d = mat2d . makeTranslation ( x , y ) ;
    // 看到 translate、rotate 和 scale 都会调用 multMatrix 方法
    this . multMatrix ( mat ) ;
}
public rotate ( angle : number = 0 ,isRadian : boolean = true): void {
    if ( isRadian === false ) {
        angle = Math2D . toRadian ( angle ) ;
    }
    let mat : mat2d = mat2d . makeRotation ( angle ) ;
    this . multMatrix ( mat ) ;
}
// 从两个向量构建旋转矩阵
public rotateFrom ( v1 : vec2 , v2 : vec2 , norm : boolean = false ) : void {
    let mat : mat2d = mat2d . makeRotationFromVectors ( v1 , v2 , norm ) ;
    this . multMatrix ( mat ) ;
```

```
    }
    public scale ( x : number = 1.0 , y : number = 1.0 ) : void {
        let mat : mat2d = mat2d . makeScale ( x , y ) ;
        this . multMatrix ( mat ) ;
    }
    public invert ( ) : mat2d {
        let ret : mat2d = new mat2d ( ) ;
        if ( mat2d . invert ( this . matrix , ret ) === false ) {
            alert ( " 堆栈顶部矩阵为奇异矩阵,无法求逆 " ) ;
            throw new Error ( " 堆栈顶部矩阵为奇异矩阵,无法求逆 " ) ;
        }
        return ret ;
    }
}
```

可以看到，上面实现的矩阵堆栈和第 4 章中的 RenderStateStack 具有类似功能，进一步可以推导出 Canvas2D 中的 CanvasRenderingContext2D 类也持有一个渲染状态堆栈，以及一个矩阵变换堆栈。在下一节将用矩阵堆栈来替换 CanvasRenderingContext2D 类中的矩阵堆栈来实现坦克 Demo 的绘制。

7.1.12　在坦克 Demo 中应用矩阵堆栈

在本节中，在 TankWithMatrix 类中实现一个新的绘制方法 drawWithMatrixStack，该方法使用矩阵堆栈进行坐标系变换，然后将变换后的坐标系传回 CanvasRenderingContext2D 上下文渲染对象，用于绘制正确的图形。

首先，在 TankWithMatrix 类中增加一个矩阵堆栈的成员变量。具体代码如下：

```
public matStatck : MatrixStack = new MatrixStack ( ) ;
```

其次，需要在 TestApplication 类中创建一个设置 CanvasRenderingContext2D 上下文渲染对象的栈顶矩阵的方法。来看一下下面的代码：

```
public setTransform ( mat : mat2d ) : void {
    if ( this . context2D === null ) {
        return ;
    }
    this . context2D . setTransform (
        mat . values [ 0 ] ,
        mat . values [ 1 ] ,
        mat . values [ 2 ] ,
        mat . values [ 3 ] ,
        mat . values [ 4 ] ,
        mat . values [ 5 ]
    ) ;
}
```

上述方法与前面实现的 transform 方法类似，它们之间的区别在于：
- setTransform 方法的作用和 MatrixStack 的 loadMatrix 一致，是将参数矩阵中各个元素的值直接复制到 CanvasRenderingContext2D 上下文渲染对象所持有的矩阵堆栈的栈顶矩阵中。
- transform 方法的作用和 MatrixStack 的 multMatrix 一致，是将当前栈顶矩阵乘以参数矩阵，因此会累积上一次的变换。

有了上面的成员变量和辅助方法后，就可以实现 drawWithMatrixStack 方法。具体代码如下：

```
public drawWithMatrixStack ( app : TestApplication ) : void {
    if ( app . context2D === null ) {
        return ;
    }

    // 整个坦克绘制 tank
    app . context2D . save ( ) ;
    this . matStack . pushMatrix ( ) ;
        // 整个坦克移动和旋转，注意局部变换的经典结合顺序（ trs : translate -> rotate -> scale ）
        /*
        app . context2D . translate ( this . pos . x , this . pos . y ) ;
        app . transform ( this . tankRotation ) ;
        app . context2D . scale ( this . scale . x , this . scale . y ) ;
        */
        this . matStack . translate ( this . pos . x , this . pos . y ) ;
        this . matStack . multMatrix ( this . tankRotation ) ;
        this . matStack . scale ( this . scale . x , this . scale . y ) ;
        app . setTransform ( this . matStack . matrix ) ;

        // 绘制坦克的底盘（矩形）
        app . context2D . save ( ) ;
        //this . matStack . pushMatrix ( ) ;       // 实际可以取消，内部并没有使用变换操作
            app . context2D . fillStyle = 'grey' ;
            app . context2D . beginPath ( ) ;
            if ( this . initYAxis ) {
                // 交换 width 和 height，这样就不需要修改 TestApplication 中的示例代码了
                app . context2D . rect ( - this . height * 0.5 , - this . width * 0.5 , this . height , this . width ) ;
            } else {
                app . context2D . rect ( - this . width * 0.5 , - this . height * 0.5 , this . width , this . height ) ;
            }
            app . context2D . fill ( ) ;
        // this . matStack . popMatrix ( ) ;       // 实际可以取消，内部并没有使用变换操作
        app . context2D . restore ( ) ;

        // 绘制炮塔 turret
        app . context2D . save ( ) ;
        this . matStack . pushMatrix ( ) ;
            // app . context2D . rotate ( this . turretRotation ) ;
            this . matStack . rotate ( this . turretRotation , true ) ;
            app . setTransform ( this . matStack . matrix ) ;
```

```
    // 椭圆炮塔 ellipse 方法
    app . context2D . fillStyle = 'red' ;
    app . context2D . beginPath ( ) ;
    if ( this . initYAxis ) {
        // 当朝着 y 轴正方向时,椭圆的 radiuX < radiuY
        app . context2D . ellipse(0 , 0 ,10 ,15 , 0 , 0 ,Math.PI * 2);
    } else {
        // 当朝着 x 轴正方向时,椭圆的 radiuX > radiuY
        app . context2D . ellipse(0 , 0 ,15 ,10 , 0 , 0 ,Math.PI * 2);
    }
    app . context2D . fill ( ) ;

    // 炮管 gun barrel (炮管)
    app . context2D . strokeStyle = 'blue' ;
    app . context2D . lineWidth = 5 ; // 炮管需要粗一点,因此 5 个单位
    app . context2D . lineCap = 'round' ; // 使用 round 方式
    app . context2D . beginPath ( ) ;
    app . context2D . moveTo ( 0 , 0 ) ;
    if ( this . initYAxis ) {
        // 当朝着 y 轴正方向时,炮管是沿着 y 轴正方向绘制的
        app . context2D . lineTo ( 0 , this . gunLength ) ;
    } else {
        // 当朝着 x 轴正方向时,炮管是沿着 x 轴正方向绘制的
        app . context2D . lineTo ( this . gunLength , 0 ) ;
    }
    app . context2D . stroke ( ) ;

    // 炮口,先将局部坐标系从当前的方向,向 x 正方向平移炮管长度个单位,此时
    局部坐标系在炮管最右侧
    if ( this . initYAxis ) {
        // 当朝着 y 轴正方向时,炮口是沿着 y 轴正方向绘制的
        // app . context2D . translate ( 0 , this . gunLength ) ;
        // app . context2D . translate ( 0 , this . gunMuzzleRadius ) ;
        this . matStack . translate ( 0 , this . gunLength ) ;
        this . matStack . translate ( 0 , this . gunMuzzleRadius ) ;
        app . setTransform ( this . matStack . matrix ) ;
    } else {
        // 当朝着 x 轴正方向时,炮口是沿着 x 轴正方向绘制的
        // app . context2D . translate ( this . gunLength , 0 ) ;
        // app . context2D . translate ( this . gunMuzzleRadius , 0 ) ;
        this . matStack . translate ( this . gunLength , 0 ) ;
        this . matStack . translate ( this . gunMuzzleRadius , 0 ) ;
        app . setTransform ( this . matStack . matrix ) ;
    }
    // 调用自己实现的 fillCircle 方法,内部使用 Canvas2D arc 绘制圆弧方法
    app . fillCircle ( 0 , 0 , 5 , 'black' ) ;
this . matStack . popMatrix ( ) ;
app . context2D . restore ( ) ;

// 绘制一个圆球,标记坦克正方向,一旦炮管旋转后,可以知道正前方在哪里
app . context2D . save ( ) ;
this . matStack . pushMatrix ( ) ;
    if ( this . initYAxis ) {
```

```
            // 当朝着 y 轴正方向时，标记坦克前方的圆球是沿着 y 轴正方向绘制的
            // app . context2D . translate ( 0 , this . height * 0.5 ) ;
            this . matStack . translate ( 0 , this . height * 0.5 ) ;
            app . setTransform ( this . matStack . matrix ) ;
        } else {
            // 当朝着 x 轴正方向时，标记坦克前方的圆球是沿着 x 轴正方向绘制的
            // app . context2D . translate ( this . width * 0.5 , 0 ) ;
            this . matStack . translate ( this . width * 0.5 , 0 ) ;
            app . setTransform ( this . matStack . matrix ) ;
        }
        app . fillCircle ( 0 , 0 , 10 , 'green' ) ;
    this . matStack . popMatrix ( ) ;
    app . context2D . restore ( ) ;

    // 坐标系是跟随整个坦克的
    if ( this . showCoord ) {
        app . context2D . save ( ) ;
        this . matStack . pushMatrix ( ) ;
            app . context2D . lineWidth = 1 ;
            app . context2D . lineCap = '' ;
            app . strokeCoord ( 0 , 0 , this . width * 1.2 , this . height
                * 1.2 ) ;
        this . matStack . popMatrix ( ) ;
        app . context2D . restore ( ) ;
    }
    this . matStack . popMatrix ( ) ;
    app . context2D . restore ( ) ;

    if ( this . showLine === false ) {
        return ;
    }

    app . context2D . save ( ) ;
    this . matStack . pushMatrix ( ) ;
        app . strokeLine ( this . pos . x , this . pos . y , app . canvas .
            width * 0.5 , app . canvas . height * 0.5 ) ;
        app . strokeLine ( this . pos . x , this . pos . y , this . target .
            x , this . target . y ) ;
    this . matStack . popMatrix ( ) ;
    app . context2D . restore ( ) ;
}
```

通过上述代码会发现：
- 凡是调用 context2D 对象的 save 和 restore 方法的地方，都增加 MatrixStack 类对应的 pushMatrix 和 popMatrix 方法。
- 凡是调用 context2D 对象的 translate、scale 和 rotate 方法的地方，都替换为 MatrixStack 类相对应的方法。
- 凡是调用 context2D 对象的 stroke 或 fill 方法前，都要先调用 setTransform 方法，将矩阵堆栈栈顶的变换后的矩阵回传给 context2D 对象。

当在 render 方法中调用坦克的 drawWithMatrix 方法后运行 Demo，会发现程序正确地运行。通过这个例子，可以了解 Canvas2D 中是如何管理坐标系的层次变换的，值得花点

时间研究该 Demo，做到知其然，知其所以然。

7.1.13　图 7.1 旋转矩阵推导图示绘制源码

下面看一下推导旋转矩阵时几何图像的生成代码。

```
// 绘制旋转矩阵推导图
let x : number = 300 ;
let y : number = 200 ;
let ptX : number = 100 ;
let ptY : number = 30 ;
let rot : number = 30 ;
if ( app . context2D !== null ) {
    // 绘制没有旋转前的 xy 坐标系
    app . context2D . save ( ) ;
        app . context2D . translate ( x , y ) ;
        app . fillText ('原点 o' , 0 , 0 , 'black' , 'center' , 'bottom' ,
'15px sans-serif') ;
        app . context2D . rotate ( 0 ) ;
        app . drawVec ( 200 , 10 , "" , "x 轴" , 1 , false , false , false ) ;
        app . context2D . rotate ( Math2D . toRadian ( 90 ) ) ;
        app . drawVec ( 200 , 10 , "" , "y 轴" , 1 , false , false , false ) ;
    app . context2D . restore ( ) ;

    // 绘制变换了 30 度后的 x'y' 坐标系
    app . context2D . save ( ) ;
        app . context2D . translate ( x , y ) ;
        app . context2D . rotate ( Math2D . toRadian ( rot ) ) ;
        app . context2D . strokeStyle = 'blue' ;

        // 绘制在 x'y' 坐标系中的某个点
        app . fillCircle ( ptX , ptY , 5 ) ;
        app . fillText ('点 p' , ptX + 5 , ptY , 'black' , 'left' , 'middle' ,
'15px sans-serif' ) ;

        // 绘制某个点的投影
        app . context2D . save ( ) ;
            app . context2D . setLineDash ( [ 2 , 2 ] ) ;
            // 绘制该点在 x' 轴上的投影
            app . strokeLine ( ptX , 0 , ptX , ptY ) ;
            app . fillText ( "x'" , ptX , 0 , 'blue' , 'center' , 'bottom' ,
'20px sans-serif') ;
            // 绘制该点在 y' 轴上的投影
            app . strokeLine ( 0 , ptY , ptX , ptY ) ;
            app . fillText ( "y'" , 0 , ptY , 'blue' , 'right' , 'middle' ,
'20px sans-serif' ) ;
        app . context2D . restore ( ) ;

        // 绘制 x'y' 坐标系
```

```
            app . drawVec ( 200 , 10 , "" , "x'轴" , 1 , false , false , false ) ;
            app . context2D . rotate ( Math2D . toRadian ( 90 ) ) ;
            app . drawVec ( 200 , 10 , "" , "y'轴" , 1 , false , false , false ) ;
        app . context2D . restore ( ) ;

        // 现在要绘制在 x'y'中定义的某个点变换到 xy 坐标系后，分别在 x 轴和 y 轴上的投影点
        let rotation : mat2d = mat2d . makeRotation ( Math2D . toRadian ( rot ) ) ;
        let pt : vec2 = Math2D . transform ( rotation , vec2 . create ( ptX ,
        ptY ) ) ;

        app . context2D . save ( ) ;
            app . context2D . translate ( x , y ) ;
            app . context2D . setLineDash ( [ 2 , 2 ] ) ;
            // 绘制该点在 x 轴上的投影，以及文字标记
            app . strokeLine ( pt . x , 0 , pt . x , pt . y ) ;
            app . fillText ( "x" , pt . x , 0 , 'black' , 'center' , 'bottom' ,
            '20px sans-serif' ) ;
            // 绘制该点在 y 轴上的投影，以及文字标记
            app . strokeLine ( 0 , pt . y , pt . x , pt . y ) ;
            app . fillText ( "y" , 0 , pt . y , 'black' , 'right' , 'middle' ,
            '20px sans-serif' ) ;
            // 不使用虚线绘制，将其设定为 [ ]
            app . context2D . setLineDash ( [ ] ) ;
            // 绘制斜边 r
            app . drawVecFromLine ( vec2 . create ( 0 , 0 ) , pt , 10 , "" , "" ,
            1 , false , true , false , true ) ;
        app . context2D . restore ( ) ;

        // 绘制角度 a
        app . context2D . save ( ) ;
            app . context2D . translate ( x , y ) ;
            app . context2D . rotate ( Math2D . toRadian ( rot * 0.5 ) ) ;
            app . context2D . translate ( 30 , 0 ) ;
            app . fillText ( 'a' , 0 , 0 , 'black' , 'center' , 'middle' , '15px
            sans-serif' ) ;
        app . context2D . restore ( ) ;

        // 绘制角度 b
        app . context2D . save ( ) ;
            let b : number = Math . atan2 ( pt . y , pt . x ) ;
                                                    // 点 p 相对 x 轴的夹角
            b = Math2D . toRadian ( rot ) + b ;
            app . context2D . translate ( x , y ) ;
            app . context2D . rotate ( b * 0.5 ) ;
            app . context2D . translate ( 40 , 0 ) ;
            app . fillText ( " b " , 0 , 0 , 'black' , 'center' , 'middle' , '15px
            sans-serif' ) ;
        app . context2D . restore ( ) ;
    }
```

7.2 贝塞尔曲线

在本节中将要关注贝塞尔曲线的相关知识，本节内容的要点如下：

- 通过一个 Demo 演示 Canvas2D 中的 quadraticCurveTo（二次贝塞曲线）和 bezierCurveTo（三次贝塞尔曲线）方法的用法。
- 通过伯恩斯坦多项式（Bernstein Polynomia）推导出二阶和三阶贝塞尔多项式。
- 使用自己实现的二次及三次贝塞尔曲线绘制方法。
- 将贝塞尔曲线生成的过程抽象成枚举器。

接下来开始贝塞尔曲线之旅吧！

7.2.1 Demo 效果

先来看一下贝赛尔曲线 Demo 的效果，具体如图 7.3 所示。

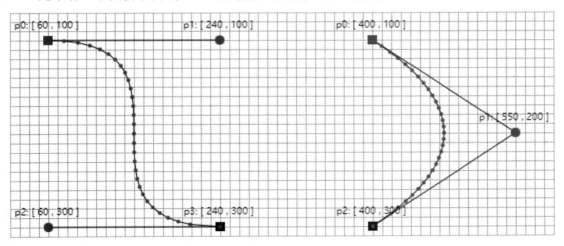

图 7.3 贝塞尔曲线初始化效果图

- 左侧显示的是三次贝塞尔曲线，右侧显示的是二次贝塞尔曲线。
- 二次贝塞尔曲线由三个点组成，其 *P*0 和 *P*2 为锚点（正方形表示），控制曲线两个端点的位置，而 *P*1 为控制点（圆形表示），用于调整曲线的曲率。
- 三次贝塞尔曲线由四个点组成，其 *P*0 和 *P*3 为锚点（正方形表示），控制曲线两个端点的位置，而 *P*1 和 *P*2 为控制点（圆形表示），用于调整曲线的曲率。
- 当使用自己实现的二次或三次贝塞尔曲线绘制方法，则会显示各个插值点（曲线上的小圆点表示）。

- 如图 7.4 所示，可以用鼠标选中锚点或控制点并拖动，可以调整曲线的位置或曲率，并且会更新各个点的坐标信息。

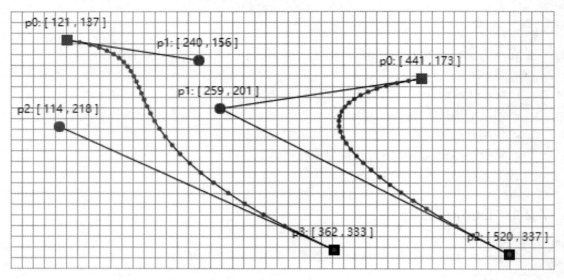

图 7.4　调整曲线端点位置和曲率

7.2.2　使用 Canvas2D 内置曲线绘制方法

首先来看一下如何使用 CanvasRenderingContext2D 上下文渲染对象内置的 quadraticCurveTo 和 bezierCurveTo 这两个方法来实现贝塞尔曲线的绘制。

来看一下这两个方法的签名。具体代码如下：

```
// 绘制二次贝塞尔曲线的方法
public quadraticCurveTo ( cpx : number , cpy : number , x : number , y : number ) : void ;
// 绘制三次贝塞尔曲线的方法
public bezierCurveTo ( cp1x : number , cp1y : number , cp2x : number , cp2y : number , x : number , y : number) :
```

参考图 7.3，来看一下上述两个方法的用法：

- 对于二次贝塞尔曲线 quadraticCurveTo 方法来说，其参数(cpx , cpy)对应的是 **P**1 点，(*x* , *y*)点对应的是 **P**2 点（图 7.3 右侧图像）。
- 对于三次贝塞尔曲线 bezierCurveTo 方法来说，其参数(cp1x , cp1y)对应的是 **P**1 点，(cpx2 , cpy2)对应的是 **P**2 点，而(*x* , *y*)对应的是 **P**3 点（图 7.2 左侧图像）。
- 对于 **P**0 点，可以使用 public moveTo (x : number , y : number) 方法来设定。

有了上述两个方法，那么来看一下如何绘制静态的二次和三次贝塞尔曲线。先来看一下二次贝塞尔曲线 QuadraticBezierCurve 类的相关绘制代码，具体如下：

```
class QuadraticBezierCurve {
    // 受保护的成员变量，能够被继承的子类访问
    protected _startAnchorPoint : vec2 ;         // 起点，相当于 p0 点
    protected _endAnchorPoint : vec2 ;           // 终点，相当于 p2 点
    protected _controlPoint0 : vec2 ;            // 控制点，相当于 p1 点

    protected _drawLine : boolean ;
                        // 是否要绘制连线，方块表示的锚点和原点表示的控制点
    protected _lineColor : string ;              // 绘制线段的颜色
    protected _lineWidth : number ;              // 绘制的线宽
    protected _radiusOrLen : number ; // 方块表示的锚点和原点表示的控制点的大小

    public constructor ( start : vec2 , control : vec2 , end : vec2 ) {
        // 初始化控制点
        this . _startAnchorPoint = start ;
        this . _endAnchorPoint = end ;
        this . _controlPoint0 = control ;
        // 初始化渲染属性
        this . _drawLine = true ;
        this . _lineColor = 'black';
        this . _lineWidth = 1 ;
        this . _radiusOrLen = 5 ;
    }

    public draw ( app : TestApplication ) {
        if ( app . context2D !== null ) {
            app . context2D . save ( ) ;
                // 设置线段的渲染属性
                app . context2D . lineWidth = this . _lineWidth ;
                app . context2D . strokeStyle = this . _lineColor ;

                // 二次贝塞尔曲线绘制的代码
                app . context2D . beginPath ( ) ;
                app . context2D . moveTo ( this . _points [ 0 ] . x , this .
                _points [ 0 ] . y ) ;
                app . context2D . quadraticCurveTo ( this . _controlPoint0 .
                x , this . _controlPoint0 . y , this . _endAnchorPoint . x ,
                this . _endAnchorPoint . y ) ;
                app . context2D . stroke ( ) ;

                // 绘制辅助的信息
                if ( this . _drawLine ) {
                    // 绘制起点 p0 到控制点 p1 的连线
                    app . strokeLine ( this . _startAnchorPoint . x , this .
                    _startAnchorPoint . y , this . _controlPoint0 . x , this .
                    _controlPoint0 . y ) ;
                    // 绘制终点 p2 到控制点 p1 的连线
                    app . strokeLine ( this . _endAnchorPoint . x , this .
                    _endAnchorPoint . y , this . _controlPoint0 . x , this .
                    _controlPoint0 . y ) ;

                    // 绘制绿色的正方形表示起点 p0
```

```
            app . fillRectWithTitle ( this . _startAnchorPoint . x -
            ( this . _radiusOrLen + 5 ) * 0.5 ,
                              this . _startAnchorPoint . y -
                              (this . _radiusOrLen + 5 ) * 0.5 ,
                              this . _radiusOrLen + 5 , this .
                              _radiusOrLen + 5 ,
                              undefined , undefined , 'green' ,
                              false ) ;
            // 绘制蓝色的正方形表示终点 p2
            app . fillRectWithTitle ( this . _endAnchorPoint . x -
            ( this . _radiusOrLen + 5 ) * 0.5 ,
                              this . _endAnchorPoint . y - ( this .
                              _radiusOrLen + 5 ) * 0.5 ,
                              this . _radiusOrLen + 5 , this .
                              _radiusOrLen + 5 ,
                              undefined , undefined , 'blue' ,
                              false ) ;
            // 绘制红色的原点表示控制点 p1
            app . fillCircle ( this . _controlPoint0 . x , this .
            _controlPoint0 . y , this . _radiusOrLen ) ;
        }

        // 绘制三个点的坐标信息，显示出当前 p0、p1 和 p2 的坐标信息
        // 有 override vec2 的 toString 方法
        app . drawCoordInfo ( 'p0:' + this . _startAnchorPoint .
        toString ( ) , this . _startAnchorPoint . x , this .
        _startAnchorPoint . y - 10 ) ;
        app . drawCoordInfo ( 'p1:' + this . _controlPoint0 . toString
        ( ) , this . _controlPoint0 . x , this . _controlPoint0 . y
        - 10 ) ;
        app . drawCoordInfo ( 'p2:' + this . _endAnchorPoint . toString
        ( ) , this . _endAnchorPoint . x , this . _endAnchorPoint .
        y - 10 ) ;
        app . context2D . restore ( ) ;
        }
    }
}
```

接下来实现三次贝塞尔曲线的相关代码。由于三次贝塞尔曲线仅比二次贝塞尔曲线多了一个控制点参数，因此直接继承自上面实现的 QuadraticBezierCurve 类，具体如下：

```
class CubeBezierCurve extends QuadraticBezierCurve {
    protected _controlPoint1 : vec2 ;

    public constructor ( start : vec2 , control0 : vec2 , control1 : vec2 ,
    end : vec2 ) {
        super ( start , control0 , end ) ;
        this . _controlPoint1 = control1 ;
    }
}
```

需要覆写（override）QuadraticBezierCurve 的 draw 方法，大部分代码都相同，仅仅是将原来的 quadraticCurveTo 方法替换为 bezierCurveTo 方法，并且增加一些绘制控制线，以及控制点的代码（因为多了一个控制点）。

当调用上面的代码，会得到如图 7.5 所示的效果。

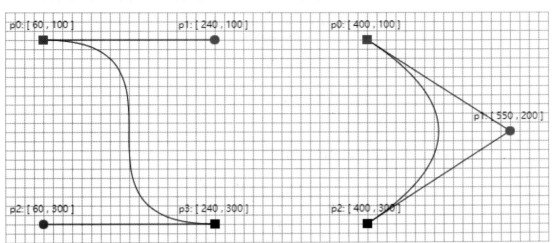

图 7.5　Canvas2D 内置贝塞尔绘制方法

图 7.5 和图 7.3 相比，缺少了贝塞尔曲线上的 30 个插值点（红色小圆点），以及无法和鼠标交互，实时调整曲线的位置和曲率。

贝塞尔曲线在游戏中的一个很重要的功能：沿着路径朝向正确地运动。

想象一下，给定一系列贝塞尔线段组成的封闭路径，让坦克沿着贝塞尔路径并且朝向正确地运行，也是一种很棒的体验。还有，例如在 3D 环境中，摄像机沿着预先设定的路径漫游。

上述这些需求的基础，就是要能够实时地计算出如图 7.3 所示的路径上的插值点（红色小圆点），这就需要知道贝塞尔多项式相关的知识，下一节使用伯恩斯坦多项式来推导二次和三次贝塞尔多项式。

7.2.3　伯恩斯坦多项式推导贝塞尔多项式

可以使用如下公式来定义 n 次（阶）贝塞尔曲线：

$$p'(t) = \sum_{i=0}^{n} P_i B_{i,n}(t) \quad \text{其中 } t \in [0,1]$$

在上述公式中，向量 P_i 表示第 i 个控制点，而 $B_{i,n}(t)$ 表示的是伯恩斯坦多项式（Bernstein Polynomial），其表达式如下：

$$B_{i,n}(t) = \frac{n!}{i!(n-i)!}(1-t)^{n-i} t^i$$

接下来使用上述公式来推导二次贝塞尔曲线的多项式，对于二次贝塞尔曲线多项式来说，$n = 2$，那么：

(1) 当 $i = 0$ 时，其伯恩斯坦多项式的值为

$$B_{0,2}(t) = \frac{2!}{0!(2-0)!}(1-t)^2 t^0$$

$$= \frac{2 \times 1}{1 \times 2 \times 1}(1-t)^2 \times 1$$

$$= \frac{2}{2}(1-t)^2$$

$$= (1-t)^2$$

(2) 当 $i = 1$ 时，其伯恩斯坦多项式的值为

$$B_{1,2}(t) = \frac{2!}{1!(2-1)!}(1-t)^1 t^1$$

$$= \frac{2 \times 1}{1 \times 1}(1-t)t$$

$$= 2(1-t)t$$

$$= 2t(1-t)$$

(3) 当 $i = 2$ 时，其伯恩斯坦多项式的值为

$$B_{2,2}(t) = \frac{2!}{2!(2-2)!}(1-t)^0 t^2$$

$$= \frac{2 \times 1}{2 \times 1} \times 1 \times t^2$$

$$= \frac{2}{2}(t)^2$$

$$= t^2$$

将上面得到的三个公式代入 $p'(t) = \sum_{i=0}^{n} P_i B_{i,n}(t)$ 公式中，得到：

$$p'(t) = (1-t)^2 p_0 + 2t(1-t)p_1 + t^2 p_2, \quad \text{其中} t \in [0,1]$$

上面的公式就是二次贝塞尔多项式的向量表示版本。

同样地，对于三次甚至更高阶次的贝塞尔曲线来说，都能使用 $p'(t) = \sum_{i=0}^{n} P_i B_{i,n}(t)$ 公式获得对应的多项式。现在直接给出三次（$n = 3$）贝塞尔多项式，具体公式如下：

$$p'(t) = (1-t)^3 p_0 + 3t(1-t)^2 p_1 + 3t^2(1-t)p_2 + t^3 p_3, \quad \text{其中} t \in [0,1]$$

既然有了二次和三次贝塞尔多项式，就在 Math2D 类中实现这些公式。具体代码如下：

```
// 二次贝塞尔曲线标量版
public static getQuadraticBezierPosition ( start : number , ctrl : number ,
end: number , t : number ) : number {
    if ( t < 0.0 || t > 1.0 ) {
        alert ( " t的取值范围必须为 [ 0 , 1 ] " ) ;
```

```
        throw new Error ( " t的取值范围必须为 [ 0 , 1 ] " ) ;
    }
    let t1 : number = 1.0 - t ;
    let t2 : number = t1 * t1 ;
    return t2 * start + 2.0 * t * t1 * ctrl + t * t * end ;
}

// 二次贝塞尔曲线向量版
public static getQuadraticBezierVector ( start : vec2 , ctrl : vec2 , end :
vec2 , t : number , result : vec2 | null = null ) : vec2 {
    if ( result === null ) result = vec2 . create ( ) ;
    result . x = Math2D . getQuadraticBezierPosition ( start . x , ctrl .
    x , end . x , t ) ;
    result . y = Math2D . getQuadraticBezierPosition ( start . y , ctrl .
    y , end . y , t ) ;
    return result ;
}

// 三次贝塞尔曲线标量版
public static getCubicBezierPosition ( start : number , ctrl0 : number ,
ctrl1 : number , end : number , t : number ) : number {
    if ( t < 0.0 || t > 1.0 ) {
        alert ( " t的取值范围必须为 [ 0 , 1 ] " ) ;
        throw new Error ( " t的取值范围必须为 [ 0 , 1 ] " ) ;
    }
    let t1 : number = ( 1.0 - t ) ;
    let t2 : number = t * t ;
    let t3 : number = t2 * t ;
    return ( t1 * t1 * t1 ) * start + 3 * t * ( t1 * t1 ) * ctrl0 + ( 3
    * t2 * t1 ) * ctrl1 + t3 * end ;
}

// 三次贝塞尔曲线向量版
public static getCubicBezierVector ( start : vec2 , ctrl0 : vec2 , ctrl1 :
vec2 , end : vec2 , t : number , result : vec2 | null = null ) : vec2 {
    if ( result === null ) result = vec2 . create ( ) ;
    result . x = Math2D . getCubicBezierPosition ( start . x , ctrl0 . x ,
    ctrl1 . x , end . x , t ) ;
    result . y = Math2D . getCubicBezierPosition ( start . y , ctrl0 . y ,
    ctrl1 . y , end . y , t ) ;
    return result ;
}
```

7.2.4 贝塞尔曲线自绘版

有了上一节推导的公式及实现的代码，就能够在 QuadraticBezierCurve 及其子类 Cubic BezierCurve 中实现自己的贝塞尔曲线绘制版本。下面先来看一下 QuadraticBezierCurve 需要增加的内容。具体代码如下：

```
// 增加如下成员变量，用来插值点相关操作
protected _drawSteps : number = 30 ;    //需要多个插值点，默认为30个
```

```typescript
protected _points !: Array < vec2 > ;    //计算出来的插值点存储在_points数组中
protected _showCurvePt : boolean = true ;  // 是否显示所有插值点，默认显示
protected _dirty : boolean = true ;    //标记变量，用来指明是否要重新计算所有插值点

// 然后实现一个虚拟方法getPosition，该方法返回 t = [ 0 , 1 ]之间的位置向量
// 继承的子类需要覆写（override）该方法
protected getPosition ( t : number ) : vec2 {
    if ( t < 0 || t > 1.0 ) {
        throw new Error ( " t的取值范围必须是[ 0 , 1 ]之间 " ) ;
    }
    // 调用推导出来的二次贝塞尔多项式的向量版本
    // 子类（CubicBezierCurve）需要override本方法，调用对应的三次贝塞尔多项式版本
    return Math2D . getQuadraticBezierVector ( this . _startAnchorPoint ,
    this . _controlPoint0 , this . _endAnchorPoint , t ) ;
}

// 私有方法
private _calcDrawPoints ( ) : void {
    if ( this . _dirty )                    //如果_dirty为true，才重新计算所有插值点
    {
        this . _points = [ ] ;              // 清空插值点数组
        this . _points . push (this . _startAnchorPoint ) ; // 第一个是起点
        let s : number = 1.0 / ( this . _drawSteps ) ;
        // 计算除第一个和最后一个锚点之外的所有插值点，将其存储到数组中去
        for ( let i = 1 ; i < this . _drawSteps - 1 ; i ++ ) {
            // 调用虚方法getPosition
            let pt : vec2 = this . getPosition ( s * i ) ;
            // 将计算出来的插值点放入数组中去
            this . _points . push ( pt ) ;
        }
        this . _points . push ( this . _endAnchorPoint ) ; // 最后一个是终点
        // 将_dirty标记设置为false
        this . _dirty = false ;
    }
}

// 让动画回调中一直调用update函数
// 而update函数内部调用_calcDrawPoints方法，该方法会根据_dirty标记决定是否重新
// 计算插值点
public update ( intervalSec : number ) : void {
    this . _calcDrawPoints ( ) ;
}

// draw方法
// 虚拟方法，子类需要override
// useMyCurveDrawFunc为true，使用自己实现的绘制贝塞尔方法，否则用Canvas2D内置方法
public draw (app : TestApplication ,useMyCurveDrawFunc : boolean = true){
    // ... 代码同7.4.2节所示
    if ( useMyCurveDrawFunc === false ) {
        app . context2D . quadraticCurveTo ( this . _controlPoint0 . x , this .
        _controlPoint0 . y , this . _endAnchorPoint . x , this .
```

```
        _endAnchorPoint . y ) ;
    } else {
        // 只需要将计算出来的插值点使用 lineTo 方法连接起来就能绘制出自己的曲线
        // 所以曲线的光滑度取决于 drawSteps 的数量，数量越多，越光滑
        for ( let i = 1 ; i < this . _points . length ; i ++ ) {
            app . context2D . lineTo ( this . _points [ i ] . x , this . _points
                [ i ] . y ) ;
        }
    }
    增加显示已经计算出来的所有插值点
    if ( this . _showCurvePt ) {
        for ( let i = 0 ; i < this . _points . length ; i ++ ) {
            app . fillCircle ( this . _points [ i ] . x , this . _points [ i ] .
                y , 2 ) ;
        }
    }
    // ... 下面的代码同 7.4.2 节所示
}
```

对于继承自 QuadraticBezierCurve 的子类 CubicBezierCurve 来说，只需要覆写（override）getPosition 方法和 draw 方法就可以了。当运行上述代码，就会得到如图 7.13 所示的效果。

7.2.5 鼠标碰撞检测和交互功能

接下来完成贝塞尔曲线 Demo 的碰撞检测和交互功能。参考图 7.2，如果要用鼠标拖动控制点，必须要先确定当前鼠标点是否在圆点或正方向内部。在第 6 章中已经实现了 Math2D 类的 isPointInRect 和 isPointInCirce 方法，继续定义一个枚举结构，用来标记碰撞检测的结果。具体代码如下：

```
enum ECurveHitType {
    NONE ,                               // 没有选中
    START_POINT ,                        // 选中起点
    END_POINT ,                          // 选中终点
    CONTROL_POINT0 ,                     // 选中控制点 0
    CONTROL_POINT1                       // 选中控制点 1，针对三次贝塞尔曲线而言
}

// 在 QuadraticBezierCurve 类中定义如下成员变量：
protected _hitType : ECurveHitType = ECurveHitType . NONE ;
                                     // 初始化时没选中
```

有了上面碰撞检测的方法和枚举后，就可以进入鼠标交互的业务逻辑。具体代码如下：

```
// 虚拟方法，子类（CubicBezierCurve 需要覆写以增加控制点 1 是否选中的情况）
protected hitTest ( pt : vec2 ) : ECurveHitType {
    if ( Math2D . isPointInCircle ( pt , this . _controlPoint0 , this .
        _radiusOrLen ) ) {
        return ECurveHitType . CONTROL_POINT0 ;     // 选中控制点 0
    } else if ( Math2D . isPointInRect (
```

```typescript
                            pt.x, pt.y,
                            this._startAnchorPoint.x - (this._radiusOrLen + 5) * 0.5,
                            this._startAnchorPoint.y - (this._radiusOrLen + 5) * 0.5,
                            this._radiusOrLen + 5, this._radiusOrLen + 5)) {
            return ECurveHitType.START_POINT;           // 选中起点
        } else if (Math2D.isPointInRect(
                            pt.x, pt.y,
                            this._endAnchorPoint.x - (this._radiusOrLen + 5) * 0.5,
                            this._endAnchorPoint.y - (this._radiusOrLen + 5) * 0.5,
                            this._radiusOrLen + 5, this._radiusOrLen + 5)) {
            return ECurveHitType.END_POINT;             // 选中终点
        } else {
            return ECurveHitType.NONE;                  // 什么都没选中
        }
    }

    // 鼠标事件处理
    public onMouseDown(evt: CanvasMouseEvent): void {
        // 每次 mouseDown 时调用 hitTest 进行碰撞检测测试
        // 将检测到的结果记录下来
        this._hitType = this.hitTest(evt.canvasPosition);
    }

    public onMouseUp(evt: CanvasMouseEvent): void {
        // 每次 mouseUp 时，将_hitType 清空为 NONE
        this._hitType = ECurveHitType.NONE;
    }

    // 虚拟方法,子类 CubicBezierCurve 需要 override 这个事件检测函数,需要增加 CONTROL_POINT1 的处理代码
    public onMouseMove(evt: CanvasMouseEvent): void {
        // 如果有选中的控制点
        if (this._hitType !== ECurveHitType.NONE) {
            switch (this._hitType) {
                case ECurveHitType.CONTROL_POINT0:
                    // 更新控制点的位置
                    this._controlPoint0.x = evt.canvasPosition.x;
                    this._controlPoint0.y = evt.canvasPosition.y;
                    this._dirty = true;
                                    // 标记_dirty,需要在 update 时重新计算插值点
                    break;

                case ECurveHitType.START_POINT:
                    this._startAnchorPoint.x = evt.canvasPosition.x;
                    this._startAnchorPoint.y = evt.canvasPosition.y;
                    this._dirty = true;
                    break;

                case ECurveHitType.END_POINT:
```

```
                    this . _endAnchorPoint . x = evt . canvasPosition . x ;
                    this . _endAnchorPoint . y = evt . canvasPosition . y ;
                    this . _dirty = true ;
                break ;
            }
        }
    }
```

类似地，在 CubicBezierCurve 中覆写（override）hitTest，以及 onMouseMove 方法后，就能以交互的方式来运行应用程序。

7.2.6 实现贝塞尔曲线枚举器

可以使用 IEnumerator＜T＞泛型接口来抽象贝塞尔曲线的生成过程，为了更加精确地控制生成过程中的迭代次数，扩展 IEnumerator＜T＞接口为 IBezierEnumerator 接口，并且实现该接口。具体代码如下：

```
export interface IBezierEnumerator extends IEnumerator < vec2 > {
    steps : number ;
}

// BezierEnumerator 类根据构造函数是否有control2，来判断当前是二次还是三次贝塞尔曲线
export class BezierEnumerator implements IBezierEnumerator {
    private _steps : number ;
    private _i : number ; // 1.0 /(this . _steps),表示每次t的增量在[0 , 1]之间
    private _startAnchorPoint : vec2 ;
    private _endAnchorPoint : vec2 ;
    private _controlPoint0 : vec2 ;
    private _controlPoint1 : vec2 | null ;
                        //如果_controlPoint1 不为 null，则说明是三次贝塞尔曲线
    private _currentIdx : number ;              // 用来标明当前迭代到哪一步

    public constructor ( start : vec2 , end : vec2 , control0 : vec2 , control1 :
    vec2 | null = null , steps : number = 30 ) {
        this . _startAnchorPoint = start ;
        this . _endAnchorPoint = end ;
        this . _controlPoint0 = control0 ;
        if ( control1 !== null ) {
            this . _controlPoint1 = control1 ;
        } else {
            this . _controlPoint1 = null ;
        }
        this . _steps = steps ;
        this . _i = 1.0 / ( this . _steps ) ;
        this . _currentIdx = -1 ;
    }

    public reset ( ) : void {
        this . _currentIdx = - 1 ;
    }
```

```typescript
public get current ( ) : vec2 {
    // 通过 this . _currentIdx * this . _i 计算出当前的 t 的数值
    if ( this . _controlPoint1 !== null ) {        //调用三次贝塞尔求值函数
        return Math2D . getCubicBezierVector ( this . _startAnchorPoint ,
            this . _controlPoint0 , this . _controlPoint1 , this .
            _endAnchorPoint , this . _currentIdx * this . _i ) ;
    } else {  // 调用二次贝塞尔求值函数
        return Math2D . getQuadraticBezierVector ( this . _startAnchor
            Point , this . _controlPoint0 , this . _endAnchorPoint , this .
            _currentIdx * this . _i ) ;
    }
}

public moveNext ( ) : boolean {
    this._currentIdx ++ ;
    return this . _currentIdx < this . _steps ;
}

public get steps ( ) : number {
    this . _i = 1.0 / ( this . _steps ) ;
    return this . _steps ;
}

// 当每次设置 steps 后，都需要重新计算所有的插值点
public set steps ( steps : number ) {
    this . _steps = steps ;
    this . reset ( ) ;
}
}

// 实现创建贝塞尔迭代器接口的工厂方法
// 由于将贝塞尔迭代器实现在 math2d.ts 文件中，因此将这些工厂方法实现在 Math2D 类中
public static createQuadraticBezierEnumerator ( start : vec2 , ctrl : vec2 ,
end : vec2  , steps : number = 30 ) : IBezierEnumerator {
    return new BezierEnumerator ( start , end , ctrl , null , steps ) ;
}

public static createCubicBezierEnumerator ( start : vec2 ,ctrl0 : vec2 ,
ctrl1 : vec2 , end : vec2  , steps : number = 30 ) : IBezierEnumerator {
    return new BezierEnumerator ( start , end , ctrl0 , ctrl1 , steps ) ;
}
```

使用 IBezierEnumerator 对象非常简单。接下来将上面的贝塞尔曲线 Demo 中计算插值点的_calcDrawPoint 方法用 IBezierEnumerator 对象来改写一下。具体代码如下：

```typescript
// this . _iter 指向 IBezierEnumerator 对象, 在类的初始化函数中使用 create 工厂
方法创建的
private _calcDrawPoints ( ) : void {
    if ( this . _dirty ) {
        this . _points = [ ] ;
        this . _iter . reset ( ) ;
        while ( this . _iter . moveNext ( ) ) {
            this . _points . push ( this . _iter . current ) ;
        }
```

```
        this . _dirty = false ;
    }
}
```

7.3 本章总结

  7.1 节主要讲解了矩阵的相关知识。首先介绍了仿射矩阵的乘法，通过仿射矩阵的乘法操作，将平移、缩放和旋转等应用都用矩阵乘法来统一操作。然后讲解了矩阵乘法的一些性质，例如矩阵乘法不符合交换律。特别要了解的是矩阵的结合律。通过结合律，可以将所有矩阵操作累积合成一个矩阵，该矩阵包含我们想要的所有变换操作。最后再进行向量与矩阵相乘，从而达到点的空间坐标系变换的效果。

  接下来介绍了单位矩阵，如果将矩阵设置为单位矩阵，就会把以前累积的矩阵效果清零，这是一个必要的操作。这里重点讲解了逆矩阵的相关知识，演示了平移、缩放矩阵，以及其逆矩阵的各种计算方式。由于旋转矩阵是矩阵变换中相对较难的一个操作，因此通过三角函数的相关知识推导出旋转矩阵，以及旋转矩阵的各种求逆方式和各自的优缺点。还推导出从两个方向向量构建旋转矩阵的算法，该算法最大的优势是不需要使用耗时的 atan2 三角函数就能直接表示朝向，为此修改了上一章的坦克 Demo。

  在第 8 章中，通过向量的有向缩放，消除了 sin 和 cos 三角函数。本章中，则使用从两个单位向量构建旋转矩阵的方式，消除 atan2 这个三角函数的使用。最后实现了 Canvas2D 中的矩阵堆栈，改写整个坦克 Demo，将原本由 Canvas2D 内置的矩阵堆栈操作替换为自己实现的矩阵堆栈，让程序正确无误地运行。

  7.2 节则关注二次和三次贝塞尔曲线的相关内容。首先通过一个 Demo 来演示 Canvas2D 中内置的二次和三次曲线的绘制方法。但是当我们想获得物体沿着路径运动的特效时，在这种情况下，需要完全地控制整个贝塞尔路径的插值过程，因此引入了伯恩斯坦多项式。通过伯恩斯坦多项式可以推导出二次和三次贝塞尔多项式，通过二次和三次贝塞尔多项式，就能完全掌控贝塞尔曲线。既然能够计算得到贝塞尔路径的所有插值点，那么只要将这些点连接起来，就能实现自己的二次和三次贝塞尔曲线的绘制方法，从而取代 Canvas2D 内置的曲线绘制方法，做到知其然，知其所以然。然后扩展了 Demo，让其支持使用鼠标的交互行为，从而可以实时地观察曲线的位置变化和曲率变化，以加深理解。最后会发现，贝塞尔插值点计算的过程就是迭代枚举的过程，因此特别适合迭代器（枚举器）设计模式，因此我们实现了贝塞尔枚举器。

  以上就是本章的所有内容。

第 4 篇
架构与实现篇

▶▶ 第 8 章　精灵系统

▶▶ 第 9 章　优美典雅的树结构

▶▶ 第 10 章　场景图系统

第 8 章 精灵系统

在本章中,要以面向接口的编程方式实现一个具备一些必要功能的、使用非场景图类型、支持精确点选、基于非立即渲染模式(保留模式),以及采取享元设计模式的精灵系统。

8.1 精灵系统的架构与接口

本节将对精灵系统的架构和接口进行详细介绍。

精灵系统的必要功能是指最小化的、可以使精灵系统运行起来的一些基础功能,例如精灵的更新、绘制、鼠标和键盘事件的分发与响应,以及为了支持鼠标事件而必须要实现的碰撞检测等相关功能。

2D 精灵系统一般可以由基于树结构的场景图(Scene Graph)和非场景图方式来实现。本章中使用非场景图的方式来实现精灵系统。在后面章节中,会再实现一个场景图模式的精灵系统,目的是为了让读者了解场景图的优点,以及面向接口编程的优势(本章中实现的 Demo 可以不做任何修改,就能运行在场景图模式中)。

精灵系统支持精确点选有两层含义。点选说明仅仅支持点与精灵的碰撞检测,本书中不涉及更加复杂的精灵之间的碰撞检测内容。精确点选是指当精灵的位置(平移)、朝向(旋转)或大小(缩放)发生改变后,仍旧能够通过例如鼠标指针单击来测试是否选中了该精灵,这部分内容涉及坐标系变换的相关知识点。

图形的渲染模式可以分为立即模式(Immediate Mode)和保留模式(Retained Mode)两种。像 Canvas2D、OpenGL 和 Direct3D 等这些底层图形 API 库都是立即模式的渲染库,它们维持一个基于当前渲染信息(渲染状态、变换矩阵、几何顶点或路径等)的状态机。每次渲染前都要设置各种渲染状态、变换矩阵,以及几何顶点或路径数据等。

渲染后,丢弃前面设置的数据,恢复到原始状态,周而复始地做着同样的工作。如果要对渲染显示后的某个图形进行操作,例如获取选中的某个图形的位置、朝向和颜色等相关信息,需要在外部设计一套机制,用来记录图形的相关渲染状态、变换信息,以及需要绘制的几何数据(保留模式)。为了满足上述的需求,特意设计了如下两个接口:

- 精灵(ISprite)接口,用来保存各种渲染状态(IRenderState 接口),以及矩阵变换信息(ITransformable 接口)。

- 形体（IShape）接口，用来抽象渲染几何顶点或路径数据，以及基于这些几何数据上的相关操作（IDrawable 接口和 IHittable 接口）。

享元设计模式（FlyWeight Design Pattern）最大的优点是能减少内存的使用。当数据量非常大时，享元设计模式的优势就会得到体现。享元模式的使用必须要在架构设计之初就要考虑进去，否则在后期重构时改动量将变得庞大。提供 ISprite 和 IShape 这两个接口，将渲染状态和变换信息放在 ISprite 接口中，而将渲染和碰撞检测使用的几何顶点或路径数据放在 IShape 接口中，就是为了支持享元模式。多个 ISprite 接口可以共享一个 IShape 接口，例如红色、旋转了 30°的矩形精灵，与黑色、旋转 10°、放大 2 倍、平移 20 个单位的矩形精灵共享一个矩形 Shape 对象，其关系如图 8.1 所示。

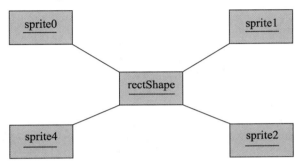

图 8.1　多个精灵共享一个形体

精灵系统的两个关键接口是 ISprite 和 IShape，但是还需要其他一些类、接口、枚举和事件等来协助完成整个精灵系统的更新、渲染、事件分发和响应流程等功能。

在此列出了整个精灵系统涉及的相关入口类及接口，具体的精灵系统静态类结构如图 8.2 所示。

接下来的内容就以图 8.2 为参考，按照自顶向下的顺序，来讲解各个类和接口之间的协作关系，然后再以自下而上的顺序来介绍实现过程。

注意：图 8.2 中的精灵系统 UML（Unified Modeling Language，统一建模语言）静态类结构图中，不同的箭头形状及实虚线有不同的含义。具体解释如下：
- 空心箭头+实线表示类或接口的泛化（Generalization，对应 TypeScript 中 extends 关键词）。
- 箭头+虚线表示类或接口的依赖关系（Dependency，例如 IDrawable 接口依赖 IRenderState 接口，则说明 IDrawable 接口中有个方法会使用 IRenderState 接口作为输入参数）。
- 箭头+实线表示单向直接关联关系（Directed Association，例如 ISprite 单向关联了 IShape 接口，则说明 ISprite 有个 IShape 类型的成员变量可以寻址到 IShape 对象，但是反过来 IShape 对象无法寻址到 ISprite 对象）。

- 实心菱形+实线表示组合关系（Composition，例如 IDispatcher 接口持有一个 ISpriteContainer 接口，并且 IDispatcher 接口掌控 ISpriteConainer 接口的生命周期，如果 IDispatcher 接口内存被释放，那么 ISpriteContainer 接口的内存也会被释放，ISpriteContainer 接口无法独立于 IDispatcher 接口单独生存）。
- 空心菱形+实线表示聚合关系（Aggregation，例如 ISpriteContainer 接口可以增、删 ISprite 接口，ISprite 接口可以独立于 ISpriteContainer 接口存在）。
- UML 中关于关联关系、聚合关系及组合关系的语义区别非常微妙，不同的架构师、开发者站在不同的角度，可能有不同的理解，因此属于比较抽象的范畴。

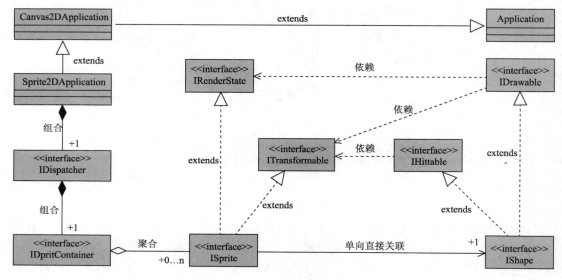

图 8.2　精灵系统静态类结构图

由于 UML 本身是一个庞大的知识体系结构，本书并不深入讲述 UML 相关内容，因此关于 UML 的更多内容，请各位读者自行查阅相关资料。

8.1.1　应用程序的入口与命令分发

根据图 8.2 所示，应用程序的入口类是 Applicaton 类及其子类。回顾第 3 章的内容，知道 Application 类的主要作用是规定整个应用程序的运行流程，包括如下几个关键功能：
- 动画循环的启动和关闭。
- 提供可以基于不同帧率进行回调的定时器。
- 需要子类实现的 update 抽象方法。
- 需要子类实现的 render 抽象方法。

- 需要子类实现的鼠标和键盘事件分发或响应抽象方法。

在 Application 基础上继承获得了 Canvas2DApplication 类，该类的作用仅仅是创建 CanvasRenderingContext2D 渲染上下文对象，并且将该对象以公开属性方式暴露出来，这样就能随时获得该上下文渲染对象进行绘图操作。

再回顾一下第 4～7 章的内容，都是从 Canvas2DApplication 类继承得到了 TestApplication 类，并且覆写了 update 虚方法进行状态更新操作，覆写 render 方法进行绘制和显示，覆写各种鼠标和键盘虚方法直接进行事件响应处理。由于只有一个物体（或精灵）的情况下，就不存在更新、渲染及事件分发的概念，直接处理单个精灵就可以解决问题。

但是从本章开始，要构造一个同时能操作多个精灵的通用系统。在这种情况下，需要有一个精灵容器，然后需要遍历精灵容器中的各个精灵，对精灵分别进行相关的操作。对于上述提到的这些操作，抽象成一个名为 IDispatcher 的接口，该接口的成员变量和成员方法如下：

```
// 进行事件、更新、绘制命令分发的接口
// 接收到分发命令的精灵都存储在 ISpriteContainer 容器中
export interface IDispatcher {
    // 只读的类型为 ISpriteContainer 的成员变量
    // 本接口中所有的dispatch开头的方法都是针对 ISpriteContainer 接口进行遍历操作
    readonly container : ISpriteContainer ;
    // 遍历 ISpriteContainer 容器，进行精灵的 update 分发
    dispatchUpdate ( msec : number , diffSec : number ) : void ;
    // 遍历 ISpriteContainer 容器，进行精灵的 render 分发
    dispatchDraw ( context : CanvasRenderingContext2D ) : void ;
    // 遍历 ISpriteContainer 容器，进行精灵的鼠标事件分发
    dispatchMouseEvent ( evt : CanvasMouseEvent ) : void ;
    // 遍历 ISpriteContainer 容器，进行精灵的键盘事件分发
    dispatchKeyEvent ( evt: CanvasKeyBoardEvent ) : void ;
}
```

有了 IDispatcher 接口后，就能实现一个名为 Sprite2DApplication 的类，该类继承自 Canvas2DApplication 类，并且引用一个 IDispatcher 接口。在 Sprite2DApplication 中覆写的虚方法中都调用 IDispatcher 接口中的 dispatch 相关方法。具体代码如下：

```
export class Sprite2DApplication extends Canvas2DApplication {
    // 声明一个受保护的类型为 IDispatcher 的成员变量
    // 下面所有的虚方法都委托调用 IDispatcher 相关的方法
    protected _dispatcher : IDispatcher ;

    // 一个方便的只读属性，返回 ISpriteContainer 容器接口
    // 可以通过该方法，和 ISprite 进行交互
    public get rootContainer ( ) : ISpriteContainer {
        return this . _dispatcher . container ;
    }

    public update ( msec : number , diff : number ): void {
        this . _dispatcher . dispatchUpdate ( msec , diff ) ;
    }
```

```typescript
public render ( ) : void {
    if ( this . context2D ) {
        // 每次都先将整个画布内容清空
        this . context2D . clearRect ( 0 , 0 , this . context2D . canvas . width , this . context2D . canvas . height ) ;
        this . _dispatcher . dispatchDraw ( this . context2D ) ;
    }
}

protected dispatchMouseDown ( evt : CanvasMouseEvent ) : void{
    // 调用基类的同名方法
    super . dispatchMouseDown ( evt ) ;
    // 事件分发
    this . _dispatcher . dispatchMouseEvent ( evt ) ;
}

protected dispatchMouseUp( evt : CanvasMouseEvent ) : void {
    super . dispatchMouseUp ( evt ) ;
    this . _dispatcher . dispatchMouseEvent ( evt ) ;
}

protected dispatchMouseMove ( evt : CanvasMouseEvent ) : void {
    super . dispatchMouseMove ( evt ) ;
    this . _dispatcher . dispatchMouseEvent ( evt ) ;
}

protected dispatchMouseDrag ( evt : CanvasMouseEvent ) : void {
    super . dispatchMouseDrag ( evt ) ;
    this . _dispatcher . dispatchMouseEvent ( evt ) ;
}

protected dispatchKeyDown ( evt : CanvasKeyBoardEvent ) : void {
    super . dispatchKeyDown ( evt ) ;
    this . _dispatcher . dispatchKeyEvent ( evt ) ;
}

protected dispatchKeyUp ( evt : CanvasKeyBoardEvent ) : void {
    super . dispatchKeyUp ( evt ) ;
    this . _dispatcher . dispatchKeyEvent ( evt ) ;
}

protected dispatchKeyPress ( evt : CanvasKeyBoardEvent ) : void {
    super . dispatchKeyPress ( evt ) ;
    this . _dispatcher . dispatchKeyEvent ( evt ) ;
}
}
```

接下来看一下与 IDispatcher 接口相关联的 ISpriteContainer 接口。IDispatcher 接口中所有的方法都是针对 ISpriteContainer 进行操作的，接收到分发命令的精灵都存储在 ISpriteContainer 容器中。下面看一下 ISpriteContainer 接口相关的成员属性和方法。具体代码如下：

```typescript
export interface ISpriteContainer {
    name : string ;                                    // 如有需要，提供一个容器名称
    // 添加一个精灵到容器中
    addSprite ( sprite : ISprite ) : ISpriteContainer ;
    // 从容器中删除一个精灵
    removeSprite ( sprite : ISprite ) : boolean ;
    // 清空整个容器
    removeAll ( includeThis : boolean ) : void ;
    // 根据精灵获取索引号，如果没找到精灵，返回-1
    getSpriteIndex ( sprite : ISprite ) : number ;
    // 根据索引号，从容器中获取精灵
    getSprite ( idx : number ) : ISprite ;
    // 获取容器中精灵的数量
    getSpriteCount ( ) : number ;
}
```

在 Sprite2DApplication 类中，可以通过使用公开的 rootContainer 只读属性寻址到 IDispatcher 接口相关联的 ISpriteContaier，然后就可以在程序运行时，通过 ISpriteContainer 的相关方法增、删、改、查 ISprite 接口。

8.1.2　IRenderState、ITransformable 和 ISprite 接口

ISprite 是整个精灵系统的核心接口，该接口主要提供如下几个方面的功能：
- 保存当前精灵的所有渲染状态，继承自 IRenderState 接口（见图 8.2）。
- 持有一个数学变换系统，继承自 ITransformable 接口（见图 8.2），用以表示平移、旋转及缩放等操作。
- 点选碰撞检测操作。
- 更新与绘制操作。
- 响应鼠标、键盘、更新及渲染事件。

来看一下 IRenderState 接口相关的内容。具体代码如下：

```typescript
export enum ERenderType {
    CUSTOM ,                    // 自定义
    STROKE ,                    // 线框渲染模式
    FILL ,                      // 填充模式
    STROKE_FILL ,               // 线框模式+填充模式
    CLIP                        // 裁剪模式
}
export interface IRenderState {
    isVisible : boolean ;                           // 是否可见
    showCoordSystem : boolean ;                     // 是否显示坐标系统
    lineWidth : number ;
    fillStyle : string | CanvasGradient | CanvasPattern ;
    strokeStyle : string | CanvasGradient | CanvasPattern ;
    renderType : ERenderType ;                      // 渲染的枚举类型
}
```

我们知道，Canvas2D 中包括更多的渲染状态，例如是否虚线绘制，以及是否有阴影等，如读者有需要，可自行添加到本接口中，这里只提供必要的，用于演示的相关信息。

接下来看一下 ITransformable 接口，该接口的相关内容如下：

```typescript
export interface ITransformable {
    x : number ;
    y : number ;
    rotation : number ;                            // 角度表示的方位
    scaleX : number ;
    scaleY : number ;

    getWorldMatrix ( ) : mat2d ;                   // 获取全局坐标系矩阵
    getLocalMatrix ( ) : mat2d ;                   // 获取局部坐标系矩阵
}
```

通过 ITransformable 接口提供的功能，就能进行坐标系变换。

除了继承自上面两个接口的属性和方法外，ISprite 自身还提供一些属性、方法和事件。具体代码如下：

```typescript
export enum EOrder {
    PREORDER,
    POSTORDER
}
// 事件回调函数签名
export type UpdateEventHandler = ( ( spr : ISprite , mesc : number, diffSec : number , travelOrder : EOrder ) => void ) ;
export type MouseEventHandler = ( ( spr : ISprite , evt : CanvasMouseEvent ) => void ) ;
export type KeyboardEventHandler = ( ( spr : ISprite , evt : CanvasKeyBoardEvent ) => void ) ;
export type RenderEventHandler = ( spr : ISprite , context : CanvasRenderingContext2D , renderOreder : EOrder ) => void ;
//接口可以扩展多个接口
//而类只能扩展一个类
export interface ISprite extends ITransformable , IRenderState {
    name : string ;                                // 当前精灵的名称
    shape : IShape ;
    // ISprite 引用一个 IShape 接口，如 draw 和 hitTest 都调用 IShape 对应的同名方法
    owner : ISpriteContainer ;                     // 双向关联，通过 owner 找到容器对象
    data : any ;           // 为了方便起见，有时需要添加一些不知道数据类型的额外数据
    hitTest ( localPt : vec2 ) : boolean ;  // 点选碰撞检测
    update ( mesc : number , diff : number , order : EOrder ) : void ;
                                                   // 更新
    draw ( context : CanvasRenderingContext2D ) : void ;          // 绘制

    // 事件处理
    mouseEvent : MouseEventHandler | null ;
    keyEvent : KeyboardEventHandler | null ;
    updateEvent : UpdateEventHandler | null ;
    renderEvent : RenderEventHandler | null ;
}
```

在 ISprite 接口中引用了一个 IShape 接口，接下来看一下 IShape 接口相关的内容。

8.1.3 IDrawable、IHittable 和 IShape 接口

IShape 接口充当渲染数据源的作用，其渲染数据以顶点或路径形式定义在局部坐标系中。IShape 和 ISprite 之间的协作关系如图 8.3 所示。

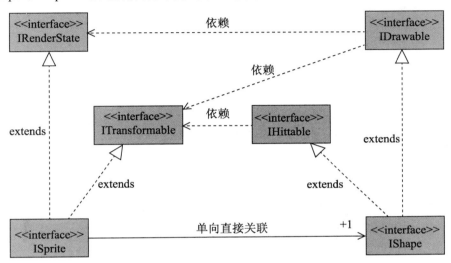

图 8.3　ISprite 和 IShape 接口之间的关系图

ISprite 接口和 IShape 接口关系的具体介绍如下。

（1）ISprite 接口扩展了 IRenderState 和 ITransformable 接口。

（2）ISprite 接口单向直接关联了（UML 中使用带方向箭头的实线表示）一个 IShape 接口，这样可以直接获取当前引用的 IShape 接口。

（3）IDrawable 依赖了（UML 中使用带方向箭头的虚线表示）ITransformable 接口和 IRenderState 接口。下面来看一下 IDrawable 接口相关的内容。具体代码如下：

```
// IDrawable 所有接口方法依赖 ITransformable、IRenderState 和 CanvasRendering
Context2D 这 3 个参数进行绘制操作
export interface IDrawable {
    // 用于 draw 之前的操作，例如渲染状态进栈、设置各个渲染状态值及设置当前变换矩阵
    beginDraw ( transformable : ITransformable , state : IRenderState ,
    context : CanvasRenderingContext2D ) : void ;
    // 用于形体的绘制操作
    draw ( transformable : ITransformable , state : IRenderState , context :
    CanvasRenderingContext2D, ) : void ;
    // 绘制后的操作，例如渲染状态恢复操作等
    endDraw ( transformable : ITransformable , state : IRenderState ,
    context : CanvasRenderingContext2D ) : void ;
}
```

（4）IHittable 接口依赖了（UML 中使用带方向箭头的虚线表示）ITransformable 接口。具体代码如下：

```
// 用于点与 IShape 的精确碰撞检测操作
// 如果选中就返回 true，否则返回 false
export interface IHittable {
    // 参数 localPt 点是相对 IShape 所在的坐标系的偏移（offset）
    // 这意味着 localPt = transform . getLocalMatrix * worldPt
    // 某些情况下可能需要获取 worldPt，可以做如下操作
    // worldPt = transform . getWorldMatrix * localPt
    // 其中*表示 Math2D . transform 方法
    hitTest ( localPt : vec2 , transform : ITransformable ) : boolean ;
}
```

（5）IShape 接口扩展了 IDrawable 接口和 IHittable 接口，并提供了额外的两个成员属性。具体代码如下：

```
export interface IShape extends IHittable , IDrawable {
    readonly type : string ;// 例如 Rect、Circle 等具有唯一性表示的字符串
    data : any ;              // 为了方便起见，有时需要添加一些不知道数据类型的额外数据
}
```

这里强调一下 UML 中的单向直接关联和依赖之间的语义区别。

ISprite 单向关联了 IShape，意味着 ISprite 接口中持有一个 IShape 类型的成员变量作为引用。而 IShape 没有反向关联一个 ISprite，意味着 IShape 接口中没有方法能够反过来查找到 ISprite 接口实例，这正是我们所需要的效果。

想像一下，IShape 是个享元类，同一个 IShape 接口的实例可以被多个 ISprite 所持有（如图 8.1 所示）。在这种情况下，如果 IShape 接口有一个反向引用指向 ISprite，就无法实现被共享的需求了。

但是享元类 IShape 需要使用 ISprite 接口中相关的渲染状态数据（IRenderState）和矩阵变换数据（ITransformable）来进行渲染和碰撞检测，可以使用成员方法参数传递的方式（UML 中的依赖关系）来实现数据传输。通过接口依赖方式，让耦合性最小化。

8.2　实现非场景图类型精灵系统

8.1 节中按照图 8.2 所示的精灵系统静态类结构图中从顶向下的顺序，介绍了精灵系统的各个接口及它们之间的关系。本节中，将以从下而上的顺序，实现前面介绍过的接口，形成一个非场景图类型的精灵系统。由于 IShape 的实现类比较多，并且涉及各种形体的碰撞检测算法及不同的渲染数据和绘制代码，因此将在下一节中介绍，并通过一个演示 Demo 来测试不同的形体对象。

8.2.1 Transform2D 辅助类

本节将实现一个有用的辅助类：Transform2D。该类封装坐标系变换的相关功能，需要使用前面章节实现的向量、矩阵及矩阵堆栈类。

首先，在 Math2D 类中声明一个静态变量，这样我们就能够以 Math2D.matStack 的方式获取一个矩阵堆栈对象。具体代码如下：

```
export class Math2D {
    public static matStack : MatrixStack = new MatrixStack ( ) ;
}
```

然后，仍旧在 math2d.ts 文件中增加并实现 Transform2D 类。具体代码如下：

```
export class Transform2D {
    public position : vec2 ;                            // 位移
    public rotation : number ;                          //方位（角度表示）
    public scale : vec2 ;                               // 缩放
    public constructor ( x : number = 0 , y : number = 0 , rotation : number
= 0 , scaleX : number = 1 , scaleY : number = 1 ) {
        this . position = new vec2( x , y ) ;
        this . rotation = rotation ;
        this . scale = new vec2( scaleX , scaleY ) ;
    }
    public toMatrix () : mat2d {
        Math2D . matStack . loadIdentity ( ) ;          // 设置矩阵栈顶矩阵归一化
        Math2D . matStack . translate ( this . position . x , this . position .
y ) ;                                                   // 先平移
        Math2D . matStack . rotate ( this . rotation , false ) ;
                // 然后旋转，最后一个参数false，表示rotation是角度而不是弧度
        Math2D . matStack . scale ( this . scale . x , this . scale . y ) ;
                //最后缩放操作
        return Math2D . matStack . matrix ;
                // 返回TRS合成后的、表示从局部到世界的变换矩阵
    }
    public toInvMatrix ( result : mat2d ) : boolean {
        // 获取局部到世界的变换矩阵
        let mat : mat2d = this . toMatrix ( ) ;
        // 对mat矩阵求逆，获得从世界到局部的变换矩阵
        return mat2d . invert ( mat , result ) ;
    }
}
```

8.2.2 ISprite 接口的实现

新建一个名为 Sprite2d.ts 文件，并在该文件中实现 Sprite2D 类。先来声明并初始化必要的成员变量。具体代码如下：

```typescript
// 继承 VS 引用
// 可以使用引用，这样可以实列化，更加灵活
// 将可变的部分，以事件方式响应
export class Sprite2D implements ISprite {
    // IRenderState 接口需要的成员属性（或变量）
    public showCoordSystem : boolean = false;
    public renderType : ERenderType = ERenderType . FILL ;
    public isVisible : boolean = true ;
    public fillStyle : string | CanvasGradient | CanvasPattern = 'white';
    public strokeStyle : string | CanvasGradient | CanvasPattern = 'black';
    public lineWidth : number = 1 ;
    // ITransformable 接口的成员属性和方法都委托到 Transform2D 来实现
    public transform : Transform2D = new Transform2D ( ) ;
    // ISprite 接口本身的成员属性
    public name : string ;
    public shape : IShape ;
    public data : any ;
    public owner ! : ISpriteContainer ;   //确保后续会被设置，当前精灵的拥有者
    // 事件回调
    public mouseEvent : MouseEventHandler | null = null ;
    public keyEvent : KeyboardEventHandler | null = null ;
    public updateEvent : UpdateEventHandler | null = null ;
    public renderEvent : RenderEventHandler | null = null ;
    // 构造函数
    public constructor ( shape: IShape , name : string ) {
        this . name = name ;
        this . shape = shape ;
    }
}
```

接下来看一下 **ITransformable** 接口的实现，该接口所有的属性和方法都是委托调用了 Transform2D 类的相关方法。具体代码如下：

```typescript
public get x ( ) : number {
    return this . transform . position . x ;
}
public set y ( y : number ) {
    this . transform . position . y = y ;
}
public get y ( ) : number {
    return this . transform . position . y ;
}
public set rotation ( rotation : number ) {
    this . transform . rotation = rotation ;
}
public get rotation ( ) : number {
    return this . transform . rotation ;
}
public set scaleX ( s : number ) {
    this . transform . scale . x = s ;
}
public get scaleX ( ) : number {
    return this . transform . scale . x ;
}
```

```
public set scaleY ( s : number ) {
    this . transform . scale . y = s ;
}
public get scaleY ( ) : number {
    return this . transform . scale . y ;
}

public getWorldMatrix ( ) : mat2d {
    return this . transform . toMatrix ( ) ;
}
// 如果矩阵求逆失败,直接抛出错误
public getLocalMatrix ( ) : mat2d {
    let src : mat2d = this . getWorldMatrix ( ) ;
    let out : mat2d = mat2d . create ( ) ;
    if ( mat2d . invert ( src , out ) ) {
        return out ;
    }else{
        alert ( "矩阵求逆失败" ) ;
        throw new Error ( "矩阵求逆失败" ) ;
    }
}
```

然后来看一下 ISprite 接口本身需要实现的成员方法。除了 update 方法外,其他两个方法都是调用 IShape 接口对应的方法。具体代码如下:

```
// 下面是 ISprite 接口成员方法的实现
public update ( mesc: number, diff:number,order: EOrder ): void {
    // 如果当前精灵有挂接 updateEvent,则触发该事件
    if ( this . updateEvent ) {
        this . updateEvent ( this , mesc , diff , order ) ;
    }
}
// 委托调用 IShape 对应的 hitTest 方法(IHittable 接口)
// 而 IShape 接口的 hitTest 方法依赖 ITransform 接口
public hitTest ( localPt : vec2 ) : boolean {
    if ( this . isVisible ) {
        //要将光标点变换到局部坐标系
        return this . shape . hitTest ( localPt , this ) ;
    } else {
        return false ;
    }
}
// 委托调用 IShape 对应的 draw 方法(IDrawable 接口)
// 而 IShape 接口的 beginDraw、draw、endDraw 方法依赖 ITransform 和 IRenderState 接口
// 如有必要同时会触发 renderEvent 事件
public draw ( context : CanvasRenderingContext2D ): void {
    if ( this . isVisible ) {
        // 渲染状态进栈
        // 然后设置渲染状态及当前变换矩阵
        this . shape . beginDraw ( this , this , context ) ;
        // 在 Draw 之前,触发 PREORDER 渲染事件
        if ( this . renderEvent !== null ) {
            this . renderEvent ( this , context , EOrder .PREORDER ) ;
```

```
        }
        // 调用主要的绘图方法
        this . shape . draw ( this , this , context ) ;
        // 在 Draw 之后，触发 POSTORDER 渲染事件
        if ( this . renderEvent !== null ) {
            this . renderEvent ( this , context , EOrder .POSTORDER ) ;
        }
        // 恢复渲染状态
        this . shape . endDraw ( this , this , context ) ;
    }
}
```

最后需要一个工厂类来创建 ISprite 接口。具体代码如下：

```
export class SpriteFactory {
    public static createSprite (shape: IShape ,name : string ) : ISprite {
        let spr : ISprite = new Sprite2D ( shape , name ) ;
        return spr ;
    }
}
```

后续要创建接口实例对象的方法，都可以放在 SpriteFactory 类中。

8.2.3　Sprite2DManager 管理类

当实现了 ISprite 接口后，需要对 ISprite 接口的各个实例对象进行管理，此时可以实现 Sprite2DManager 类，该类实现 ISpriteContainer 接口，对 ISprite 进行增、删、改、查相关操作。将 Sprite2DManger 类实现在一个名为 sprite2DSystem.ts 文件中。具体代码如下：

```
export class Sprite2DManager implements ISpriteContainer {
    public name : string = 'sprite2dManager' ;
    private _sprites : ISprite [ ] = [ ] ; // 数组存放 ISprite 接口的实列对象
    public addSprite ( sprite : ISprite ) : ISpriteContainer {
        sprite . owner = this ;
        this . _sprites . push ( sprite ) ;
        return this ;
    }
    public removeSpriteAt ( idx : number ) : void {
        // 数组的 splice 方法：从 idx 开始，删除 1 个元素
        this . _sprites . splice( idx , 1 ) ;
    }

    public removeSprite ( sprite : ISprite ) : boolean {
        // 根据 sprite 查找索引号
        let idx = this . getSpriteIndex ( sprite ) ;
        if ( idx != -1 ) {
            // 如果找到，就删除，并返回 true
            this . removeSpriteAt ( idx ) ;
            return true ;
        }
        // 如果没找到，返回 false
        return false ;
```

```
    }
    public removeAll ( ) : void {
        // 我们有 4 种数组情况的方法，具体来看一下
        /*
        // 第 1 种方式，调用我们自己实现的 removeSpriteAt 方法
        for ( let i = this . _sprites . length - 1 ; i >= 0 ; i -- ) {
            this . removeSpriteAt ( i ) ;
        }
        */

        /*
        // 第 2 种方式，调用数组的 pop 方法
        for ( let i = this . _sprites . length - 1 ; i >= 0 ; i -- ) {
            this . _sprites . pop ( ) ;
        }
        */

        /*
        // 第 3 种方式，直接将数组的 length 设置为 0
        this . _sprites . length = 0
        */

        // 目前我们使用第 4 种方式，重新生成一个新数组赋值给 this._sprites 变量
this . _sprites = [ ] ; }
    public getSprite ( idx : number ) : ISprite {
        if ( idx < 0 || idx > this . _sprites . length - 1 ) {
            throw new Error ( "参数 idx 越界!!" ) ;
        }
        return this . _sprites [ idx ] ;
    }
    public getSpriteCount ( ) : number {
        return this . _sprites . length ;
    }
    public getSpriteIndex ( sprite: ISprite ): number {
        for ( let i = 0 ; i < this . _sprites . length ; i++ ) {
            if ( this . _sprites [ i ] === sprite ) {
                return i ;
            }
        }
        return -1 ;
    }
}
```

由于现在实现的是一个线性的、非场景图管理类型的精灵系统，不涉及复杂的树结构操作，因此可以直接将 IDispatcher 接口相关内容由 Sprite2DManager 来实现。下面来看一下具体实现：

```
// Sprite2DManager 类实现 IDispatcher 接口
export class Sprite2DManager implements  IDispatcher  {
    // IDispatcher 接口实现
    // 由_dragSprite 接受 drag 事件的处理
    // 也就是说，drag 事件都是发送到_dragSprite 精灵的
    private _dragSprite : ISprite | undefined = undefined ;
```

```
public get container ( ) : ISpriteContainer {
    return this ;
}
// 分发 update 事件
public dispatchUpdate ( msec : number , diff : number ) : void {
    // 从前到后遍历精灵数组，触发 PREORDER updateEvent
    for ( let i = 0 ; i < this . _sprites . length ; i ++ ) {
        this . _sprites [ i ] . update ( msec , diff , EOrder . PREORDER ) ;
    }
    // 从后到前遍历精灵数组，触发 POSTORDER updateEvent
    for ( let i = this . _sprites . length -1 ; i >= 0 ; i -- ) {
        this . _sprites [ i ] . update ( msec , diff , EOrder . POSTORDER ) ;
    }
}
// 分发 draw 命令
public dispatchDraw ( context : CanvasRenderingContext2D ) : void {
    // 从前到后遍历精灵数组，调用 draw 方法
    // 绘制的顺序是先添加的精灵先绘制，后添加的精灵后绘制
    for ( let i = 0 ; i < this . _sprites . length ; i++ ) {
        this . _sprites [ i ] . draw ( context ) ;
    }
}

  // 分发键盘事件，采取最简单的方式
  // 遍历每个精灵，凡是有键盘处理事件的都触发该事件
  public dispatchKeyEvent ( evt : CanvasKeyBoardEvent ) : void {
    let spr: ISprite ;
    for ( let i = 0 ; i < this . _sprites . length ; i++ ) {
        spr = this . _sprites [ i ] ;
        if ( spr . keyEvent ) {
            spr . keyEvent ( spr, evt ) ;
        }
    }
  }
// 分发鼠标事件
public dispatchMouseEvent ( evt : CanvasMouseEvent ) : void {
    // 每次按下鼠标时，将_dragSprite 设置为当前鼠标指针下面的那个精灵
    // 每次释放鼠标时，将_dragSprite 设置为 undefined
    if ( evt . type === EInputEventType . MOUSEUP ) {
        this . _dragSprite = undefined ;
    } else if (evt . type === EInputEventType . MOUSEDRAG) {
        // 触发 drag 事件，由_dragSprite 接受并处理
        if ( this . _dragSprite !== undefined ) {
            if ( this . _dragSprite . mouseEvent !== null ) {
                this . _dragSprite . mouseEvent ( this . _dragSprite ,
                evt ) ;
                // 一旦处理完成，不再继续分发，直接退出方法
                return ;
            }
        }
    }
    let spr : ISprite ;
    // 遍历精灵数组，注意是从后向前的顺序遍历
```

```typescript
        // 绘制时，以由前往后方式遍历，这样可以保证绘制的深度正确性
        // 绘制鼠标事件时，以由后往前方式遍历，确保上面的精灵能先接收到事件
        for ( let i = this . _sprites . length - 1 ; i >= 0 ; i-- ) {
            // 获取当前精灵
            spr = this . _sprites [ i ] ;
            // 获取当前精灵的局部矩阵
            let mat : mat2d | null = spr . getLocalMatrix ( ) ;
            // 将全局表示的 canvasPosition 点变换到相对当前精灵局部坐标系的表示：
            localPosition
            Math2D . transform ( mat , evt . canvasPosition , evt . localPosition ) ;
            // 要测试的点和精灵必须在同一个坐标系中，切记
            // 如果碰撞检测测试成功，说明选中该精灵
            if ( spr . hitTest ( evt . localPosition ) ) {
                evt . hasLocalPosition = true ;
                // 鼠标按下并且有点选选中的精灵时，记录下来
                // 后续的 drag 事件都是发送到该精灵上的
                if ( evt . type === EInputEventType . MOUSEDOWN) {
                    this . _dragSprite = spr ;
                }
                // 如果有选中的精灵，并且有事件处理程序，则立刻触发该事件处理程序
                if ( spr . mouseEvent ) {
                    spr . mouseEvent ( spr , evt ) ;
                    // 一旦处理完成，不再继续分发，直接退出方法
                    return ;
                }
            }
        }
    }
}
```

下面来看一下，如何使用 Sprite2DManager 类。

在 8.1.1 节中，实现了 Sprite2DApplication 类，该类中持有一个 IDispatcher 接口。当时的实现代码并没有提供构造函数，在此处实现 Sprite2DApplication 类的构造函数，创建 Sprite2DManager 类（实现了 IDispatcher 接口）的实例。具体代码如下：

```typescript
public constructor ( canvas : HTMLCanvasElement ) {
    super ( canvas ) ;
    this . _dispatcher = new Sprite2DManager ( ) ;
}
```

这样就将整个系统的运行流程全部串联起来了，Sprite2DApplication 启动动画循环后，会不断地通过 IDispatcher 分发更新、渲染命令，以及鼠标和键盘事件。

8.3　IShape 形体系统

本节将实现一个用于演示 Demo 的基础形体系统，如图 8.4 所示。

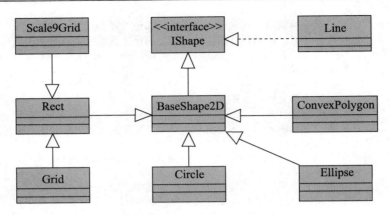

图 8.4　IShape 接口及实现类

8.3.1　线段 Line 类

参考图 8.4 会发现，Line 类直接实现了 IShape 接口，而其他类都是从 BaseShape2D 类继承而来的。这是因为在几何中，我们可以分为点、线、面、体等基础形体表示，其中我们设计的 BaseShape2D 类表示面的概念，其特点是第一个顶点和最后一个顶点重合，从而形成一个封闭性的面（能计算出面积），最小的面是三角形，需要 3 个顶点表示。而两点就可以确定一条直线了，与我们 BaseShape2D 类的设计目的不同。顺便提一下，对于点的表示，我们可以使用 BaseShape2D 的子类 Circle 来模拟，而对于体，则属于 3D 中的概念，本书不涉及 3D 相关内容。

下面来看一下 Line 的实现代码，具体如下：

```
export class Line implements IShape {
    // 线段起点和终点
    public start : vec2 ;
    public end : vec2 ;
    public data : any ;
    // t 的取值范围在 0～1 之间，用来控制线段的原点位于线段上某一处
    public constructor ( len : number = 10 , t : number = 0 ) {
        if ( t < 0.0 || t > 1.0 ) {
            alert ( "参数 t 必须处于 [ 0 , 1 ]之间!!" ) ;
            throw new Error ( "参数 t 必须处于 [ 0 , 1 ]之间!!" ) ;
        }
        this . start = vec2 . create ( - len * t , 0 ) ;
        this . end = vec2 . create ( len * ( 1.0 - t ) , 0 ) ;
        this . data = undefined ;
    }
    // override IHittable 接口的 hitTest 方法
    public hitTest ( localPt : vec2 ,transform :ITransformable): boolean {
        return Math2D . isPointOnLineSegment ( localPt , this . start , this .
        end ) ;
    }
```

```
    // 做 3 件事
    public beginDraw ( transformable : ITransformable , state : IRenderState ,
    context : CanvasRenderingContext2D ): void {
        // 1．渲染状态进栈
        context . save ( ) ;
        // 2．设置当前渲染状态
        context . lineWidth = state . lineWidth ;
        context . strokeStyle = state . strokeStyle ;
        // 3．设置当前变换矩阵
        let mat : mat2d = transformable . getWorldMatrix ( ) ;
        context . setTransform ( mat . values [ 0 ] , mat . values [ 1 ] ,
        mat . values [ 2 ] , mat . values [ 3 ] , mat . values [ 4 ] , mat .
        values [ 5 ] ) ;
    }
    public draw ( transformable : ITransformable , state : IRenderState ,
    context : CanvasRenderingContext2D ) : void {
        context . beginPath ( ) ;
        context . moveTo ( this . start . x , this . start . y ) ;
        context . lineTo ( this . end . x , this . end . y ) ;
        context . stroke ( ) ;
    }
    // 恢复渲染状态堆栈
    public endDraw ( transformable: ITransformable, state : IRenderState ,
    context: CanvasRenderingContext2D ): void {
        context.restore();
    }
    //子类必须覆写（override）type 属性，返回当前子类的实际类型
    public get type (): string {
        return "Line";
    }
}
```

上面的代码中，IShape 接口的实现类中的 hitTest 方法中直接调用了第 6 章实现的 Math2D . isPointOnLineSegment 静态方法，作为一个惯例，将下面形体类中所有的点选方法都作为静态函数实现在 Math2D 类中。

其次，所有 IShape 的子类（图 8.4 中的所有实现类）都实现在一个名为 shape.ts 的文件中。

最后，所有的公开接口（图 8.2 所示的所有接口，以及一些辅助的枚举结构和事件回调函数签名）都声明在一个名为 interface.ts 的文件中，该文件还包括一个名为 SpriteFactory 的工厂类，用来创建各种接口。在 8.2.2 节中，实现了 createSprite 方法用来生产 ISprite 接口，现在继续在 SpriteFactory 类中生产表示 Line 的 IShape 接口。具体代码如下：

```
// 通过两个点获得一条直线
public static createLine ( start : vec2 , end : vec2 ) : IShape {
    let line : Line = new Line ( ) ;
    line . start = start ;
    line . end = end ;
    return line ;
}
```

```
// 通过线段长度和[0, 1]之间的t获得一条与x轴方向平行的、原点在该线段任意一点的直线
public static createXLine ( len : number = 10 ,t : number = 0 ):IShape {
    return new Line ( len , t ) ;
}
```

上面的代码通过两种方式获得线段，在本章 Demo 中会演示 createXLine 方法的使用，在后面章节中会使用 createLine 方法。

8.3.2　BaseShape2D 抽象基类

将一些子类共享的代码在 BaseShape2D 抽象基类中实现，BaseShape2D 本身扩展了 IShape 接口。具体代码如下：

```
// 在class之前使用 abstract 关键字表示当前类为抽象类
export abstract class BaseShape2D implements IShape {
    // 在属性前使用 abstract 关键字表示一个抽象属性
    // 子类必须override，返回当前子类的实际类型字符串
    public abstract get  type (): string ;
    // 在方法前面使用 abstract 关键字表示一个抽象方法
    // IHittable 接口方法,子类必须覆写本方法
    public abstract hitTest ( localPt : vec2 , transform : ITransformable ) : boolean ;
    // 2D 局部坐标的绘制
    public axisXStyle : string | CanvasGradient | CanvasPattern;
    public axisYStyle : string | CanvasGradient | CanvasPattern;
    public axisLineWidth : number;
    public axisLength : number ;
    // 接口成员属性实现
    public data : any ;
    public constructor ( ) {
        this . axisXStyle = "rgba( 255 , 0 , 0 , 128 ) " ;
        this . axisYStyle = "rgba( 0 , 255 , 0 , 128 ) " ;
        this . axisLineWidth = 1 ;
        this . axisLength = 100 ;
        this . data = undefined ;
    }
    // 用于绘制坐标系 x 轴和 y 轴,以利于 debug 显示
    protected drawLine ( ctx : CanvasRenderingContext2D , style : string | CanvasGradient | CanvasPattern , isAxisX : boolean = true ) {
        ctx .save ( ) ;
        ctx . strokeStyle = style ;
        ctx . lineWidth = this . axisLineWidth ;
        ctx . beginPath ( ) ;
        ctx . moveTo( 0 , 0 ) ;
        if ( isAxisX ) {
            ctx . lineTo ( this . axisLength , 0 ) ;
        } else {
            ctx.lineTo( 0 , this . axisLength ) ;
        }
        ctx . stroke ( ) ;
        ctx . restore ( ) ;
```

```typescript
    }
    public beginDraw ( transformable: ITransformable, state : IRenderState ,
    context: CanvasRenderingContext2D ): void {
        // 渲染状态入堆栈
        context . save ( ) ;
        // 设置渲染状态
        context . lineWidth = state . lineWidth ;
        context . strokeStyle = state . strokeStyle ;
        context . fillStyle = state . fillStyle ;
        // 通过ITransformable接口获得当前IShape的全局矩阵
        // 然后设置变换矩阵
        let mat : mat2d = transformable . getWorldMatrix ( ) ;
        context . setTransform ( mat . values [ 0 ] , mat . values [ 1 ] ,
        mat.values [ 2 ] , mat . values [ 3 ] , mat . values [ 4 ] , mat .
        values [ 5 ] ) ;
    }
    // 子类必须覆写（override）draw方法，如有需要（大部分情况下），则调用基类方法
    (super . draw ( ) )
    // 具体绘制过程在draw中实现，基类实现调用fill、stroke、STROKE_FILL和clip命令
    public draw ( transformable: ITransformable, state : IRenderState ,
    context: CanvasRenderingContext2D, ): void {
        //调用绘制命令
        if ( state.renderType === ERenderType.STROKE ) {
            context.stroke();
        } else if ( state.renderType === ERenderType.FILL ) {
            context.fill();
        } else if ( state.renderType === ERenderType.STROKE_FILL ) {
            context.stroke();
            context.fill();
        } else if ( state . renderType === ERenderType . CLIP ) {
            context . clip ( ) ;
        }
    }

    public endDraw ( transformable : ITransformable , state : IRenderState ,
    context : CanvasRenderingContext2D ) : void {
        // 非裁剪状态下，并且showCoordSystem为true的情况下，绘制出坐标系
        if ( state . renderType !== ERenderType . CLIP ) {
            if ( state . showCoordSystem ) {
                this . drawLine ( context , this . axisXStyle , true ) ;
                this . drawLine ( context , this . axisYStyle , false ) ;
            }
        }
        // 渲染状态出栈
        context . restore ( ) ;
    }
}
```

通过上述代码可以知道，BaseShape2D 提供了一个局部坐标系绘制的功能（在 endDraw 方法中），并且抽象了子类的实现流程。大部分情况下，BaseShape2D 的子类要做的是实现一个抽象属性 type、一个抽象方法 hitTest，以及覆写 draw 方法，除了 ERenderType . CUSTOM 类型之外的其他绘制命令、都需要在 draw 方法中调用基类的同名方法（super . draw）。

8.3.3 Rect 类和 Grid 类

下面直接来看 Rect 的代码，具体如下：

```typescript
// 继承自 BaseShape2D
export class Rect extends BaseShape2D {
    // 位置与尺寸
    public width : number ;
    public height : number ;
    public x : number ;
    public y : number ;
    // 使用 u、v 控制矩形的原点偏移，其中 u、v 的取值范围为[0，1]之间，表示原点相对矩形左上角的偏移比例
    public constructor ( w: number = 1, h: number = 1 , u : number = 0 , v : number = 0  ) {
        super ( ) ;
        this . width = w ;
        this . height = h ;
        this . x = - this . width * u ;
        this . y = - this . height * v ;
    }
    // 覆写基类的 type 属性
    public get type ( ): string {
        return "Rect";
    }
    // 覆写基类 hitTest 方法，调用前面实现的 isPointInRect 方法
    public hitTest(localPt : vec2 , transform : ITransformable): boolean {
        return Math2D . isPointInRect ( localPt . x , localPt . y , this . x , this . y , this . width , this . height ) ;
    }
    // 复写基类 draw 方法，最后调用 super . draw 方法用来显示当前绘制的形体
    public draw ( transformable: ITransformable, state : IRenderState , context: CanvasRenderingContext2D ): void {
        //使用 MoveTo，lineTo 及 closePath 渲染命令生成封闭路径
        context.beginPath();
        context.moveTo( this . x , this . y );
        context.lineTo( this . x + this . width , this . y );
        context.lineTo( this . x + this . width , this . y + this . height );
        context.lineTo( this . x , this . y + this . height );
        //3 条没有封闭的线，但是不调用 closePath，可以填充，但是 stroke 描边时缺少一条线段
        //此时要么调用 closePath，要么增加一条线段
        context.closePath();
        //调用基类绘制命令
        super . draw ( transformable, state , context ) ;
    }
}
```

Rect 定义在局部坐标系中，默认情况下原点位于左上角，可以通过 u、v 参数来控制原点位置，前面章节已作详解。其他代码比较简单，请大家自行阅读。

接下来看一下 Grid 类，该类继承自 u、v 设置为[0 , 0]的 Rect 类，主要作为 Demo 的背景使用。具体代码如下：

```
export class Grid extends Rect {
    public xStep: number;
    public yStep: number;
    public constructor ( w : number = 10, h : number = 10 ,xStep :number = 10, yStep : number = 10 ) {
        super( w , h , 0 , 0 ) ;
        this . xStep = xStep ;
        this . yStep = yStep ;
    }
    public draw ( transformable : ITransformable , state : IRenderState , context : CanvasRenderingContext2D ) : void {
        state . renderType = ERenderType . CUSTOM ;
        context . fillRect ( 0 , 0 , this . width , this . height ) ;
        context . beginPath ( ) ;
        for ( var i = this . xStep + 0.5 ; i < this . width ; i += this . xStep )
        {
            context . moveTo ( i , 0 ) ;
            context . lineTo ( i , this . height ) ;
        }
        context . stroke ( ) ;
        context . beginPath ( ) ;
        for ( var i = this . yStep + 0.5 ; i < this . height ; i += this . yStep )
        {
            context . moveTo ( 0 , i );
            context . lineTo ( this . width , i ) ;
        }
        context . stroke ( ) ;
    }
    public get type ( ) : string {
        return "Grid";
    }
}
```

8.3.4　Circle 类和 Ellipse 类

先来看一下表示圆的 Circle 类，该类只需要一个半径参数，就能定义圆心在局部坐标系原点的一个圆。具体代码如下：

```
export class Circle extends BaseShape2D {
    public radius : number ;   //半径
    public constructor ( radius: number = 1 ) {
        super ( ) ;
        this . radius = radius;
    }
    // 调用 Math2D 类的 isPointInCircle 静态方法
    public hitTest(localPt : vec2 , transform :ITransformable ): boolean {
        return Math2D . isPointInCircle ( localPt , vec2 . create ( 0 , 0 ),
```

```
        this.radius ) ;
    }
    public draw ( transformable : ITransformable , state : IRenderState ,
    context : CanvasRenderingContext2D ): void {
        context.beginPath();
        context.arc( 0, 0, this.radius, 0.0, Math.PI * 2.0, true );
        //必须放在最后,调用绘制命令
        super.draw( transformable, state , context );
    }
    public get type (): string {
        return "Circle";
    }
}
```

实际上,Circle 类是椭圆 Ellipse 类的特殊形式,即 Ellipse 类的 x 轴半径和 y 轴半径相等时,表示的就是 Circle。下面来看 Ellipse 类的相关代码,具体如下:

```
export class Ellipse extends BaseShape2D {
    public radiusX : number ;
    public radiusY : number ;
    // 椭圆圆心在局部坐标系的[ 0 , 0 ]位置处
    public constructor ( radiusX : number = 10 , radiusY : number = 10 ) {
        super ( ) ;
        this . radiusX = radiusX ;
        this . radiusY = radiusY ;
    }
    public hitTest(localPt : vec2 ,transform : ITransformable ): boolean {
        let isHitted : boolean = Math2D . isPointInEllipse ( localPt . x ,
        localPt . y , 0 , 0 , this . radiusX, this . radiusY ) ;
        return isHitted;
    }
    public draw ( transform : ITransformable , state : IRenderState , context :
    CanvasRenderingContext2D ) : void {
        context . beginPath ( ) ;
        context . ellipse ( 0 , 0 , this . radiusX , this . radiusY , 0 ,
        0 , Math.PI * 2 ) ;
        super . draw ( transform , state , context ) ;
    }
    public get type ( ) : string {
        return "Ellipse" ;
    }
}
```

8.3.5　ConvexPolygon 类

下面是凸多边形 ConvexPolygon 实现内容,代码如下:

```
export class ConvexPolygon extends BaseShape2D {
    public points : vec2 [ ] ;
    public constructor ( points : vec2 [ ] ) {
        if ( points . length < 3 ) {
            alert ( "多边形顶点必须大于 3 或等于 3!!")
            new Error ( "多边形顶点必须大于 3 或等于 3!!") ;
```

```
        }
        if ( Math2D . isConvex ( points ) === false ) {
            alert ( "当前多边形不是凸多边形!!" ) ;
            new Error ( "当前多边形不是凸多边形!!" ) ;
        }
        super ( ) ;
        this . points = points ;
    }
    public hitTest(localPt :vec2 ,transform : ITransformable ) : boolean {
        for ( let i : number = 2 ; i < this . points . length ; i ++ ) {
            if ( Math2D . isPointInTriangle ( localPt , this . points [ 0 ] ,
            this . points [ i - 1 ] , this . points [ i ] ) ) {
                return true ;
            }
        }
        return false ;
    }
    public draw ( transformable: ITransformable, state : IRenderState ,
    context: CanvasRenderingContext2D ): void {
        context . beginPath ( ) ;
        context . moveTo ( this . points[ 0 ] . x ,this . points [ 0 ] . y ) ;
        for ( let i = 1 ; i < this . points . length ; i ++ ) {
            context . lineTo (this . points[ i ].x ,this . points[ i ] . y ) ;
        }
        context . closePath ( ) ;
        super . draw ( transformable , state , context ) ;
    }
    public get type ( ) : string {
        return "Polygon" ;
    }
}
```

8.3.6 Scale9Grid 类

下面先来看一下 Scale9Grid（九宫缩放）的演示效果，如图 8.5 所示。

图 8.5 Scale9Grid 效果图

从图 8.5 中可以看到，左侧表示源图，为了更好地理解九宫缩放，在内部使用两横、两纵 4 条线段将其分隔为 9 个区域；而右侧图像（目标图）是将左侧图像的 9 个区域按照一定规则映射（绘制）后的效果，具体的映射规则如下：

- 源图的四个角（左上、右上、右下、左下）按原始大小（不缩放）直接映射（绘制）到目标图对应的四个角上。

- 源图的左边和右边仅以纵向缩放的方式映射（绘制）到目标图对应区域。
- 源图的上边和下边仅以横向缩放的方式映射（绘制）到目标图对应区域。
- 源图的中间部分以纵横缩放的方式映射（绘制）到目标图的对应区域。

为了更方便地实现上述需求，先提供两个辅助类。其中第一个是 Inset 类，将其定义在 math2d.ts 文件中。具体代码如下：

```typescript
export class Inset {
    public values : Float32Array ;
    public constructor ( l : number = 0 , t : number = 0 , r : number = 0 , b : number = 0 ) {
        this . values = new Float32Array ( [ l , t , r , b ] ) ;
    }
    public get leftMargin ( ) : number {
        return this . values [ 0 ] ;
    }
    public set leftMargin ( value : number ) {
        this . values [ 0 ] = value ;
    }
    public get topMargin ( ) : number {
        return this . values [ 1 ] ;
    }
    public set topMargin ( value : number ) {
        this . values [ 1 ] = value ;
    }
    public get rightMargin ( ) : number {
        return this . values [ 2 ] ;
    }
    public set rightMargin ( value : number ) {
        this . values [ 2 ] = value ;
    }
    public get bottomMargin ( ) : number {
        return this . values [ 3 ] ;
    }
    public set bottomMargin ( value : number ) {
        this . values [ 3 ] = value ;
    }
    // 静态 create 方法
    public static create ( l : number = 0 , t : number = 0 , r : number = 0 , b : number = 0 ) : Inset {
        return new Inset ( l , t , r , b ) ;
    }
}
```

很简单的 Inset 类，其 4 个 margin 和 CSS 中的外边距含义相同。

另外一个类是使用原点+尺寸表示的 Rectangle。具体代码如下：

```typescript
export class Size {
    public values : Float32Array ;                    // 使用 float32Array

    public constructor ( w : number = 1 , h : number = 1 ) {
        this . values = new Float32Array ( [ w , h ] ) ;
    }
    set width ( value : number ) { this . values [ 0 ] = value ; }
```

```
        get width () : number { return this.values[0]; }
        set height ( value : number ) { this.values[1] = value; }
        get height () : number { return this.values[1]; }
        // 静态 create 方法
        public static create ( w : number = 1 , h : number = 1 ) : Size {
            return new Size ( w , h );
        }
    }
    export class Rectangle {
        public origin : vec2 ;
        public size : Size ;
        public constructor ( orign : vec2 = new vec2 () , size : Size = new Size ( 1 , 1 ) ) {
            this.origin = orign ;
            this.size =  size ;
        }
        public isEmpty () : boolean {
            let area : number = this.size.width * this.size.height ;
            if ( Math2D.isEquals ( area , 0 ) === true ) {
                return true ;
            } else {
                return false ;
            }
        }
        // 静态 create 方法
        public static create ( x : number = 0 , y : number = 0 , w : number = 1 , h : number = 1 ) : Rectangle {
            let origin : vec2 = new vec2 ( x , y ) ;
            let size : Size = new Size ( w , h ) ;
            return new Rectangle ( origin , size ) ;
        }
    }
```

有了上面几个辅助类，就可以实现 Scale9Grid 类了。具体代码如下：

```
export enum EImageFillType {
    NONE ,                              // 没有任何效果
    STRETCH ,                           // 拉伸模式
    REPEAT ,                            // x 和 y 重复填充模式
    REPEAT_X ,                          // x 方向重复填充模式
    REPEAT_Y                            // y 方向重复填充模式
}

export class Scale9Data {
    // 设计为私有，只能由构造函数传入？
    public image : HTMLImageElement ;

    // 私有成员变量
    private _inset : Inset ;

    // 使用 set 属性
    public set inset ( value : Inset ) {
        this._inset = value ;
    }
```

```typescript
    public get leftMargin ( ) : number {
        return this . _inset . leftMargin ;
    }

    public get rightMargin ( ) : number {
        return this . _inset . rightMargin ;
    }

    public get topMargin ( ) : number {
        return this . _inset . topMargin ;
    }

    public get bottomMargin ( ) : number {
        return this . _inset . bottomMargin ;
    }

    public constructor ( image : HTMLImageElement , inset : Inset ) {
        this . image = image ;
        // 不能在构造函数中直接使用 this . inset = inset ;
        this . _inset = inset ;
    }
}

export class Scale9Grid extends Rect {

    public data : Scale9Data ;
    public srcRects ! : Rectangle [ ] ;
    public destRects ! : Rectangle [ ] ;

    public get type ( ) : string {
        return "Scale9Grid" ;
    }

    public constructor ( data : Scale9Data , width : number , height : number , u : number , v : number ) {
        super ( width , height , u , v ) ;
        this . data = data ;
        this . _calcDestRects ( ) ;
    }

    // 关键方法，计算出源图和目标图中 9 个区块的坐标和尺寸
    // 计算规则请参考 8.3.6 节讲解的九宫计算的相关规则
    // 1. 计算出 4 个角 rectangle
    // 2. 计算出 4 条边 rectangle
    // 3. 计算出中心 content rectangle
    private _calcDestRects ( ) : void {
        this . destRects = [ ] ;
        this . srcRects = [ ] ;

        let rc : Rectangle ;

        // 左上角
        rc = new Rectangle ( ) ;              // 源 Rect
```

```
rc . origin = vec2 . create ( 0 , 0 ) ;
rc . size = Size . create ( this . data . leftMargin , this . data .
topMargin ) ;
this . srcRects . push ( rc ) ;

rc = new Rectangle ( ) ; // 目标 Rect
rc . origin = vec2 . create ( this . x , this . y ) ;
rc . size = Size . create ( this . data . leftMargin , this . data .
topMargin ) ;
this . destRects . push ( rc ) ;

// 右上角
rc = new Rectangle ( ) ;
rc . origin = vec2 . create ( this . data . image . width - this .
data . rightMargin , 0 ) ;
rc . size = Size . create ( this . data . rightMargin , this . data .
topMargin ) ;
this . srcRects . push ( rc ) ;

rc = new Rectangle ( ) ;
rc . origin = vec2 . create ( this . right - this . data . rightMargin ,
this . y ) ;
rc . size = Size . create ( this . data . rightMargin , this . data .
topMargin ) ;
this . destRects . push ( rc ) ;

// 右下角
rc = new Rectangle ( ) ;
rc . origin = vec2 . create ( this . data . image . width - this .
data . rightMargin , this . data . image . height - this . data .
bottomMargin ) ;
rc . size = Size . create ( this . data . rightMargin , this . data .
bottomMargin ) ;
this . srcRects . push ( rc ) ;

rc = new Rectangle ( ) ;
rc . origin = vec2 . create ( this . right - this . data . rightMargin ,
this . bottom - this . data . bottomMargin ) ;
rc . size = Size . create ( this . data . rightMargin , this . data .
bottomMargin ) ;
this . destRects . push ( rc ) ;

// 左下角
rc = new Rectangle ( ) ;
rc . origin = vec2 . create ( 0 , this . data . image . height - this .
data . bottomMargin ) ;
rc . size = Size . create ( this . data . leftMargin , this . data .
bottomMargin ) ;
this . srcRects . push ( rc ) ;

rc = new Rectangle ( ) ;
rc . origin = vec2 . create ( this . x , this . bottom - this . data .
bottomMargin ) ;
rc . size = Size . create ( this . data . leftMargin , this . data .
```

```
bottomMargin ) ;
this . destRects . push ( rc ) ;

// 左边
rc = new Rectangle ( ) ;
rc . origin = vec2 . create ( 0 , this . data . topMargin ) ;
rc . size = Size . create ( this . data . leftMargin , this .data .
image . height - this . data . topMargin - this . data . bottomMargin ) ;
this . srcRects . push ( rc ) ;

rc = new Rectangle ( ) ;
rc . origin = vec2 . create ( this . x , this . y + this . data .
topMargin ) ;
rc . size = Size . create ( this . data . leftMargin , this . height
- this . data . topMargin - this . data . bottomMargin ) ;
this . destRects . push ( rc ) ;

// 上边
rc = new Rectangle ( ) ;
rc . origin = vec2 . create ( this . data . leftMargin , 0 ) ;
rc . size = Size . create ( this . data . image . width - this . data .
leftMargin - this . data .rightMargin , this . data . topMargin ) ;
this . srcRects . push ( rc ) ;

rc = new Rectangle ( ) ;
rc . origin = vec2 . create ( this . x + this . data . leftMargin ,
this . y ) ;
rc . size = Size . create ( this . width - this . data . leftMargin
- this . data . rightMargin , this . data . topMargin ) ;
this . destRects . push ( rc ) ;

// 右边
rc = new Rectangle ( ) ;
rc . origin = vec2 . create ( this . data . image . width - this .data .
rightMargin , this .data . topMargin ) ;
rc . size = Size . create ( this . data . rightMargin , this . data .
image . height - this . data . topMargin - this . data . bottomMargin ) ;
this . srcRects . push ( rc ) ;

rc = new Rectangle ( ) ;
rc . origin = vec2 . create ( this . right - this . data . rightMargin ,
this . y + this . data . topMargin ) ;
rc . size = Size . create ( this . data . rightMargin , this . height
- this .data . topMargin - this . data . bottomMargin ) ;
this . destRects . push ( rc ) ;

// 下边
rc = new Rectangle ( ) ;
rc . origin = vec2 . create ( this . data . leftMargin , this . data .
image . height - this .data . bottomMargin ) ;
rc . size = Size . create ( this . data . image . width - this . data .
leftMargin - this . data . rightMargin , this . data . bottomMargin ) ;
this . srcRects . push ( rc ) ;
```

```typescript
        rc = new Rectangle ( ) ;
        rc . origin = vec2 . create ( this . x + this . data . leftMargin ,
        this . bottom - this . data . bottomMargin ) ;
        rc . size = Size . create ( this . width - this . data . leftMargin
        - this . data . rightMargin , this . data . bottomMargin ) ;
        this . destRects . push ( rc ) ;

        // 中心
        rc = new Rectangle ( ) ;
        rc . origin = vec2 . create ( this . data . leftMargin , this . data .
        topMargin ) ;
        rc . size = Size . create ( this . data . image . width - this . data .
        leftMargin - this . data . rightMargin , this . data . image . height
        - this .data . topMargin - this .data . bottomMargin ) ;
        this . srcRects . push ( rc ) ;

        rc = new Rectangle ( ) ;
        rc . origin = vec2 . create ( this . x + this . data . leftMargin ,
        this . y + this . data . topMargin ) ;
        rc . size = Size . create ( this . width - this . data . leftMargin
        - this . data . rightMargin , this . height - this .data . topMargin
        - this .data . bottomMargin ) ;
        this . destRects . push ( rc ) ;
    }

    private _drawImage ( context : CanvasRenderingContext2D , img :
    HTMLImageElement | HTMLCanvasElement , destRect : Rectangle , srcRect :
    Rectangle ,fillType :EImageFillType = EImageFillType .STRETCH ):boolean {
        // 该方法实现请参考 4.3.3 节实现加强版的 drawImage 方法中的内容
    }

    // 遍历绘制 9 个区块
    public draw ( transformable : ITransformable , state : IRenderState ,
    context : CanvasRenderingContext2D ) : void {
        for ( let i : number = 0 ; i < this . srcRects . length ; i ++ ) {
            this . _drawImage ( context , this . data . image , this . destRects
            [ i ] , this . srcRects [ i ] , EImageFillType . STRETCH ) ;
        }
    }
}
```

8.3.7 SpriteFactory 生产 IShape 产品

前面已经在 SpriteFactory 类中生产了 ISprite 接口和 Line 类型的 IShape 接口。现在继续遵循上述模式生产已经实现的各种 IShape。具体代码如下：

```typescript
public static createGrid ( w : number , h : number , xStep : number = 10,
yStep : number = 10 ) : IShape {
    return new Grid ( w , h , xStep , yStep ) ;
}
public static createCircle ( radius : number ) : IShape {
    return new Circle ( radius ) ;
}
```

```
public static createRect ( w : number , h : number , u : number = 0 , v :
number = 0 ) : IShape {
    return new Rect ( w , h , u , v ) ;
}
public static createEllipse (radiusX : number , radiusY : number ): IShape {
    return new Ellipse ( radiusX , radiusY ) ;
}
public static createPolygon ( points : vec2 [ ] ) : IShape {
    if ( points . length < 3 ) {
        throw new Error ( "多边形顶点数量必须大于或等于3!!!") ;
    }
    return new ConvexPolygon ( points ) ;
}
public static createScale9Grid ( data : Scale9Data , width : number ,
height : number , u : number = 0 , v : number = 0 ) : IShape {
    return new Scale9Grid ( data , width , height ,u , v ) ;
}
```

到此为止，非场景图类型的精灵系统全部完成。下一节将通过一个 Demo 来测试一下成果。

8.4 精灵系统测试 Demo

本节中将实现一个名为 IShape_Event_Hittest_Draw_Test_Demo 的演示程序，下面先来看一下演示效果，具体如图 8.6 所示。

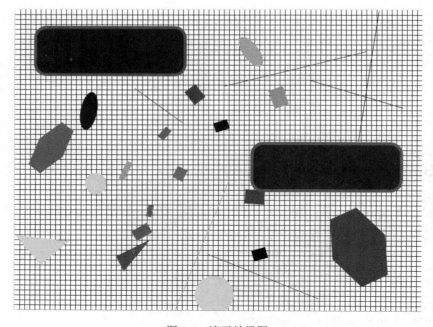

图 8.6 演示效果图

需求描述：
- 当单击鼠标右键后，在单击处产生一个使用不同 IShape 对象的不同缩放值不同方位、具有随机颜色的精灵。
- 除了 IShape 类型为 Line、Circle 和 Scale9Grid 之外的所有精灵都自动旋转。
- 能够通过鼠标左键按下时点选精灵，当选中精灵时，呈现红色，并且可以拖动到任意位置。

需求就是这么简单，基本涵盖了精灵系统现存的各个功能点，下一节将介绍实现过程。

8.4.1 Demo 的运行流程

先来看一下 IShape_Event_Hittest_Draw_Test_Demo 演示的成员变量及构造函数。具体代码如下：

```
class IShape_Event_Hittest_Draw_Test_Demo {
    // 引用一个 Sprite2DApplication 对象
    private _app : Sprite2DApplication ;

    // 用于九宫缩放的图像对象
    private _image : HTMLImageElement ;
    // 将添加的 IShape 存储在该数组中
    private _shapes : IShape [ ] = [ ] ;
    // 用来追踪添加的 IShape 的索引号
    private _idx : number ;
    // 下面的变量用来记录选中精灵原本的颜色
    // 需求是当选中某个精灵后，该精灵变成红色
    // 当鼠标被释放后，要恢复该精灵原本的颜色
    private _lastColor : string | CanvasGradient | CanvasPattern ;
                               // 追踪上一次的颜色
    public constructor ( app : Sprite2DApplication ) {
        this . _lastColor = 'red' ;
        this . _app = app ;
        this . _idx = 0 ;         // 这个变量表示当前的坦克精灵位于哪条曲线段上面，
                                  // 初始化时，将坦克精灵设置在第一条曲线段上

        // 创建 HTMLImageElement 对象
        this . _image = document . createElement ( 'img' ) as HTMLImageElement ;
        // 设置要载入的图片 URL 路径
        this . _image . src = "./data/scale9.png" ;
        // 使用箭头函数后，this 指向当前类
        this . _image . onload = ( ev : Event ) : void => {
            // 创建享元类 IShape 接口
            this . _createShapes ( ) ;
            // 创建 grid 及初始化
            this . createSprites ( ) ;
```

```
            // 启动 Apllication 的动画循环
            this._app.start();
        }
    }
}
```

通过演示程序的构造函数，可以知道整个 Demo 的运行流程：首先载入九宫缩放使用图片，在图片全部载入完成后，通过 createShapes 方法创建所有需要用到的享元 IShape 对象，然后根据创建的 IShape 对象来创建各个精灵（createSprites 方法），最后调用 Application 类的 start 方法，启动动画循环。

8.4.2　创建各种 IShape 对象

接下来看一下 createShapes 方法，该方法创建目前已经实现的所有 IShape 对象。具体代码如下：

```
private _createShapes() : void {
    // 清空容器
    this._shapes = [];
    // 创建一条原点在起点的线段
    this._shapes.push(SpriteFactory.createLine(vec2.create(0, 0), vec2.create(100, 0)));
    // 创建一条原点在中心的线段
    this._shapes.push(SpriteFactory.createXLine(100, 0.5));
    // 创建原点在不同位置的 Rect
    this._shapes.push(SpriteFactory.createRect(10, 10));
    this._shapes.push(SpriteFactory.createRect(10, 10, 0.5, 0.5));
    this._shapes.push(SpriteFactory.createRect(10, 10, 0.5, 0));
    this._shapes.push(SpriteFactory.createRect(10, 10, 0, 0.5));
    this._shapes.push(SpriteFactory.createRect(10, 10, -0.5, 0.5));
    // 创建圆与椭圆
    this._shapes.push(SpriteFactory.createCircle(10));
    this._shapes.push(SpriteFactory.createEllipse(10, 15));
    let points : vec2[];
    // 创建三角形来测试
    points = [vec2.create(0, 0), vec2.create(20, 0), vec2.create(20, 20)];
    this._shapes.push(SpriteFactory.createPolygon(points));
    // 创建六边形来测试
    points = [vec2.create(20, 0),
        vec2.create(10, 20),
        vec2.create(-10, 20),
        vec2.create(-20, 0),
        vec2.create(-10, -20),
        vec2.create(10, -20)];
    this._shapes.push(SpriteFactory.createPolygon(points));
    // 创建九宫缩放 IShape
    let data : Scale9Data = new Scale9Data(this._image, new Inset(30, 30, 30, 30));
```

```
    // 创建原点位于九宫缩放图像中心的精灵
    this . _shapes . push ( SpriteFactory . createScale9Grid ( data , 300 ,
    100 , 0.5 , 0.5 ) ) ;
}
```

createShapes 方法非常简单，仅仅是创建已经实现过的 IShape 对象，如果还有其他需求，可以从 BaseShape2D 继承，实现自己想要的 IShape 实现类。

8.4.3　创建网格精灵和事件处理函数

有了上述的 IShape 数据列表后，可以通过 createSprites 方法来创建相应的各种 ISprite 对象了。具体代码如下：

```
private createSprites ( ) : void {
    // 创建与画布相同尺寸的背景 grid 形体和精灵
    let grid : IShape = SpriteFactory . createGrid ( this . _app . canvas .
    width , this . _app . canvas . height ) ;
    let gridSprite : ISprite = SpriteFactory . createSprite ( grid , 'grid' ) ;
    // 白色填充，黑色网格线
    gridSprite . fillStyle = "white" ;
    gridSprite . strokeStyle = 'black';
    // 将精灵添加到 ISpriteContainer 中
    this . _app . rootContainer . addSprite ( gridSprite ) ;
    // 设置精灵的鼠标事件处理函数
    gridSprite . mouseEvent =( s : ISprite ,evt :CanvasMouseEvent):void => {
        // 如果_shapes 数组为空，说明还没创建好形体
        if ( this . _shapes . length === 0 ) {
            return ;
        }
        // gridSprite 仅对鼠标右键单击做出响应，不响应鼠标中键和左键
        if ( evt . button === 2 )
        {
            // gridSprite 仅仅响应鼠标右键的 MouseUp 事件
            if ( evt . type === EInputEventType . MOUSEUP ) {
                // 使用%取模操作，获得周期性索引 [ 0 , 1 , 2 … , 0 , 1 , 2 , … ]
                this . _idx = this . _idx % this . _shapes . length ;
                // 获得索引指向的享元 IShape 对象
                // 然后生成使用该 IShape 的精灵
                let sprite : ISprite = SpriteFactory . createSprite ( this .
                _shapes [ this . _idx ] ) ;
                // 让精灵位于鼠标单击处的位置
                sprite . x = evt . canvasPosition . x ;
                sprite . y = evt . canvasPosition . y ;
                // 对于非 Scale9Grid 类型的 IShape 对象，则随机获得-180°～180°
                之间的方位
                // 对于 Scale9Grid 类来说，旋转后会在 9 个区域内的各自拼接处形成一条
                可见的线段，目前暂时没有解决方案，因此直接不让 Scale9Grid 进行旋转操作
                if ( sprite . shape . type !== "Scale9Grid" ) {
```

```
                sprite . rotation = Math2D . random ( -180 , 180 ) ;
            }
            // default 情况下，设置精灵的渲染类型为 FILL
            sprite . renderType = ERenderType . FILL ;
            // 如果是线段，为了点选测试方便，不让其每次更新时都旋转
            if ( this . _shapes [ this . _idx ] . type === 'Line' ) {
                                                            // 线段
                // 其渲染类型为描边
                sprite . renderType = ERenderType . STROKE ;
                // 仅 x 轴方向随机缩放 1～2 倍
                sprite . scaleX = Math2D . random ( 1 , 2 ) ;
                                                            // 仅 x 轴缩放
                // 随机选择描边颜色
                sprite . strokeStyle = Canvas2DUtil . Colors [ Math . floor
                (Math2D .random (3 ,Canvas2DUtil . Colors . length - 1))] ;
            } else {
                // 随机选择填充颜色
                sprite . fillStyle = Canvas2DUtil . Colors [ Math . floor
                ( Math2D . random (3 ,Canvas2DUtil.Colors . length - 1))] ;
                // 对于圆圈来说，等比缩放，否则大部分情况下会变成椭圆
                if ( this . _shapes [ this . _idx ] . type === 'Circle' )
                {                                           // 圆圈等比缩放
                    let scale : number = Math2D . random ( 1 , 3 ) ;
                    sprite . scaleX = scale ;
                    sprite . scaleY = scale ;
                } else if ( this . _shapes [ this . _idx ] . type !==
                'Scale9Grid' ) {
                    // 对于非 Scale9Grid 来说，随机非等比缩放 1～3 倍
                    sprite . scaleX = Math2D . random ( 1 , 3 ) ;
                    sprite . scaleY = Math2D . random ( 1 , 3 ) ;
                }
            }
            // 对于每个精灵，都挂接两个事件回调函数
            sprite . mouseEvent = this . mouseEventHandler . bind ( this ) ;
            sprite . updateEvent = this . updateEventHandler . bind
            ( this ) ;
            // 将其新创建的精灵放入 ISpriteContainer 容器中
            // 这样就能自动进行命令分发操作
            this . _app . rootContainer . addSprite ( sprite ) ;
            // idx 指向下一个要创建的 IShape 索引
            this . _idx ++ ;
        }
      }
   }
}
```

上述代码中主要做了两件事，即创建一个网格作为背景的精灵，并以该网格精灵作为鼠标右键 MOUSEUP 事件的接收方，当在该网格精灵上单击鼠标右键并弹起后，会根据 _shapes 数组中的内容，周而复始地创建各种形体的精灵。

8.4.4 非网格精灵的事件处理函数

在 createSprites 方法中创建的精灵都绑定了（bind）了两个事件处理函数，本节将介绍这两个事件函数的内容。先来看一下 UpdateEvent 事件处理函数。具体代码如下：

```
private updateEventHandler ( s : ISprite , mesc : number , diffSec : number ,
order : EOrder ) : void {
    // update 分发时, 先从前到后, 然后再从后到前触发两次 update 事件
    // 仅仅处理从前到后的 update 事件
    if ( EOrder . POSTORDER ) {
        return ;
    }
    if ( s . shape . type !== 'Circle' && s . shape . type !== 'Line' &&
    s . shape . type !== 'Scale9Grid')
    {
        s . rotation += 100 * diffSec ;
    }
}
```

上述代码很简单，对于 Circle 类型的精灵来说，因为圆的旋转毫无意义，根本无法分辨出一个圆是否在旋转，因此就不让它进行旋转了。

而对于 Line 精灵来说，由于其线宽仅有 1 个像素，如果一直旋转，则很难选用鼠标精确选中该精灵，因此也在 updateEvent 事件处理函数中过滤掉 Line 精灵。

对于 Scale9Grid 精灵来说，一旦发生旋转后会导致九宫格拼接处出现虚线效果，如图 8.7 所示。

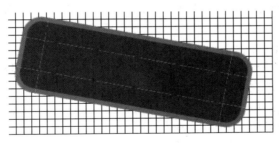

图 8.7 旋转导致九宫拼接处出现虚线效果

上图出现的这种问题，到目前也没有较好的解决方案，因此我们采取最简单的方式，在 updateEvent 中直接过滤掉该类型的精灵。

接下来看一下鼠标事件处理代码，具体如下：

```
private mouseEventHandler (s : ISprite ,evt :CanvasMouseEvent ) : void {
    // 这里 button 必须为 0, 否则出问题
    if ( evt . button === 0 ) {
        // 左键 down 事件处理
        if ( evt . type === EInputEventType . MOUSEDOWN ) {
            // 对于 Line 类型的精灵来说, 选中后变成红色, 并且线宽设置为 10 个单位
```

```
                if ( s . shape . type === "Line" ) {
                    this . _lastColor = s . strokeStyle ;   // 记录选中前精灵的颜色
                    s . strokeStyle = 'red' ;
                    s . lineWidth = 10 ;
                } else {   // 非 Line 类型精灵，则选中变成红色
                    this . _lastColor = s . fillStyle ;
                    s . fillStyle = 'red' ;
                }
            } // 左键 up 事件处理
            else if ( evt . type === EInputEventType .MOUSEUP ) {
                // 对于 Line 类型精灵来说，up 后恢复到线宽为 1，原来的颜色
                if ( s . shape . type === "Line" ) {
                    s . lineWidth = 1 ;
                    s . strokeStyle = this . _lastColor ;
                } else {   // 恢复到原来的颜色
                    s . fillStyle = this . _lastColor ;
                }
            } // 左键 drag 事件处理
            else if ( evt . type === EInputEventType . MOUSEDRAG ) {
                // 精灵跟随鼠标位置移动
                s . x = evt . canvasPosition . x ;
                s . y = evt . canvasPosition . y ;
            }
        }
    }
```

8.4.5 Demo 的入口代码

到目前为止，已经完成了整个 Demo 的源码分析。要让这个 Demo 运行起来，需要做如下步骤：

```
//获取 Canvas 对象
let canvas : HTMLCanvasElement | null = document . getElementById ('canvas')
as HTMLCanvasElement ;
// 根据 Canvas 对象创建 App 对象
let app : Sprite2DApplication = new Sprite2DApplication ( canvas ) ;
// 通过 App 对象创建测试 Demo，Demo 内部会调用 App 的 start 方法进行动画循环状态
new IShape_Event_Hittest_Draw_Test_Demo ( app ) ;
```

当使用 F5 快捷键启动 Debug 模式，就能按照需求的描述来运行应用程序了。

8.5 本章总结

在本章的 8.1 节中，以自上向下的顺序介绍了精灵系统的整体架构及如下一些主要接口：
- IDispatcher 接口用来分发精灵的更新、绘制和事件等相关操作。
- ISpriteContainer 接口用来增、删、改、查以及管理精灵。

- ISprite 接口扩展自 IRenderState 和 ITransformable 接口，ISprite 接口具有一个数学变换系统，并且保存当前的所有渲染状态（保留模式），能够进行点选、更新、绘制操作，以及对鼠标、键盘和渲染事件做出必要的响应。
- IShape 接口扩展自 IDrawable 和 IHittable 接口，充当渲染数据源的作用，通过 ISprite 和 IShaper 接口的配合，实现了享元设计模式。

在 8.2 节中，以自下而上的顺序实现了精灵系统中除 IShape 之外的所有接口。在 8.3 节中，则实现了 IShape 接口的所有子类，除了 Scale9Grid 子类外，其他基本形体子类在前面章节中都有所涉及。最后的 8.4 节中通过一个 Demo 演示来测试精灵系统的各个功能。

第 9 章 优美典雅的树结构

在日常生活中，经常会遇到树的概念。例如，在计算机科学中，最熟悉的树的用法就是文件系统。文件被放在目录（文件夹）中，而这些目录（文件夹）又被递归定义为子目录（子文件夹）；公司组织架构也是经典的树结构应用；在编译器或解释器中，将所有文本字符串解析成抽象语法树，然后进行后续操作。

对于游戏开发或计算机图形编程来说，树结构可能也是应用最广泛的数据结构之一。可以使用树结构来表示场景图渲染系统、角色动画中的骨骼系统、具有层次性的空间分割系统或者是层次包围体系统等，可以说树结构无处不在。

由于树结构是如此的重要，并且在后续章节中，将要用到树结构来实现场景图渲染系统，因此在本章中会详细地介绍树结构的相关概念，以及各种操作，为后续章节打下坚实的基础。

9.1 树的数据结构

关于树结构的相关内容，将会分为三个部分来介绍。本节中将会讲解树结构中节点的创建、删除，以及各种层次关系的查询操作。然后在下一节中介绍树结构的各种遍历算法。最后介绍树结构如何进行序列化和反序列化操作。

9.1.1 树结构简介

先来看一下树结构的表现形式，具体如图 9.1 所示。

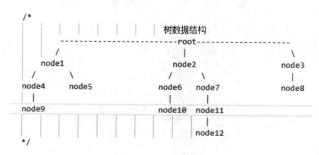

图 9.1 树数据结构

根据图 9.1，可以得知以下内容。

（1）就一颗树（Tree）而言，有且只有一个根节点（Root Node），这里用 root 指代根节点。

（2）对于根节点 root 来说，它有三个直接的儿子节点（Child Nodes），它们分别是 [node1，node2，node3]。依此类推，node1 的直接儿子节点为 [node4，node5]。

（3）以 node1 节点为例，它的父亲节点（Parent Node）是根节点 root。同样地，如果是 node10，它的父亲节点就是 node6。比较特别的是，根节点 root 没有父亲节点。

（4）可以将某个节点以下的所有节点称为该节点的子孙节点（Descendant Nodes），例如以 node2 节点为例，它的直接儿子节点为 [node6，node7]，它的子孙节点是 [node6，node7，node10，node11，node12]。

（5）以 node10 节点为例，[node6，node2，root] 节点都是其祖先节点（Ancestor Nodes）。

（6）以 node2 节点为例，它有两个儿子节点 [node6，node7]，那么 node6 和 node7 之间是兄弟节点（Sibling Nodes）的关系，它们有共同的父亲节点 node2。

（7）对于 [node9，node10，node12，node8] 这 4 个节点来说，它们没有儿子节点，因此可以把它们看作叶子节点（Leaf Nodes）。

（8）关于节点的深度（Depth）和高度（Height）比较抽象，下面罗列一些比较典型的节点的相关信息来了解一下。

- 根节点 root 的深度为 0（root 没有祖先节点），高度为 4（root～node12 之间的节点个数）。
- 节点 node1 的深度为 1（node1～root 之间的节点个数），高度为 2（node1～node9 之间的节点个数）。
- 节点 node12 的深度为 4（node12～root 之间的节点个数），高度为 0（因为 node12 是没有儿子节点的叶子节点）。

通过上述三个典型节点的高度和深度相关信息，可以知道：

- 约定当前节点为第 0 层。
- 节点的深度是从当前节点到根节点的层数（从下到上）。
- 节点的高度是从当前节点到最大层数的叶子节点之间的距离（从上到下）。

要在计算机中表示出上面描述的树的层次性结构，一般要使用在内存中方便寻址到父亲节点和所有儿子节点的存储方式，因此可以使用如下方式存储树节点：

```
export class TreeNode < T > {
    // 父子节点的存储方式：
    private _parent : TreeNode < T > | undefined ;
    // 指向当前节点的父亲节点，如果 this 是根节点，则指向 undefined，因为根节点肯定没有父亲节点
    private _children : Array< TreeNode < T > > | undefined ;
    // 数组中保存所有直接儿子节点，如果是叶子节点，则 _children 为 undefined 或 _children . length = 0
```

```
    public name : string ;
      // 当前节点的名称，有利于 debug 时信息输出或按名字获取节点（集合）操作
    public data : T | undefined ;
      // 一个泛型对象，指向一个需要依附到当前节点的对象
}
```

上述的存储结构，可以非常方便、高效地获取当前节点的父亲节点和所有儿子节点，而且使用泛型方式，可以给当前节点附加各种类型的对象，灵活易用。

9.1.2 树节点添加时的要点

假设要给当前的某个节点添加一个儿子节点，第一个反应是：这是一个很简单的操作，只要将一个儿子节点添加到当前节点的_children 数组中去，然后设置儿子节点的父亲节点_parent 为当前节点。

但是添加儿子节点真的这么简答吗？答案是：未必！

参考图 9.1，假设当前节点为 node4，它的祖先为[node1 , root]，然后来看一下添加节点的情况：

- 如果 node4 要添加一个新创建的节点，例如名为 node13，则上面描述的算法没有任何问题。
- 如果在 node4 中要添加 node1 或 root 节点作为儿子节点，由于 node1 和 root 已经是 node4 的祖先节点了，这会导致循环引用，从而让整个程序崩溃。因此在为某个节点添加子节点时，必须检查要添加的子节点是否是当前节点的祖先节点。
- 考虑另外一种情况，假设当前节点仍旧是 node4，要添加 node5 作为子节点。此时 node5 并不是 node4 的祖先节点，但是 node5 已经有父节点 node1。在这种情况下，选择的处理方式是将 node5 先从 node1 的_children 列表中移除，然后重新添加到 node4 中作为儿子节点。

通过上述分析，可以将要添加的子节点分三种情况处理：

- 首先判断要添加的儿子节点如果是当前节点的祖先节点，则不需要做任何处理，直接退出操作即可。
- 然后判断要添加的儿子节点是否有父亲节点，如果有父亲节点，需要先将儿子节点从父亲节点中移除，再添加到当前节点中。
- 如果要添加的子节点是新创建的节点，没有父亲节点，也不会有循环引用，则使用标准处理方式。

9.1.3 树节点 isDescendantOf 和 remove 方法的实现

要实现上述的子节点添加算法，需要先实现一些必要的辅助方法。

首先实现一个名为 isDescendantOf 的方法，该方法用来判断一个要添加的子节点是否

是当前节点的祖先节点。具体代码如下：

```
/**
 *
 * @param ancestor { TreeNode < T > | undefined } 用于测试参数 ancestor 是 this 节点的祖先
 * @returns { boolean } 如果 this 节点是 ancestor 的子孙节点，返回 true，否则，返回 false
 */
public isDescendantOf ( ancestor : TreeNode < T >|undefined ): boolean {
    //undefined 值检查
    if ( ancestor === undefined ) {
        return false ;
    }
    //从当前节点的父亲节点开始向上遍历
    let node : TreeNode < T > | undefined = this . _parent ;
    for ( let node : TreeNode < T > | undefined = this . _parent ; node !== undefined ; node = node . _parent ) {
        //如果当前节点的祖先等于 ancestor，说明当前节点是 ancestor 的子孙，返回 true
        if ( node === ancestor ) {
            return true ;
        }
    }
    //否则遍历完成，说明当前节点不是 ancestor 的子孙，返回 false
    return false ;
}
```

然后来看移除某个节点的源码，代码如下：

```
/**
 *
 * @param index { number } 要移除的子节点的索引号
 * @returns { TreeNode<T> | undefined } 如果移除成功，返回的就是参数 child 节点，否则，为 undefined
 */
public removeChildAt ( index : number ): TreeNode < T > | undefined {
    //由于使用延迟初始化，必须要进行 undefined 值检查
    if ( this . _children === undefined ) {
        return undefined ;
    }
    //根据索引从_children 数组中获取节点
    let child : TreeNode<T> | undefined = this . getChildAt ( index ) ;
    //索引可能会越界，这是在 getChildAt 函数中处理的
    //如果索引越界了，getChildAt 函数返回 undefined
    //因此必须要进行 undefined 值检查
    if ( child === undefined ) {
        return undefined;
    }
    /*
    TypeScript / js splice 方法：向/从数组中添加/删除项目，然后返回被删除的项目
    参数：
    index 必需。整数，规定添加/删除项目的位置，使用负数可从数组结尾处规定位置
    howmany 必需。要删除的项目数量。如果设置为 0，则不会删除项目
```

```
        item1 , ... , itemX 可选。向数组添加的新项目
        这里使用了 index 和 howmany 这两个参数,含义是:将 index 处的元素删除
        */
        this . _children . splice( index , 1 ) ;        // 从子节点列表中移除
        child . _parent = undefined ;         // 将子节点的父亲节点设置为 undefined
        return child ;
}
```

上述 removeChildAt 方法是通过索引来定位要删除的子节点。有时候已经获得了一个 TreeNode < T >类型的引用,此时需要删除的是该引用,那么可以实现一个更加方便的方法。具体代码如下:

```
/**
 *
 * @param child { TreeNode < T > | undefined } 要移除的子节点
 * @returns { TreeNode < T > | undefined } 如果移除成功,返回的就是参数 child 节点,否则为 null
 */
public removeChild (child : TreeNode < T > |undefined ) : TreeNode < T > | undefined {
    // 参数为 undefined 的处理
    if ( child == undefined ) {
        return undefined;
    }
    // 如果当前节点是叶子节点的处理
    if ( this._children === undefined ) {
        return undefined;
    }
    // 由于使用数组线性存储方式,从索引查找元素是最快的
    // 但是从元素查找索引,必须遍历整个数组
    let index : number = -1 ;
    for ( let i = 0 ; i < this . _children . length ; i++ ) {
        if ( this . getChildAt ( i ) === child ) {
            index = i; // 找到要删除的子节点,记录索引
            break ;
        }
    }
    //没有找到索引
    if ( index === -1 ) {
        return undefined ;
    }
    //找到要移除的子节点的索引,那么就调用 removeChildAt 方法
    return this . removeChildAt ( index ) ;
}
```

remove 的第三个版本,该方法将 this 节点从父亲节点中删除。具体代码如下:

```
/**
 *
 * 将 this 节点从父节点中移除
 * @returns { TreeNode < T > | undefined } 如果移除成功,返回的就是 this 节点
 */
public remove ( ) : TreeNode<T> | undefined {
```

```ts
    if ( this . _parent !== undefined ) {
       return this . _parent . removeChild ( this ) ;
    }
    return undefined ;
 }
```

9.1.4 实现添加树节点方法

有了 isDescendantOf 方法和 remove 相关方法,就能实现添加子节点的方法了。具体代码如下:

```ts
/**
 *
 * @param child { TreeNode < T > } 要添加的子节点
 * @param index { number } 要添加到的索引位置
 * @returns { TreeNode < T > | undefined } 如果添加子节点成功,返回true,否则,返回false
 */
public addChildAt (child : TreeNode<T> ,index : number ):TreeNode < T > | undefined {
    // 第一种情况:要添加的子节点是当前节点的祖先的判断
    // 换句话说就是,当前节点已经是child节点的子孙节点,这样会循环引用,那就直接退出方法
    if ( this . isDescendantOf ( child ) ) {
       return undefined ;
    }
    //延迟初始化的处理
    if ( this . _children === undefined ) {
       //有两种方式初始化数组,笔者推荐[ ]方式,可以少写一些代码
       this . _children = [ ] ;
       // this._children = new Array<TreeNode<T>>();
    }
    //索引越界检查
    if ( index >= 0 && index <= this . _children . length ) {
       if ( child . _parent !== undefined ) {
          //第二种情况:要添加的节点是有父亲节点的,需要从父亲节点中remove掉
          child . _parent . removeChild ( child ) ;
       }
       //第三种情况: 要添加的节点不是当前节点的祖先并且也没有父亲节点(新节点或已从父亲移除)
       //设置父亲节点并添加到_children中
       child . _parent = this ;
       this . _children . splice ( index , 0 , child ) ;
       return child ;
    }
    else {
       return undefined ;
    }
 }
 /**
  * 简便方法,在子列表最后添加一个儿子节点
  * @param child { TreeNode < T > } 要添加的子节点
```

```
 * @returns { TreeNode < T > | undefined} 如果添加子节点成功，返回 true，否则，
返回 false
 */
public addChild ( child: TreeNode < T > ) : TreeNode<T> | undefined {
    if ( this . _children === undefined ) {
        this . _children = [ ] ;
    }
    //在列表最后添加一个节点
    return this . addChildAt ( child , this . _children . length ) ;
}
```

到此为止，实现了树结构的两个重要的操作，即树节点的添加和移除操作。这样就能使用编程方式生成一棵节点树，也能删除某个节点，甚至可以删除整棵节点树。

下面来看一下树节点的构造函数。具体代码如下：

```
/**
 * @param data { T | undefined } 设置要创建的树节点上依附的数据 T，默认为 undefined
 * @param parent { parent: TreeNode < T > | undefined = undefined } 设置
要创建的树节点的父亲节点，默认为 undefined
 * @param name { string } 设置要创建的树节点的名称，默认为空字符串
 */
public constructor(data:T | undefined = undefined,parent:TreeNode < T > |
undefined = undefined , name : string = "" ) {
    this . _parent = parent ;
    this . _children = undefined ;
    this . _name = name ;
    this . data = data ;
    // 如果有父亲节点，则将 this 节点添加到父亲节点的儿子列表中
    if ( this . _parent !== undefined ) {
        this . _parent . addChild ( this ) ;
    }
}
```

9.1.5　树结构的层次关系查询操作

在 9.1 节中，涉及树的很多层次关系相关的术语，例如根节点、父亲节点、儿子节点和兄弟节点等，那么本节中，将从代码的角度实现这些关系的查询。

（1）获取当前节点的父亲节点和儿子节点的相关信息和操作。具体代码如下：

```
public get parent () : TreeNode < T > | undefined {
    return this . _parent ;
}
// 从当前节点中获取索引指向的儿子节点
public getChildAt ( index: number ) : TreeNode < T > | undefined {
    if ( this . _children === undefined ) {
        return undefined ;
    }
    if ( index < 0 || index >= this . _children . length ) {
        return undefined ;
    }
```

```
      return this . _children [ index ] ;
   }
   // 获取当前节点的儿子个数
   public get childCount ( ) : number {
      if ( this . _children !== undefined ) {
         return this . _children . length;
      }
      else {
         return 0 ;
      }
   }
   // 判断当前节点是否有儿子节点
   public hasChild ( ) : boolean {
      return this . _children !== undefined && this . _children . length > 0 ;
   }
```

由于实现的树结构本身使用父亲节点和儿子数组的存储方式,因此获取这些属性的代码比较简单。

(2) 从当前节点获取根节点,以及获取当前节点的深度。具体代码如下:

```
public get root ( ) : TreeNode < T > | undefined {
   let curr: TreeNode < T > | undefined = this ;
   // 从 this 开始,一直向上遍历
   while ( curr !== undefined && curr . parent !== undefined ) {
      curr = curr . parent ;
   }
   // 返回 root 节点
   return curr ;
}
public get depth (): number {
   let curr : TreeNode < T > | undefined = this ;
   let level : number = 0 ;
   while ( curr !== undefined && curr . parent !== undefined ) {
      curr = curr . parent ;
      level++ ;
   }
    return level ;
}
```

可以看到,获取当前节点的根节点和获取当前节点的深度代码基本相同。

(3) 获取当前节点的 firstChild 和 lastChild。具体代码如下:

```
// 获取当前节点的第一个儿子节点
public get firstChild ( ) : TreeNode < T > | undefined {
   // 如果当前节点存在儿子节点,返回第一个儿子节点,否则,返回 undefined
   if ( this. _children !== undefined && this. _children . length > 0 ) {
      return this . _children [ 0 ] ;
   } else {
      return undefined ;
   }
}
// 获取当前节点的最后一个儿子节点
public get lastChild ( ) : TreeNode < T > | undefined {
   if ( this. _children !== undefined && this. _children. length > 0 ){
```

```
            return this._children[this._children.length - 1];
        } else {
            return undefined;
        }
    }
```

如图 9.1 所示，节点 node1 的 firstChild 是节点 node4，其 lastChild 是节点 node5。而节点 node4 的 firstChild 和 lastChild 都是节点 node9。

（4）获取当前节点左兄弟（prevSibling）和右兄弟（nextSibling）的操作。具体代码如下：

```
    public get nextSibling(): TreeNode<T> | undefined {
        // 没有父亲节点，肯定没有兄弟节点
        if (this._parent === undefined) {
            return undefined;
        }
        // 只有当前节点的父亲节点的儿子节点的数量大于 1 才说明有兄弟节点
        // 如果只有 1，就是 this 啦，不可能有兄弟节点
        if (this._parent._children !== undefined && this._parent._children.length > 1) {
            // 此时只说明可能有兄弟节点，还要知道是否有右兄弟节点
            // 先要知道当前节点在父亲节点的子节点列表中的索引号
            let idx: number = -1;
            for (let i = 0; i < this._parent._children.length; i++) {
                if (this === this._parent._children[i]) {
                    idx = i;
                    break;
                }
            }
            // idx 肯定不为-1
            // 如果 idx 不是父亲节点子节点列表中最后一个，说明有 nextSibling
            if (idx !== this._parent._children.length - 1) {
                return this._parent._children[idx + 1];
            } else {
                // 说明当前节点在父亲节点的子列表的最后一个，不可能有右兄弟节点了
                return undefined;
            }
        } else {
            return undefined;
        }
    }

    public get prevSibling(): TreeNode<T> | undefined {
        // 没有父亲节点，肯定没有兄弟节点
        if (this._parent === undefined) {
            return undefined;
        }
        // 只有当前节点的父亲节点的儿子节点的数量大于 1 才说明有兄弟节点
        // 如果只有 1，就是 this 啦，不可能有兄弟节点
        if (this._parent._children !== undefined && this._parent._children.length > 1) {
            // 此时只说明可能有兄弟节点，还要知道是否有右兄弟节点
```

```
        // 先要知道当前节点在父亲节点的子节点列表中的索引号
        let idx: number = - 1 ;
        for ( let i = 0 ; i < this . _parent . _children . length; i++ ) {
            if ( this === this . _parent . _children [ i ] ) {
                idx = i ;
                break ;
            }
        }
        // idx 肯定不为-1
        // 如果 idx 不是父亲节点子节点列表中最前一个的话,说明有 nextSibling
        if ( idx !== 0 ) {
            return this . _parent . _children [ idx - 1 ] ;
        } else {
            // 说明当前节点在父亲的子列表的最前一个,不可能有左兄弟了
            return undefined ;
        }
    } else {
        return undefined ;
    }
}
```

如图 9.1 所示,节点 node4 的 prevSibling 为 undefined,而它的 nextSibling 为节点 node5。同样地,节点 node5 的 prevSibling 为节点 node4,而它的 nextSibling 为 undefined。

(5) 获取当前节点最左侧(mostLeft)和最右侧(mostRight)的子孙节点。具体代码如下:

```
public get mostLeft ( ) : TreeNode < T > | undefined {
    let node : TreeNode < T > | undefined = this ;
    // 以深度优先方式,不断调用 firstChild,直到最后一个 firstChild,肯定是最左子孙节点
    while ( true ) {
        let subNode : TreeNode < T > | undefined = undefined ;
        if ( node !== undefined ) {
            // 调用 firstChild 只读属性
            subNode = node . firstChild ;
        }
        if ( subNode === undefined ) {
            break ;
        }
        node = subNode ;
    }
    return node ;
}
public get mostRight ( ) : TreeNode < T > | undefined {
    let node : TreeNode<T> | undefined = this ;
    // 以深度优先方式,不断调用 lastChild,直到最后一个 lastChild,肯定是最右侧子孙节点
    while ( true ) {
        let subNode : TreeNode < T > | undefined = undefined ;
        if ( node !== undefined ) {
            // 调用 lastChild 只读属性
            subNode = node . lastChild ;
        }
```

```
            if ( subNode === undefined ) {
                break ;
            }
            node = subNode ;
        }
        return node ;
    }
```

以图 9.1 为例，根节点 root 的最左侧子孙节点是 node7，最右侧子孙节点是 node8。同样地，如果当前节点是 node2，那么它的最左侧子孙节点是 node10，最右侧子孙节点是 node12。

会看到，通过上述这些只读属性或方法，就能查询树节点各种层次关系。而我们能唯一改变树节点层次关系的操作，只能是 addChild 和 removeChild 系列方法。

9.2　树数据结构的遍历

在上一节中了解了树节点的添加和移除操作，并实现了树节点各种层次查询操作，本节将关注树结构的遍历操作。很多重要的算法都是建立在树节点的遍历操作上，因此会介绍各种递归和非递归的遍历算法。

9.2.1　树结构遍历顺序

首先按照是广度优先（层次遍历）还是深度优先、是从上到下还是从下到上，以及是从左到右还是从右到左这三种遍历顺序来排列组合成如下 8 种形式的遍历算法，即：
- 广度优先（层次）、从上到下（先根）、从左到右的遍历算法。
- 广度优先（层次）、从上到下（先根）、从右到左的遍历算法。
- 广度优先（层次）、从下到上（后根）、从左到右的遍历算法。
- 广度优先（层次）、从下到上（后根）、从右到左的遍历算法。
- 深度优先、从上到下（先根）、从左到右的遍历算法。
- 深度优先、从上到下（先根）、从右到左的遍历算法。
- 深度优先、从下到上（后根）、从左到右的遍历算法。
- 深度优先、从下到上（后根）、从右到左的遍历算法。

接下来看一下广度优先（Breadth First）的四个遍历算法，参考图 9.1，会得到如下结果。

（1）广度优先（层次）、从上到下（先根）、从左到右遍历后得到如下所示的列表：

[root，node1，node2，node3，node4，node5，node6，node7，node8，node9，node10，node11，node12]。

如果将上面的列表以从右到左的方式来阅读，会发现这是广度优先（层次）、从下到上（后根）、从右到左的遍历顺序。

（2）广度优先（层次）、从上到下（先根）、从右到左的遍历后得到如下的列表：

[root，node3，node2，node1，node8，node7，node6，node5，node4，node11，node10，node9，node12]。

同样地，将上面的列表以从右到左的方式阅读，就是广度优先（层次）、从下到上（后根）、从左到右的遍历顺序。

再来看一下深度优先（Depth First）的四个遍历算法，具体结果如下所述。

（1）深度优先、从上到下（先根）、从左到右遍历后得到如下的列表：

[root，node1，node4，node9，node5，node2，node6，node10，node7，node11，node12，node3，node8]。

以相反的顺序读取上述列表，会得到深度优先、从下到上（后根）、从右到左的遍历顺序。

（2）深度优先、从上到下（先根）、从右到左遍历后得到如下的列表：

[root，node3，node8，node2，node7，node11，node12，node6，node10，node1，node5，node4，node9]。

同样地，以相反的顺序读取上面的列表，将会得到深度优先、从下到上（后根）、从左到右的遍历顺序。

综上所述，具体结果如下所述。

（1）排列组合，一共有八种遍历策略，如图 9.2 所示。

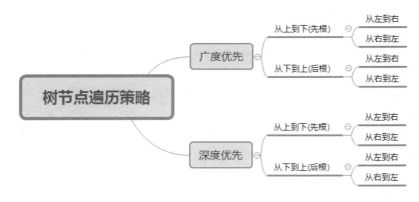

图 9.2 树节点遍历策略

（2）得到的遍历关系如下所述。
- 结论 1：广度或深度先根左右遍历的逆为广度或深度后根右左遍历。
- 结论 2：广度或深度先根右左遍历的逆为广度或深度后根左右遍历。

9.2.2 树结构线性遍历算法

先来看一下广度优先（层次）、从上到下（先根）、从左到右的非递归遍历算法，该算法需要使用队列数据结构。还是以如图 9.1 所示的树结构为例，来看一下算法的具体流程：

（1）初始化时，将 root 节点入队列，此时队列中元素有[root]。

（2）将 root 节点出队列，然后将 root 节点的所有儿子节点入队列，此时队列中的元素有[node1，node2，node3]。

（3）将 node1 节点出队列，然后将 node1 的儿子节点入队列，此时队列中的元素有[node2，node3，node4，node5]。

（4）将 node2 节点出队列，然后将 node2 的儿子节点入队列，此时队列中的元素有[node3，node4，node5，node6，node7]。

（5）不停地重复上述步骤，一直到队列中没有任何元素（也就是队列为空），表示遍历结束。

综上所述，得到的遍历关系如下。

- 结论 3：节点出队列的顺序就是遍历的顺序，上面的算法的遍历顺序是广度优先（层次）、从上到下（先根）、从左到右。
- 结论 4：如果将所有子节点以相反的方式入队列，那么将得到广度优先（层次）、从上到下（先根）、从右到左的遍历顺序。

再来看一下深度优先、从上到下（先根）、从右到左的非递归遍历算法，该算法需要使用栈数据结构来实现，还是以如图 9.1 所示的树结构为例，来看一下算法的具体流程：

（1）初始化时，将 root 节点入栈，此时栈中元素有[root]。

（2）将 root 节点出栈，然后将 root 节点的所有儿子节点入栈，此时栈中的元素有[node1，node2，node3]。

（3）将 node3 节点出栈，然后将 node3 的儿子节点入栈，此时栈中的元素有[node1，node2，node8]。

（4）将 node8 节点出栈，由于 node8 没有儿子节点，则没有任何节点入栈，此时栈中的元素有[node1，node2]。

（5）不停地重复上述步骤，一直到栈中没有任何元素（也就是栈为空），表示遍历结束。

综上所述，得到的遍历关系如下：

- 结论 5：出栈的顺序就是遍历的顺序，上面的算法的遍历顺序是深度优先、从上到下（先根）、从右到左。
- 结论 6：如果将所有子节点以相反的方式入栈，那么将得到深度优先、从上到下（先根）、从左到右的遍历顺序。

综合广度优先和深度优先的遍历算法，我们会发现这些算法的流程是完全相同的。

9.2.3 树结构遍历枚举器

在 2.3.4 节中曾经定义过一个泛型枚举器：IEnumerator < T >接口，而树结构的非递归遍历很适合使用枚举器模式，因此树结构枚举器将实现 IEnumerator < T >，从而用统一的迭代模式进行树结构遍历。

如果要实现全部的树结构遍历，需要实现八个枚举器，但是根据上述的描述与 6 个结论，会发现如下几个策略。

（1）广度优先（层次）遍历使用队列数据结构，深度优先使用栈数据结构，它们除了使用的容器对象不同外，所有的算法流程没有区别，因此将队列结构和栈接口适配成统一的 IAdapter 接口，以泛型的方式传入枚举器，那么就可以将八个枚举器缩减为 4 个枚举器：
- 从上到下（先根）_从左到右_枚举器 < 容器适配器 IAdapter > 。
- 从上到下（先根）_从右到左_枚举器 < 容器适配器 IAdapter > 。
- 从下到上（后根）_从左到右_枚举器 < 容器适配器 IAdapter > 。
- 从下到上（后根）_从右到左_枚举器 < 容器适配器 IAdapter > 。

（2）如果将从左到右和从右到左的算法使用一个回调函数 Indexer 来表示，以泛型的方式传入枚举器，那么又可以将四个枚举器减少到两个枚举器，如下：
- 从上到下（先根）_枚举器 < 容器适配器 IAdapter , Indexer > 。
- 从下到上（后根）_枚举器 < 容器适配器 IAdapter , Indexer > 。

（3）根据上一节的结论 1 和结论 2，知道从上到下（先根）遍历的逆就是后根遍历，那么只要实现从上到下（先根）枚举器，然后将从上到下（先根）枚举器包装一下，就可以成为从下到上（后根）枚举器。因此还是实现两个枚举器，但是后根枚举器依赖于先根枚举器。

9.2.4　树结构枚举器的实现

根据上一节的介绍，先来定义 IAdapter 接口，该接口用来适配栈和队列的数据结构，将它们的操作统一成一种调用形式，很经典的适配器设计模式，来看一下该接口的定义。具体代码如下：

```
export interface IAdapter < T > {
    add ( t : T ) : void ;                  // 将 t 入队列或堆栈
    remove ( ) : T | undefined ;            // 弹出队列或堆栈顶部的元素
    clear ( ) : void ;                      // 清空队列或堆栈，用于重用
    //属性
    length : number ;                       // 当前队列或堆栈的元素个数
    isEmpty : boolean ;                     // 判断当前队列或堆栈是否为空
}
```

关于如何实现 IAdapter 接口，在本章的最后一节来实现，目前只要知道该接口的签名方法，就可以以面向接口的编程方式来推进源码解析。

然后，声明 Indexer 函数签名并实现两个具体的回调方法。具体代码如下：

```
//回调函数类型定义
export type Indexer = ( len : number , idx : number ) => number ;
// 实现获取从左到右的索引号
```

```typescript
export function IndexerL2R ( len : number , idx : number ) : number {
    return idx ;
}
// 实现获取从右到左的索引号
export function IndexerR2L ( len : number , idx : number ) : number {
    return ( len - idx - 1 ) ;
}
```

有了上面的 IAdapter 接口和 Indexer 回调函数类型签名后，就可以从泛型枚举器接口继承实现先根枚举器。具体代码如下：

```typescript
/**
 * NodeT2BEnumerator 枚举器实现了 IEnumerator < TreeNode < T > > 泛型接口
 * 泛型参数 :   T 表示树节点中附加的数据的类型
 *            IdxFun 泛型参数必须是 Indexer 类型
 *            Adapter 泛型参数必须是 IAdapter < TreeNode < T > > 类型
 *            TypeScript 中可以使用 extends 进行泛型的类型限定
 */
export class NodeT2BEnumerator < T , IdxFunc extends Indexer , Adapter extends IAdapter < TreeNode < T > > > implements IEnumerator < TreeNode < T > > {
    private _node : TreeNode < T > | undefined ;     //头节点，指向输入的根节点
    private _adapter ! : IAdapter < TreeNode < T > > ;
            // 枚举器内部持有一个队列或堆栈的适配器，用于存储遍历的元素，指向泛型参数
    private _currNode ! : TreeNode < T > | undefined ;
                                                     //当前正在操作的节点类型
    private _indexer ! : IdxFunc ;
            // 当前的 Indexer，用于选择从左到右还是从右到左遍历，指向泛型参数

    /**
     * 构造函数
     * @param node { TreeNode < T > | undefined } 要遍历的树结构的根节点
     * @param func { IdxFunc } IdxFunc extends Indexer 必须是 Indexer 类型的回调函数
     * @param adapter { Adapter } 必须是实现 IAdapter < TreeNode < T > > 接口的类，该类必须要实现无参数的构造函数(new ( ) => Adapter )
     */
    public constructor ( node : TreeNode < T > | undefined , func : IdxFunc , adapter : new ( ) => Adapter ) {
        // 必须要有根节点，否则无法遍历
        if ( node === undefined ) {
            return ;
        }
        this . _node = node ;                   //头节点，指向输入的根节点
        this . _indexer = func ;                //设置回调函数
        this . _adapter = new adapter ( ) ;     //调用 new 回调函数

        this . _adapter . add ( this . _node ) ;
                                                //初始化时将根节点放入堆栈或队列中去
        this . _currNode = undefined ;          //设定当前 node 为 undefined
    }
```

```typescript
/**
 * 实现接口方法,将枚举器设置为初始化状态,调用 reset 函数后,可以重用枚举器
 */
public reset ( ) : void {
    if ( this . _node === undefined ) {
        return ;
    }
    this . _currNode = undefined;
    this . _adapter . clear ( ) ;
    this . _adapter . add ( this . _node ) ;
}

/**
 * 实现接口函数 moveNext , 返回 false 表示枚举结束,否则返回 true
 */
public moveNext ( ) : boolean {

    //当队列或者栈中没有任何元素,说明遍历已经全部完成了,返回 false
    if ( this . _adapter . isEmpty ) {
        return false ;
    }

    //弹出头或尾部元素,依赖于 adapter 是 stack 还是 queue
    this . _currNode = this . _adapter . remove ( ) ;

    // 如果当前的节点不为 undefined
    if ( this . _currNode != undefined ) {
        // 获取当前的节点的儿子节点个数
        let len : number = this . _currNode . childCount ;
        // 遍历所有的儿子节点
        for ( let i = 0 ; i < len ; i++ ) {
            // 儿子节点是从左到右,还是从右到左进入队列或堆栈
            // 注意, _indexer 是在这里调用的
            let childIdx : number = this . _indexer ( len , i ) ;
            let child : TreeNode < T > | undefined = this . _currNode .
            getChildAt ( childIdx ) ;
            if ( child !== undefined ) {
                this . _adapter . add ( child ) ;
            }
        }
    }

    return true ;
}

/**
 * 实现接口放 current,用于返回当前正在枚举的节点 TreeNode < T > | undefined
 */
public get current ( ) : TreeNode < T > | undefined {
    return this . _currNode ;
}
}
```

接下来实现后根枚举器。后根枚举器也实现了 IEnumerator < TreeNode < T > > 泛型接

口并在内部使用先根枚举器。具体代码如下：

```typescript
export class NodeB2TEnumerator < T > implements IEnumerator < TreeNode < T > > {
    private _iter : IEnumerator < TreeNode < T > > ;  // 持有一个枚举器接口
    private _arr ! : Array < TreeNode < T > | undefined > ;
                                            //声明一个数组对象
    private _arrIdx ! : number ;            // 当前的数组索引

    /**
     *
     * @param iter { IEnumerator < TreeNode < T > > } 指向树结构的先根迭代器
     */
    public constructor ( iter : IEnumerator < TreeNode < T > > ) {
        this . _iter = iter ;              // 指向先根迭代器
        this . reset ( ) ;                 // 调用reset，填充数组内容及_arrIdx
    }

    /**
     * 实现接口方法，将枚举器设置为初始化状态，调用 reset 函数后，可以重用枚举器
     * 关键方法
     */
    public reset ( ) : void {
        this . _arr = [ ] ; // 清空数组

        // 调用先根枚举器，将结果全部存入数组
        while ( this . _iter . moveNext ( ) ) {
            this . _arr . push ( this . _iter . current ) ;
        }
        // 设置_arrIdx 为数组的 length
        // 因为后根遍历是先根遍历的逆操作，所以是从数组尾部向顶部的遍历
        this . _arrIdx = this . _arr . length ;
    }

    /**
     * 实现接口放 current，用于返回当前正在枚举的节点 TreeNode < T > | undefined
     */
    public get current ( ) : TreeNode < T > | undefined {
        // 数组越界检查
        if ( this . _arrIdx >= this . _arr . length ) {
            return undefined ;
        } else {
            // 从数组中获取当前节点
            return this . _arr [ this . _arrIdx ] ;
        }
    }

    /**
     * 实现接口函数 moveNext，返回 false，表示枚举结束，否则，返回 true
     */
```

```typescript
    public moveNext ( ) : boolean {
        this . _arrIdx --;
        return ( this . _arrIdx >= 0 && this . _arrIdx < this . _arr . length ) ;
    }
}
```

最后来实现 NodeEnumeratorFactory 的工厂类，使用上面的两个枚举器，生产出 8 种产品。具体代码如下：

```typescript
export class NodeEnumeratorFactory {
    // 创建深度优先( stack )、从左到右 ( IndexerR2L )、从上到下的枚举器
    public static create_df_l2r_t2b_iter < T > ( node : TreeNode < T > |
undefined ) : IEnumerator < TreeNode < T > > {
        let iter : IEnumerator < TreeNode < T > > = new NodeT2BEnumerator
        ( node , IndexerR2L , Stack ) ;
        return iter ;
    }
    // 创建深度优先( stack )、从右到左( IndexerL2R )、从上到下的枚举器
    public static create_df_r2l_t2b_iter < T > ( node : TreeNode < T > |
undefined ) : IEnumerator < TreeNode < T > > {
        let iter : IEnumerator < TreeNode < T > > = new NodeT2BEnumerator
        ( node , IndexerL2R , Stack ) ;
        return iter ;
    }

    // 创建广度优先( Queue )、从左到右( IndexerL2R )、从上到下的枚举器
    public static create_bf_l2r_t2b_iter < T > ( node : TreeNode < T > |
undefined ) : IEnumerator < TreeNode < T > > {
        let iter : IEnumerator < TreeNode < T > > = new NodeT2BEnumerator
        ( node , IndexerL2R , Queue ) ;
        return iter ;
    }

    // 创建广度优先( Queue )、从右到左( IndexerR2L )、从上到下的枚举器
    public static create_bf_r2l_t2b_iter < T > ( node : TreeNode < T > |
undefined ) : IEnumerator < TreeNode < T > > {
        let iter: IEnumerator < TreeNode < T > > = new NodeT2BEnumerator
        ( node , IndexerR2L , Queue ) ;
        return iter ;
    }

    // 上面都是从上到下(先根)遍历
    // 下面都是从下到上(后根)遍历，是对上面的从上到下(先根)枚举器的包装

    // 创建深度优先、从左到右、从下到上的枚举器
    public static create_df_l2r_b2t_iter < T > ( node : TreeNode < T > |
undefined ) : IEnumerator < TreeNode < T > > {
        //向上转型，自动(向下转型，需要as或< >手动)
        let iter : IEnumerator < TreeNode < T > > = new NodeB2TEnumerator
        < T > ( NodeEnumeratorFactory . create_df_r2l_t2b_iter ( node ) ) ;
        return iter ;
    }
```

```typescript
    // 创建深度优先、从右到左、从下到上的枚举器
    public static create_df_r2l_b2t_iter < T > ( node : TreeNode<T> |
undefined ) : IEnumerator < TreeNode < T > > {
        let iter : IEnumerator < TreeNode < T > > = new NodeB2TEnumerator
        < T > ( NodeEnumeratorFactory . create_df_l2r_t2b_iter ( node ) ) ;
        return iter ;
    }

    // 创建广度优先、从左到右、从下到上的枚举器
    public static create_bf_l2r_b2t_iter < T > ( node : TreeNode < T > |
undefined ) : IEnumerator < TreeNode < T > > {
        let iter: IEnumerator < TreeNode < T > > = new NodeB2TEnumerator
        < T > ( NodeEnumeratorFactory.create_bf_r2l_t2b_iter( node ) ) ;
        return iter ;
    }

    // 创建广度优先、从右到左、从下到上的枚举器
    public static create_bf_r2l_b2t_iter < T > ( node : TreeNode < T > |
undefined ) : IEnumerator < TreeNode < T > > {
        let iter : IEnumerator < TreeNode < T > > = new NodeB2TEnumerator
        < T > ( NodeEnumeratorFactory . create_bf_l2r_t2b_iter ( node ) ) ;
        return iter ;
    }
}
```

要注意的一点是，NodeEnumeratorFactory 类都是静态工厂模板方法，并且每个静态工厂模板方法返回的是 IEnumerator < TreeNode < T >>类型的接口，而不是具体的两个实现类：NodeT2BEnumerator 和 NodeB2TEnumerator。以这种面向接口的编程模型，既能减少类之间的耦合性，又能将复杂的操作丢给实现者，将实现细节隐藏起来，对于调用者来说，使用起来很方便。下一节将介绍如何使用上面的 8 个枚举器。

9.2.5　测试树结构枚举器

本节将测试上一节介绍的 8 个迭代器的用法和输出结果。在测试前，来看一下泛型类 TreeNode < T >的两种用法。

第 1 种用法，直接使用 TreeNode < T >的方式，例如将泛型参数设置为 number 类型。具体代码如下：

```typescript
// 生成一个名为 root 的节点，其父亲是 undefined，节点附加的对象类似 number，其值为 0
let root : TreeNode < number >  = new TreeNode < number > ( 0 , undefined ,
" root " ) ;
// 生成一个名为 node1 的节点，其父亲是 root 节点，节点附加的对象类似 number，其值为 1
let node1 : TreeNode < number > = new TreeNode < number > ( 1 , root , "
node1 " ) ;
// 生成一个名为 node2 的节点，其父亲是 root 节点，节点附加的对象类似 number，其值为 2
let node2 : TreeNode < number > = new TreeNode < number > ( 2 , nnode, "
node2 " ) ;
```

第 2 种，可以使用继承的方式来使用树节点类，还是以附加 number 类型为例子。具

体代码如下：

```
// NumberNode 类继承自泛型参数为 number 的 TreeNode 类
// 这样就自动会设置 TreeNode 的 data 成员变量的类型为 number
class NumberNode extends TreeNode < number > { }
export class TreeNodeTest {
    // 创建一棵如图 9.1 所示结构一致的树结构
    public static createTree ( ) : NumberNode {
        let root : NumberNode = new NumberNode ( 0 , undefined , " root " ) ;
        let node1 : NumberNode = new NumberNode ( 1 , root , " node1 " ) ;
        let node2 : NumberNode = new NumberNode ( 2 , root , " node2 " ) ;
        let node3 : NumberNode = new NumberNode ( 3 , root , " node3 " ) ;
        let node4 : NumberNode = new NumberNode ( 4 , node1 , " node4 " ) ;
        let node5 : NumberNode = new NumberNode ( 5 , node1 , " node5 " ) ;
        let node6 : NumberNode = new NumberNode ( 6 , node2 , " node6 " ) ;
        let node7 : NumberNode = new NumberNode ( 7 , node2 , " node7 " ) ;
        let node8 : NumberNode = new NumberNode ( 8 , node3 , " node8 " ) ;
        let node9 : NumberNode = new NumberNode ( 9 , node4 , " node9 " ) ;
        let node10 : NumberNode = new NumberNode ( 10 , node6 , " node10 " ) ;
        let node11 : NumberNode = new NumberNode ( 11 , node7 , " node11 " ) ;
        let node12 : NumberNode = new NumberNode ( 12 ,node11 ," node12 " ) ;
        return root ;
    }
}
```

后面将使用 NumberNode 类进行演示。为了能够层次化地输出使用 createTree 方法创建的树结构，在 TreeNode < T >类中增加一个名为 repeatString 的公开方法。具体代码如下：

```
/**
 * 将一个字符串输出 n 次
 * @param target { string } 要重复输出的字符串
 * @param n { number } 要输出多少次
 */
public repeatString ( target: string, n: number ): string {
    let total: string = "";
    for ( let i = 0 ; i < n ; i ++) {
        total += target ;
    }
    return total;
}
```

有了上面实现的 createTree 方法，以及 TreeNode < T >的 repeatString 方法和 depth 属性，结合各个枚举器，就能格式化输出节点相关的内容。下面来看深度优先、从上到下、从左到右的遍历输出代码和结果，具体源码如下：

```
let root : NumberNode = TreeNodeTest . createTree ( ) ;
let iter : IEnumerator < TreeNode < number > > ;    // IEnumerator 枚举器
let current : TreeNode < number > | undefined = undefined ;
                                            //枚举器持有的当前的节点
```

```
console.log("1、depthFirst_left2rihgt_top2bottom_enumerator");
// 下面的代码应该输出的结果为：[ root, node1, node4, node9, node5, node2,
node6, node10, node7, node11, node12, node3, node8 ]
iter = NodeEnumeratorFactory.create_df_l2r_t2b_iter<number>(root);
while(iter.moveNext()){
    current = iter.current;
    if(current !== undefined){
        // 根据当前的depth获得缩进字符串（下面使用空格字符），然后和节点名合成当前节
        点输出路径
        console.log(current.repeatString(" ", current.depth * 4)
        + current.name);
    }
}
```

运行上述代码，会获得如图9.3所示的结果。

depthFirst_left2rihgt_top2bottom_enumerator
root
node1
node4
node9
node5
node2
node6
node10
node7
node11
node12
node3
node8

图9.3 枚举器输出结果

根据如图9.3所示可以发现，从上到下阅读输出结果，的确就是深度优先、从上到下、从左到右的节点遍历顺序。

接下来调用其他7个枚举器。具体代码如下：

```
// 辅助方法，根据输入的枚举器，线性输出节点内容
public static outputNodesInfo(iter: IEnumerator<TreeNode<number>>): string {
    let output: string[] = [];
    let current: TreeNode<number> | undefined = undefined;
    while(iter.moveNext()){
        current = iter.current;
        if(current !== undefined){
            output.push(current.name);
        }
```

```typescript
    }
    return ( " 实际输出：[" + output.join(",") + "]" );
}

// 下面的代码线性输出所有遍历结果，验证迭代的正确性
console.log( " 2、depthFirst_right2left_top2bottom_enumerator " );
console.log( " 应该输出：[ root , node3 , node8 , node2 , node7 , node11 ,
node12 , node6 , node10 , node1 , node5 , node4 , node9 ] " );
iter = NodeEnumeratorFactory.create_df_r2l_t2b_iter<number>( root );
console.log( TreeNodeTest.outputNodesInfo( iter ) );

console.log( " 3、depthFirst_left2right_bottom2top_enumerator " );
iter = NodeEnumeratorFactory.create_df_l2r_b2t_iter<number>( root );
console.log( " 应该输出：[ node9 , node4 , node5 , node1 , node10 , node6 ,
node12 , node11 , node7 , node2 , node8 , node3 , root ] " );
console.log( TreeNodeTest.outputNodesInfo( iter ) );

console.log( " 4、depthFirst_right2left_bottom2top_enumerator " );
iter = NodeEnumeratorFactory.create_df_r2l_b2t_iter<number>( root );
console.log( " 应该输出：[ node8 , node3 , node12 , node11 , node7 , node10 ,
node6 , node2 , node5 , node9 , node4 , node1 , root ] " );
console.log( TreeNodeTest.outputNodesInfo( iter ) );

console.log( " 5、breadthFirst_left2right_top2bottom_enumerator " );
iter = NodeEnumeratorFactory.create_bf_l2r_t2b_iter<number>( root );
console.log( " 应该输出：[ root , node1 , node2 , node3 , node4 , node5 ,
node6 , node7 , node8 , node9 , node10 , node11 , node12 ] " );
console.log( TreeNodeTest.outputNodesInfo( iter ) );

console.log( " 6、breadthFirst_rihgt2left_top2bottom_enumerator " );
iter = NodeEnumeratorFactory.create_bf_r2l_t2b_iter<number>( root );
console.log( " 应该输出：[ root , node3 , node2 , node1 , node8 , node7 ,
node6 , node5 , node4 , node11 , node10 , node9 , node12 ] " );
console.log( TreeNodeTest.outputNodesInfo( iter ) );

console.log( " 7、breadthFirst_left2right_bottom2top_enumerator " );
iter = NodeEnumeratorFactory.create_bf_l2r_b2t_iter<number>( root );
console.log( " 应该输出：[ node12 , node9 , node10 , node11 , node4 , node5 ,
node6 , node7 , node8 , node1 , node2 , node3 , root ] " );
console.log( TreeNodeTest.outputNodesInfo( iter ) );

console.log( " 8、breadthFirst_right2left_bottom2top_enumerator " );
iter = NodeEnumeratorFactory.create_bf_r2l_b2t_iter<number>( root );
console.log( " 应该输出：[ node12 , node11 , node10 , node9 , node8 , node7 ,
node6 , node5 , node4 , node3 , node2 , node1 , root ] " );
console.log( TreeNodeTest.outputNodesInfo( iter ) );
```

调用上述代码后，会获得如图9.4所示的遍历结果。

```
2、depthFirst_right2left_top2bottom_enumerator
应该输出：[ root , node3 , node8 , node2 , node7 , node11 , node12 , node6 , node10 , node1 , node5 , node4 , node9 ]
实际输出：[ root , node3 , node8 , node2 , node7 , node11 , node12 , node6 , node10 , node1 , node5 , node4 , node9 ]
3、depthFirst_left2right_bottom2top_enumerator
应该输出：[ node9 , node4 , node5 , node1 , node10 , node6 , node12 , node11 , node7 , node2 , node8 , node3 , root ]
实际输出：[ node9 , node4 , node5 , node1 , node10 , node6 , node12 , node11 , node7 , node2 , node8 , node3 , root ]
4、depthFirst_right2left_bottom2top_enumerator
应该输出：[ node8 , node3 , node12 , node11 , node7 , node10 , node6 , node2 , node5 , node9 , node4 , node1 , root ]
实际输出：[ node8 , node3 , node12 , node11 , node7 , node10 , node6 , node2 , node5 , node9 , node4 , node1 , root ]
5、breadthFirst_left2right_top2bottom_enumerator
应该输出：[ root , node1 , node2 , node3 , node4 , node5 , node6 , node7 , node8 , node9 , node10 , node11 , node12 ]
实际输出：[ root , node1 , node2 , node3 , node4 , node5 , node6 , node7 , node8 , node9 , node10 , node11 , node12 ]
6、breadthFirst_rihgt2left_top2bottom_enumerator
应该输出：[ root , node3 , node2 , node1 , node8 , node7 , node6 , node5 , node4 , node11 , node10 , node9 , node12 ]
实际输出：[ root , node3 , node2 , node1 , node8 , node7 , node6 , node5 , node4 , node11 , node10 , node9 , node12 ]
7、breadthFirst_left2right_bottom2top_enumerator
应该输出：[ node12 , node9 , node10 , node11 , node4 , node5 , node6 , node7 , node8 , node1 , node2 , node3 , root ]
实际输出：[ node12 , node9 , node10 , node11 , node4 , node5 , node6 , node7 , node8 , node1 , node2 , node3 , root ]
8、breadthFirst_right2left_bottom2top_enumerator
应该输出：[ node12 , node11 , node10 , node9 , node8 , node7 , node6 , node5 , node4 , node3 , node2 , node1 , root ]
实际输出：[ node12 , node11 , node10 , node9 , node8 , node7 , node6 , node5 , node4 , node3 , node2 , node1 , root ]
```

图 9.4 枚举器遍历

9.2.6 深度优先的递归遍历

本节介绍树结构深度优先的递归遍历算法。在 TreeNode < T > 中增加一个名为 visit 的方法，该方法以深度优先的方式递归遍历其子孙节点。

首先来看一下 visit 方法的原型或签名，具体如下：

```
public visit ( preOrderFunc : NodeCallback < T > | null = null ,
postOrderFunc : NodeCallback < T > | null = null , indexFunc : Indexer =
IndexerL2R ) : void
```

可以看到，preOrederFunc 和 postOrderFunc 这两个参数，表示的是在递归遍历的过程中，在先根访问和后根访问的时机点时需要调用的回调函数，该回调函数的签名如下：

```
//回调函数类型名，需要加 < T >        参数类型           返回类型
export type NodeCallback < T > = ( node : TreeNode < T > ) => void ;
```

至于 indexFunc 回调的类型，在上一节中有详解，默认为 IndexerL2R 方法。接下来看一下 visit 方法的实现代码，代码如下：

```
public visit ( preOrderFunc : NodeCallback < T > | null = null ,
postOrderFunc : NodeCallback < T > | null = null , indexFunc : Indexer =
IndexerL2R ) : void {
    // 在子节点递归调用 visit 之前，触发先根（前序）回调
    // 注意前序回调的时间点是在此处！！！！
    if ( preOrderFunc !== null ) {
        preOrderFunc ( this ) ;
    }
```

```typescript
// 遍历所有子节点
let arr : Array < TreeNode < T > > | undefined = this . _children ;
if ( arr !== undefined ) {
    for ( let i : number = 0 ; i < arr . length ; i++ ) {
        // 根据 indexFunc 选取左右遍历还是右左遍历
        let child : TreeNode < T > | undefined = this . getChildAt
        ( indexFunc ( arr . length , i ) ) ;
        if ( child !== undefined ) {
            // 递归调用 visit
            child . visit ( preOrderFunc , postOrderFunc , indexFunc ) ;
        }
    }
}

// 在这个时机点触发 postOrderFunc 回调
// 注意后根（后序）回调的时间点是在此处！！！
if ( postOrderFunc !== null ) {
    postOrderFunc ( this ) ;
}
}
```

来看一下调用上述方法的测试代码，代码如下：

```typescript
function printNodeInfo ( node : NumberNode ) : void {
    console . log ( node . name ) ;
}

let root : NumberNode = TreeNodeTest . createTree ( ) ;
root . visit ( printNodeInfo , null , IndexerL2R ) ;
                        // 深度优先、从上到下（先根）、从左到右遍历
root . visit ( printNodeInfo , null , IndexerR2L ) ;
                        // 深度优先、从上到下（先根）、从右到左遍历
root . visit ( null , printNodeInfo , IndexerL2R ) ;
                        // 深度优先、从下到上（后根）、从左到右遍历
root . visit ( null , printNodeInfo , IndexerR2L ) ;
                        // 深度优先、从下到上（后根）、从右到左遍历
root . visit ( printNodeInfo , printNodeInfo , IndexerL2R ) ;
                        // 先根和后根回调同时触发
root . visit ( printNodeInfo , printNodeInfo , IndexerR2L ) ;
                        // 先根和后根回调同时触发
```

可以看出，调用上述测试代码，就能获得正确的结果。

9.2.7 使用儿子兄弟方式递归遍历算法

为了方便进行树节点的层次查询，一般情况下有两种存储方式：
- 父亲节点+儿子列表方式，现在使用的都是这种存储方式。
- 父亲节点+儿子兄弟形式，如下面所示的结构：

```typescript
export class LinkTreeNode<T> {
    private _parent: LinkTreeNode<T> | undefined; // 指向当前节点的父亲节点
```

```
    private _firstChild: LinkTreeNode<T> | undefined;
                        // 指向当前节点的第一个儿子节点
    private _lastChild: LinkTreeNode<T> | undefined;
                        // 指向当前节点的最后一个儿子节点
    private _nextSibling: LinkTreeNode<T> | undefined;
                        // 指向当前节点的下一个兄弟节点
    private _prevSibling: LinkTreeNode<T> | undefined;
                        // 指向当前节点的上一个兄弟节点

    public name: string = '';
            // 当前节点的名称, 有利于 debug 时信息输出或按名字获取节点(集合)操作
    public data: T | undefined;
            // 一个泛型对象, 指向一个你需要依附到当前节点的对象
}
```

上面的树节点使用了 _firstChild 和 _nextSibling 成员变量用来快速地进行子节点从左到右的遍历，而使用 _lastChild 和 _prevSibling 成员变量用来快速地进行子节点从右到左的遍历。

本书仅使用父亲节点+儿子列表的存储方式（TreeNode < T > 类），但是为了让大家了解父亲节点+儿子兄弟存储方式的遍历流程，在 TreeNode < T > 类中模拟实现了如下四个只读查询属性（实现代码可以参阅 9.1.5 节相关源码）：

- public get firstChild ( ) : TreeNode < T > | undefined ;
- public get nextSibling ( ) : TreeNode < T > | undefined ;
- public get lastChild ( ) : TreeNode < T > | undefined ;
- public get prevSibling ( ) : TreeNode < T > | undefined ;

现在用上面这些方法来进行深度优先的递归遍历操作。具体代码如下：

```
public visitForward ( preOrderFunc : NodeCallback < T > | null = null ,
postOrderFunc : NodeCallback < T > | null = null ) : void {
    // 先根（前序）遍历时 preOrderFunc 触发的时机点
    if ( preOrderFunc ) {
        preOrderFunc ( this ) ;
    }
    let node : TreeNode < T > | undefined = this . firstChild ;
    while ( node !== undefined ) {
        node . visitForward ( preOrderFunc , postOrderFunc ) ;
        node = node . nextSibling ;
    }
    // 后根（后序）遍历时 postOrderFunc 触发的时机点
    if ( postOrderFunc ) {
        postOrderFunc ( this ) ;
    }
}

public visitBackward ( preOrderFunc : NodeCallback < T > | null = null ,
postOrderFunc : NodeCallback < T > | null = null ) : void {
    // 先根（前序）遍历时 preOrderFunc 触发的时机点
```

```
        if ( preOrderFunc ) {
            preOrderFunc ( this ) ;
        }
        let node : TreeNode < T > | undefined = this . lastChild ;
        while ( node !== undefined ) {
            node . visitBackward ( preOrderFunc , postOrderFunc ) ;
            node = node . prevSibling ;
        }
        // 后根（后序）遍历时 postOrderFunc 触发的时机点
        if ( postOrderFunc ) {
            postOrderFunc ( this ) ;
        }
    }

// 测试代码如下：
let root : NumberNode = TreeNodeTest . createTree ( ) ;

root . visitForward ( printNodeInfo , null ) ;
                            // 深度优先、从上到下 (先根)、从左到右遍历
root . visitForward ( null , printNodeInfo ) ;
                            // 深度优先、从下到上 (后根)、从左到右遍历
root . visitForward ( printNodeInfo , printNodeInfo ) ;
                            // 深度优先、从左到右同时遍历

root . visitBackward ( printNodeInfo , null ) ;
                            // 深度优先、从上到下 (先根)、从右到左遍历
root . visitBackward ( null , printNodeInfo ) ;
                            // 深度优先、从下到上 (后根)、从右到左遍历
root . visitBackward ( printNodeInfo , printNodeInfo ) ;
                            // 深度优先、从右到左同时遍历
```

9.2.8 儿子兄弟方式非递归遍历算法

到目前为止，对树结构实现了如下几种方式的遍历算法：
- 使用队列结构实现了非递归方式的、广度优先的四种枚举器。
- 使用栈结构实现了非递归方式的、深度优先的四种枚举器。
- 实现了深度优先、儿子列表方式的递归遍历。
- 实现了深度优先、儿子兄弟方式递归遍历。

接下来，看一下如何使用儿子兄弟方式实现非递归的、不需要使用栈结构的、深度优先的各种遍历算法。

为了更方便和更好地了解相关算法，提供一张节点图，如图9.5所示。

此处需要强调的是，参考图9.5，这张图和前面的图9.1相比，node1 节点下增加了一个节点 node6，并且调整了 node6 后整个节点的索引号，之所以要调整，是为了演示下面第（2）点的情况。

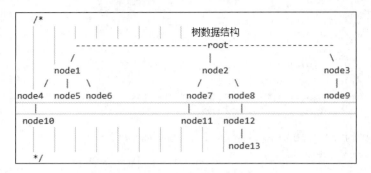

图 9.5 树结构遍历参考图

接下来了解一下非递归、无须栈结构的先根遍历算法的要点：

（1）如果当前的节点有左儿子节点，则返回左儿子节点。例如节点 node4 有左儿子节点 node10，则返回 node10。

（2）如果当前的节点没有左儿子节点，但是有右兄弟节点，则返回右兄弟节点。例如节点 node5 没有 firstChild，但是有 nextSibling（节点 node6），则返回 node6。

（3）如果当前的节点（例如 node6、node13 及 node9）既没有左儿子节点，也没有右兄弟节点，那就要向上回溯找祖先节点，看看祖先节点有没有右兄弟节点：

- 祖先节点有右兄弟节点的情况，例如 node13 节点一直向上查找到祖先节点 node2 有右兄弟节点 node3，就返回 node2 的右兄弟节点 node3。
- 祖先节点没有右兄弟的情况，例如 node9 的 parent 为 node3，而 node3 没有右兄弟节点，这种情况下，继续向上查找到 root 节点，此时 root 节点既没有父亲节点，也没有右兄弟节点，那么在这种情况下，说明全部遍历完成，就返回 undefined。

了解了上面的情况后，就能实现深度优先的、从上到下的、从左到右的、非递归的、不需要栈结构辅助的遍历方法 moveNext。具体代码如下：

```
// 深度优先、从上到下、从左到右、非递归的遍历算法
// 总体来说就是： 先儿子节点、后兄弟节点、最后祖先节点的遍历顺序
public moveNext ( ) : TreeNode < T > | undefined {
    // 如果有左儿子节点，则返回儿子节点
    let ret: TreeNode < T > | undefined = this . firstChild ;
    if ( ret !== undefined ) {
        return ret ;
    }
    // 如果没有儿子节点，但是有兄弟节点，则返回当前节点的右兄弟节点
    ret = this . nextSibling ;
    if ( ret !== undefined ) {
        return ret ;
    }
    // 当前节点既没有左儿子节点，也没有右兄弟节点
    ret = this ;
    // 一直回溯查找有右兄弟节点的祖先节点
    while ( ret !== undefined && ret . nextSibling === undefined ) {
        ret = ret . parent ;
```

```
        }
        // 如果祖先节点有右兄弟节点,则返回右兄弟节点
        if ( ret !== undefined ) {
            return ret . nextSibling ;
        }
        // 否则表示遍历结束
        return undefined ;
    }
```

同样地,在上述代码中,只要将 firstChild 替换为 lastChild,并且将 nextSibling 替换为 prevSibling,就能实现深度优先的、从上到下的、从右到左的、非递归的、不需要栈结构辅助的遍历方法 movePrev。具体代码如下:

```
    public movePrev ( ) : TreeNode<T> | undefined {
        // 如果有右儿子节点,则返回儿子节点
        let ret: TreeNode < T > | undefined = this . lastChild ;
        if ( ret !== undefined ) {
            return ret ;
        }
        // 如果没有儿子节点,但是有兄弟节点,则返回当前节点的左兄弟节点
        ret = this . prevSibling ;
        if ( ret !== undefined ) {
            return ret ;
        }
        // 当前节点既没有右儿子节点,也没有左兄弟节点
        ret = this ;
        while ( ret !== undefined && ret . prevSibling === undefined ) {
            ret = ret . parent ;
        }
        // 如果父亲有左兄弟节点,则返回左兄弟节点
        if ( ret !== undefined ) {
            return ret . prevSibling ;
        }
        // 否则表示遍历结束
        return undefined ;
    }
```

有了上面的两个方法,就能很方便地进行深度优先、从上到下(先根前序)的、从左到右(moveNext)或从右到左(movePrev)的遍历,遍历用法如下代码所示。

```
for ( let n : TreeNode < number > | undefined = root ; n !== undefined ; n = n . moveNext ( ) ) {
    // 缩进层次输出节点名
    console . log ( "moveNext : " + n . repeatString ( ' ' , n . depth * 4 )
    + n . name ) ;
}

for ( let n : TreeNode < number > | undefined = root ; n !== undefined ; n = n . movePrev ( ) ) {
    // 缩进层次输出节点名
    console . log ( "movePrev : " + n . repeatString ( ' ' , n . depth * 4 )
    + n . name ) ;
}
```

接下来看一下深度优先、从下到上（后根后序）的、从左到右，以及从右到左的非递归、无须栈结构辅助的遍历算法。还是以图 9.5 所示为下列，来了解一下从左到右遍历算法，具体流程如下：

（1）使用已经实现了的 mostLeft 方法获取最左侧的节点（node10）。

（2）从当前节点开始遍历，此时 node10 为当前节点，需要考虑如下两种情况：

- 如果当前节点（node10）没有右兄弟节点，则返回当前节点的父亲节点（node4）。
- 如果当前节点（假设为 node7）有右兄弟节点（node8），那么需要找到其右兄弟节点（node8）最左侧的子孙节点（node13）并将其返回。

（3）不断重复第（2）步操作，一直到返回树遍历结束，此时返回 undefined。

上述算法也适合从右到左的遍历，只要将查询右兄弟节点换成查询左兄弟节点，把查询左儿子节点换成查询右儿子节点就可以了。来看一下实现代码，具体如下：

```
public moveNextPost ( ) : TreeNode<T> | undefined {
    // 如果当前节点没有右兄弟节点，则返回当前节点的父亲节点
    let next : TreeNode<T> | undefined = this . nextSibling ;
    if ( next === undefined ){
        return this . parent ;
    }
    // 如果当前节点存在右兄弟节点，则获取当前节点的右兄弟节点的最左边儿子节点
    let first : TreeNode<T> | undefined = undefined ;
    while ( next !== undefined && ( first = next . firstChild ) ) {
        next = first;
    }
    return next;
}
public movePrevPost ( ) : TreeNode<T> | undefined {
    // 如果当前节点没有左兄弟节点，则返回当前节点的父亲节点
    let prev : TreeNode<T> | undefined = this . prevSibling ;
    if ( prev === undefined ) {
        return this . parent ;
    }
    // 如果当前节点存在左兄弟节点，则获取当前节点的左兄弟节点的最右边儿子节点
    let last : TreeNode<T> | undefined = undefined ;
    while ( prev !== undefined && ( last = prev . lastChild ) ) {
        prev = last;
    }
    return prev;
}
```

有了上面的 moveNextPost 和 movePrevPost 方法后，可以使用这两个方式来进行深度优先、从下到上（后根后序）的、从左到右、以及从右到左的非递归、无须栈结构辅助的遍历输出。具体代码如下：

```
for( let n : TreeNode < number > | undefined = root . mostLeft ; n !== undefined ; n = n . moveNextPost ( ) ) {
    console . log ( "moveNextPost : " + n . repeatString ( ' ' , n . depth * 4 ) + n . name ) ;
}
for ( let n : TreeNode < number > | undefined = root . mostRight ; n !==
```

```
undefined ; n = n . movePrevPost ( ) ) {
    console . log ( "movePrevPost : " + n . repeatString ( ' ' , n . depth
    * 4 ) + n . name ) ;
}
```

需要注意的一点是，在使用 moveNextPost 函数之前，需要通过 mostLeft 获得整棵树最左侧的节点，从最左侧节点开始调用 moveNextPost 函数进行遍历。该规则同样适合 movePrevPost，只是将 mostLeft 替换为 mostRight，大家可以参考上述代码加深了解。

当调用 moveNext、movePrev、moveNextPost 和 movePrevPost 这 4 个深度优先的遍历算法后，会获得如图 9.6 所示的结果。

```
moveNext     : root                    movePrev     : root
moveNext     :     node1               movePrev     :     node3
moveNext     :         node4           movePrev     :         node8
moveNext     :             node9       movePrev     :         node2
moveNext     :         node5           movePrev     :             node7
moveNext     :     node2               movePrev     :             node11
moveNext     :         node6           movePrev     :                 node12
moveNext     :         node10          movePrev     :         node6
moveNext     :         node7           movePrev     :         node10
moveNext     :             node11      movePrev     :     node1
moveNext     :             node12      movePrev     :         node5
moveNext     :     node3               movePrev     :         node4
moveNext     :         node8           movePrev     :             node9
moveNextPost :             node9       movePrevPost :         node8
moveNextPost :         node4           movePrevPost :     node3
moveNextPost :         node5           movePrevPost :                 node12
moveNextPost :     node1               movePrevPost :             node11
moveNextPost :         node10          movePrevPost :             node7
moveNextPost :         node6           movePrevPost :         node10
moveNextPost :             node12      movePrevPost :         node6
moveNextPost :             node11      movePrevPost :         node2
moveNextPost :         node7           movePrevPost :         node5
moveNextPost :     node2               movePrevPost :             node9
moveNextPost :         node8           movePrevPost :         node4
moveNextPost :     node3               movePrevPost :     node1
moveNextPost : root                    movePrevPost : root
```

图 9.6　非递归深度优先遍历

9.3　树数据结构的序列化与反序列化

关于树结构中的节点增加和删除操作、各种层次关系查询操作，以及深度或广度优先的各种遍历算法，在前面章节已经详细了解过了。接下来介绍如何将树结构进行序列化和反序列化操作。

所谓序列化是指将某个对象将其全部或部分成员变量转换为可以存储或传输的数据的过程。而反序列化就是上述过程的相反操作。

9.3.1 树节点自引用特性导致序列化错误

实际上，TypeScript / JavaScript 内置了非常方便的 JSON 序列化功能，可以通过 JSON 类的静态方法进行序列化和反序列化操作：

- **stringify** 将当前的对象及其所有属性序列化成 JSON 格式字符串，这样就可以将其存储到服务器上或者传输给其他用户。
- **parser** 将某个符合 JSON 格式的字符串反序列化成 JavaScript 对象。

但是使用 JSON 类的静态方法也会出现一些问题，如果在 TreeNode＜T＞对象上使用如下代码进行序列化操作：

```
let root : NumberNode = TreeNodeTest . createTree ( ) ;
JSON . stringify ( root ) ;
```

会得到树结构序列化循环引用错误的结果，如图 9.7 所示。

```
⊗ Error: (SystemJS) Converting circular structure to JSON
    TypeError: Converting circular structure to JSON
        at JSON.stringify (<anonymous>)
        at Function.TreeNodeTest.testTreeNode (http://localhost:3000/treeNodeTest.ts!transp
iled:149:26)
        at execute (http://localhost:3000/treeNodeTest.ts!transpiled:274:26)
    Error loading http://localhost:3000/treeNodeTest.ts
        at JSON.stringify (<anonymous>)
        at Function.TreeNodeTest.testTreeNode (http://localhost:3000/treeNodeTest.ts!transp
iled:149:26)
        at execute (http://localhost:3000/treeNodeTest.ts!transpiled:274:26)
    Error loading http://localhost:3000/treeNodeTest.ts
```

图 9.7 树结构序列化循环引用错误

这是因为树数据结构具有自引用的特征，树结构中的每个节点都含有指向其父亲的节点和儿子列表节点，每个节点又以同样的方式递归的引用。

在这种情况下，stringify 会导致解析成 JSON 的过程中发生循环引用的错误。

9.3.2 树节点的序列化和反序列化操作

在 9.2.7 节中曾经提到过父亲节点+儿子列表和父亲节点+儿子兄弟这两种存储方式。其实这两种方式更加适合在内存中表示树节点，但是并不适合在文件中的存储，因为存储的内容过多，并且有循环引用的问题。

那么如何最小序列化存储内容，并且解决循环引用的问题呢？

可以通过以下三个步骤来解决。

（1）分离内存表示的树节点和序列化存储表示的树节点，例如现在需要将树节点的层次关系（必须）和节点名称（可选）序列化成 JSON 字符串，那么可以定义如下序列化存储结构：

```
export class NodeData {
    // 节点的父亲索引号，节点必须要序列化的成员变量，否则无法表示出树节点的层次性
    // parentIdx 的数据类型是 number，这样就能正确地序列化成 JSON 字符串
    public parentIdx : number ;
    public name : string ;   // 节点名称，可选的成员变量
    public constructor ( name : string , parentIdx : number ) {
        this . name = name ;
        this . parentIdx = parentIdx ;
    }
}
```

（2）以深度优先，从上到下（先根前序）的方式将树节点输出到数组中，并且以相应的顺序构建 NodeData 数组，设定 NodeData 的名称，初始化时，将 parentIdx 设定为-1，表示无父亲节点。具体代码如下：

```
public static convertTreeToJsonString < T > ( node : TreeNode < T > ) : string {
    let nodes : Array < TreeNode < T > > = [ ] ;
                              // 深度优先、从上到下（先根前序）保存树节点
    let datas : Array < NodeData > = [ ] ;
    for (let n : TreeNode < T > | undefined = node ; n !== undefined ; n = n . moveNext ( ) ) {
        // NodeData 和 node 在数组中的顺序是一一对应的，此时可以确定存储名称，但是
        层次关系还不正确，需要后续处理，因此设定为-1
        datas . push ( new NodeData ( n . name , -1 ) ) ;
        nodes . push ( n ) ;
    }
```

（3）扫描节点数组，查询当前节点的父亲节点在深度优先、从上到下（先根前序）顺序存储的数组中的索引号，将该索引号赋值给对应位置的 NodeData 对象的 parentIdx 属性。具体代码如下：

```
    // 接上面的代码
    for ( let i : number = 0 ; i < datas . length ; i ++ ) {
        // 获取当前节点的 parent
        let parent : TreeNode < T > | undefined = nodes [ i ] . parent ;
        // 如果当前节点的父亲节点为 undefined，则肯定是根节点，根节点的父亲节点为-1
        if ( parent === undefined ) {
            datas [ i ] . parentIdx = -1 ;
        } else {
            // 查找当前节点的 parent 在深度优先的数组中的索引号
            for ( let j : number = 0 ; j < datas . length ; j ++ ) {
                // 也可以用名称比较，更好的方式用地址比较
                // if ( parent . name === nodes [ j ] . name )
                if ( parent === nodes [ j ] )
                {
```

```
                    datas[i].parentIdx = j;
                }
            }
        }
        // 使用stringify方法进行object到string的序列化操作
        // 因为使用NodeData存储树节点的层次关系,而且parentIdx是number类型,指向的
        // 是深度优先、从上到下(先根前序)顺序方式存储的数组中,因此解决了自引用的问题
        return JSON.stringify(datas);
    }
```

说到底,之所以树节点会导致 JSON 序列化出错,是因为 JSON 序列化只能是例如 string 或 number 等基本数据类型,而 TreeNode<T>中例如 parent 成员变量无法转换为 string 或 number 等基本类型,从而产生错误。上面所做的就是将 parent 成员变量从引用类型(指针指向内存地址)转换成指向数组中某个元素的索引号,这样,索引号是 number 类型,不会导致循环引以,可以被序列化,而且索引指向的元素具有确定性。

从上述代码可以看到,表示树节点层次关系的存储方式还有第三种,即仅以父亲节点表示,所以再罗列一下树节点的存储方式。

- 父亲节点。
- 父亲节点+儿子列表形式。
- 父亲节点+儿子兄弟形式。

下面来看一下如何将 NodeData 数组表示的内容反序列化成父亲节点+儿子列表的形式。具体代码如下:

```
public static convertJsonStringToTree<T>(json:string):TreeNode
<T>|undefined {
    // 首先使用JSON.parse方法,将json字符串反序列化成Array对象(datas)
    let datas:[] = JSON.parse(json);
    let data!:NodeData;
    let nodes:TreeNode<T>[] = [];
    // 根据NodeData列表生成节点数组
    for(let i:number = 0; i < datas.length; i++) {
        // 将datas中每个元素都转型为NodeData对象
        data = datas[i] as NodeData;
        // 如果当前的NodeData的parentidx为-1,表示根节点
        // 实际上,datas是深度优先,从上到下(先根前序)顺序存储的
        // 因此datas[0]肯定是根节点
        if(data.parentIdx === -1) {
            nodes.push(new TreeNode<T>(undefined, undefined, data.
            name));
        }
        else { // 不是-1,说明有父亲节点
            // 利用了深度优先,从上到下(先根前序)顺序存储的nodes数组的特点
            // 当前节点的父亲节点总是已经存在nodes中了
            // 在先根存储的数组中,父亲节点总是在儿子节点的前面,因此序列化如果是后根
            遍历存储,那么下面的代码就会崩溃,因为后根存储的数组最大特点是,儿子节点在
            父亲节点的前面,此时nodes[data.parentIdx]返回的是undefined,
```

            在形成树结构遍历时，导致程序崩溃
            nodes . push (new TreeNode < T > (undefined , nodes [data . parentIdx] , data . name)) ;
        }
    }
    // 返回反序列化中的根节点
    return nodes [0] ;
}
```

序列化代码必须要使用深度优先、从上到下（先根前序）顺序的原因，在反序列化方法的注释中有说明，请大家注意。

## 9.4　队列与栈的实现

在 9.2.4 节中，声明了 IAdapter 接口，该接口用来适配栈和队列的数据结构，将它们的所有操作都统一成一种调用形式。到目前为止都是以面向接口编程的方式在调用 IAdapter 的接口方法。下面将实现该接口。具体代码如下：

```
export abstract class AdapterBase < T > implements IAdapter < T > {
    protected _arr : Array < T > ;

    public constructor () {
        this . _arr = new Array < T > ( ) ;
    }

    public add (t : T) : void {
        this . _arr . push (t) ;
    }

    // 子类实现抽象的 remove 方法
    public abstract remove ( ) : T | undefined ;

    public get length ( ) : number {
        return this . _arr . length ;
    }

    public get isEmpty ( ) : boolean {
        return this . _arr . length <= 0 ;
    }

    public clear ( ) : void {
        this . _arr = new Array < T > ( ) ;
    }

```typescript
    public toString ( ) : string {
        return this . _arr . toString ( ) ;
    }
}

export class Stack < T > extends AdapterBase < T > {
    public remove ( ) : T | undefined {
        if ( this . _arr . length > 0 )
            return this . _arr . pop ( ) ;
        else
            return undefined ;
    }
}

export class Queue < T > extends AdapterBase < T > {
    public remove ( ) : T | undefined {
        if ( this . _arr . length > 0 )
            return this . _arr . shift ( ) ;
        else
            return undefined ;
    }
}
```

9.5 本章总结

本章主要关注数据结构方面的相关知识。

在 9.1 节中，讲解了一些树节点相关的术语，并且介绍了如何在树节点添加子节点时防止循环引用的方法。随后实现了节点的增加和删除操作，以及在节点层次关系查询时经常用到的一些操作方法。

在 9.2 节中，主要介绍了树节点的各种遍历算法。首先介绍了通用树结构的 8 种遍历顺序：

- 广度优先（层次），从上到下（先根），从左到右的遍历算法。
- 广度优先（层次），从上到下（先根），从右到左的遍历算法
- 广度优先（层次），从下到上（后根），从左到右的遍历算法。
- 广度优先（层次），从下到上（后根），从右到左的遍历算法。
- 深度优先、从上到下（先根），从左到右的遍历算法。
- 深度优先、从上到下（先根），从右到左的遍历算法。
- 深度优先、从下到上（后根），从左到右的遍历算法。
- 深度优先、从下到上（后根），从右到左的遍历算法。

然后使用辅助的队列数据结构和栈数据结构,分别实现了广度优先和深度优先的 8 个遍历枚举器。

接着使用递归的方式实现了深度优先的 4 种树遍历算法。

最后使用儿子兄弟的方式实现了深度优先且不使用栈结构的遍历算法。

在 9.3 节中,介绍了树结构由于循环引用的关系,无法直接使用 JSON 类的静态方法来进行序列化和反序列化。将自引用的成员变量 parent 转换成数组中的索引表示,从而解决了树结构的序列化和反序列化的问题。

由于在树结构的枚举器遍历中,一直在使用 IAdapter 接口,而该接口的两个实现子类 Queue 和 Stack 尚未实现,因此在 9.4 节中实现了这两个类。

以上就是本章所讲的内容。

第 10 章 场景图系统

在第 8 章中，我们以面向接口的编程方式实现了一个具有必要功能（更新、重绘、事件分发和响应）的、使用非场景图类型的、支持精确点选的、基于非立即渲染模式（保留模式）的、采取享元设计模式的精灵系统。而在第 9 章中，则实现了一个基于泛型的通用树数据结构。

本章将第 9 章中实现的树的数据结构应用到第 8 章中实现的精灵系统中，从而以面向接口的编程方式实现一个具有必要功能（更新、重绘、裁剪及事件分发和响应）的、使用场景图类型的、支持精确点选的、基于非立即渲染模式（保留模式）的、采取享元设计模式的，并兼容第 8 章实现的非场景图类型的精灵系统。

10.1 实现场景图精灵系统

在本节中，先通过一个简单的例子来看一下非场景图精灵系统的不足之处，然后来了解一下场景图的特点，接着对精灵系统接口增加一些必要的、基于树结构层次的操作方法，最后实现场景图精灵系统。在实现场景图精灵系统时，会发现仅需要重写 IDispatcher 和 ISpriteContainer 接口的实现类，并不需要修改 ISprite 接口及 IShape 接口实现类。

10.1.1 非场景图精灵系统的不足之处

在第 5 章中，实现了一个坦克 Demo。本节将通过这个简化版的坦克 Demo 来了解一下非场景图精灵系统的不足之处。首先来构建初始化的效果图，具体代码如下：

```
class SimpleTankTest {
    private _app : Sprite2DApplication ;
    private _tankSprite : ISprite ;
    private _turretSprite : ISprite ;
    public constructor ( app : Sprite2DApplication ) {
        this . _app = app ;

        this . _tankSprite = SpriteFactory . createSprite ( SpriteFactory .
        createRect ( 80 , 50 , 0.5 , 0.5 ) ) ;
        this . _tankSprite . fillStyle = 'blue' ;
        this . _tankSprite . keyEvent = this . keyEvent . bind ( this ) ;
```

```
            // _turretSprite的位置必须永远和_tankSprite一样
            this . _turretSprite = SpriteFactory . createSprite ( SpriteFactory .
            createXLine ( 100 ) ) ;
            this . _turretSprite . strokeStyle = 'red' ;
            this . _turretSprite . lineWidth = 5 ;
            this . _turretSprite . keyEvent = this . keyEvent . bind ( this ) ;
            this . _app . rootContainer . addSprite ( this . _tankSprite ) ;
            this . _app . rootContainer . addSprite ( this . _turretSprite ) ;
            this . _app . start ( ) ;
        }
    }
```

当运行上述代码后,会获得如图 10.1 所示的效果图。
下面来实现键盘事件处理方法,该方法主要作用如下:

- 按 A / Q 键让整个坦克分别顺时针和逆时针旋转 2°。
- 按 D / E 键单让炮管分别顺时针和逆时针旋转 5°。

图 10.1 坦克效果图

- 按 W / S 键让整个坦克沿着坦克当前的朝向前进和后退。

实现上述需求的代码如下:

```
    private keyEvent ( spr : ISprite , evt : CanvasKeyBoardEvent ) : void {
        if ( evt . type === EInputEventType . KEYPRESS ) {
            if ( evt . key === 'a'){
                this . _tankSprite . rotation += 2 ;
                this . _turretSprite . rotation += 2 ;
            } else if ( evt . key === 'q') {
                this . _tankSprite . rotation -= 2 ;
                this . _turretSprite . rotation -= 2 ;
            } else if ( evt . key === 'd' ) {
                this . _turretSprite . rotation += 5 ;
            } else if ( evt . key === 'e') {
                this . _turretSprite . rotation -= 5 ;
            } else if ( evt . key === 'w' ) {
                let forward : vec2 = this . _tankSprite . getWorldMatrix ( ) .
                xAxis ;
                this . _tankSprite . x += forward . x * 3 ;
                this . _tankSprite . y += forward . y * 3 ;
                this . _turretSprite . x += forward . x * 3 ;
                this . _turretSprite . y += forward . y * 3 ;
            } else if ( evt . key === 's' ) {
                let forward : vec2 = this . _tankSprite . getWorldMatrix ( ) .
                xAxis ;
                this . _tankSprite . x -= forward . x * 3 ;
                this . _tankSprite . y -= forward . y * 3 ;
                this . _turretSprite . x -= forward . x * 3 ;
                this . _turretSprite . y -= forward . y * 3 ;
            }
        }
    }
}
```

代码很简单,目的是为了演示层级运动的特点,即_tankSprite 精灵移动时需要让_turretSprite 也跟着一起移动,_tankSprite 精灵旋转时_turretSprite 也跟着一起旋转,但是_turretSprite 自己旋转时,不能影响到_tankSprite。

可以看到，非场景图精灵系统缺乏一个关键的层次关系寻址系统，导致基于基本形体层级组合而成的对象（坦克由矩形精灵和线段精灵组合而成）无法得到有效控制。上面代码中，使用硬编码的方式，在 _tankSprite 发生移动或旋转时，强制 _turretSprite 做一致的变换操作。问题是，当层级关系非常复杂时，如图 10.2 所示，不可能每次都手动去控制解决各个层级精灵之间的坐标系变换数据同步问题，在后续章节中就是来解决这个缺陷。

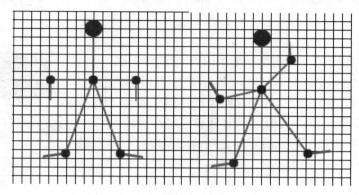

图 10.2　骨骼层级精灵

> 注意：上述代码中，需要关注一个有用的数学操作，即 this._tankSprite.getWorldMatrix().xAxis，这句代码的作用是获得_tankSprite 的全局（世界）变换矩阵，然后从全局（世界）变换矩阵中抽取 x 轴向量（之所以抽取 x 轴向量是因为坦克初始化时朝着局部坐标系的 x 轴正方向的，再次强调初始化朝向的重要性），由于没有使用缩放，因此 x 轴向量是单位方向向量，表示坦克在全局坐标系中的朝向。关于 xAxis 属性实现的源码在第 7 章矩阵相关章节。

接着使用了第 6 章中介绍向量的加减法和向量的缩放相关的知识点，让坦克朝着当前正确的朝向，根据不同的键盘值前后运动，可以看到，不需要使用任何三角函数，坦克就能做快速的有向运动。

前面章节都是朝着鼠标位置运动，这个 Demo 使用键盘控制朝向运动，游戏或 UI 开发中，方向控制基本也是使用这两种常用方法，正好在此一并介绍。

10.1.2　树结构场景图系统

上一节讲解了非场景图精灵系统潜在的不足之处，本节来详细讲解上一节中的图 10.2 所示的细节。如图 10.3 所示，将图 10.2 中的骨架以树结构的方式呈现出来。

如果通过树结构来建模整个骨骼场景系统，就能解决上一节的不足之处（即非场景图精灵缺乏一个关键的层次关系寻址系统）。而通过树结构，我们就可以获得层次相对运动的效果，如果旋转躯干骨骼，那么整个骨架都会跟着旋转；如果旋转左腿骨骼，则左腿骨骼旋转的同时，左脚骨骼也会跟着一起旋转；如果旋转左脚骨骼，只会影响自身，不会干扰到其父亲骨骼。

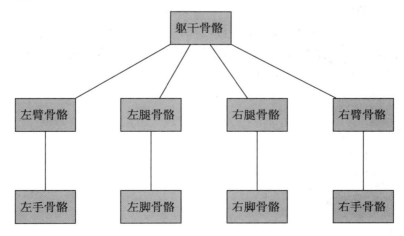

图 10.3　骨骼精灵树状图

父亲节点影响子孙节点的行为，但是子孙节点不会干扰父亲节点的行为，这是需要的效果。这些操作都不需要硬编码，通过应用第 9 章的树数据结构和前面章节的向量及矩阵相关知识，能够构建一个树结构层次与变换系统（场景图），就能很容易实现上述需求。

另外，通过树结构构建层次与变换系统，还可以做更多的事情：

- 可以使用深度优先、从上到下（先根／前序）、从左到右的遍历顺序来绘制所有精灵，之所以使用这种顺序，是因为可以完美处理精灵的遮挡关系，先绘制背景，再绘制前景。
- 可以使用深度优先、从下到上（后根／后序）的遍历方式来删除所有的树节点，析构内存，这种方式总是先从叶子节点（从左到右或从右到左）开始删除，逐步往上删除。
- 在本书中，为简单起见采用了最方便的方式，即深度优先、从下到上（后根／后序）、从右到左的方式来进行鼠标命中检测，返回选中的精灵。这种遍历顺序和渲染时候使用的深度正好相反。
- 可以从当前精灵节点一直向上遍历其所有父亲精灵节点，从而计算出局部-世界变换矩阵，以及对其求逆获得世界-局部的变换矩阵，在后续代码中，会有大量矩阵变换操作。
- 还可以在目前简单的事件分发和处理系统上添加对冒泡事件系统的支持，对于 HTML 开发人员来说，冒泡事件应该是非常熟悉的，该事件系统也是建立在树结构的基础上的。

通过树结构构建层次与变换系统，还有更多的优点，可以在具体应用中发现并分析。

10.1.3　矩阵堆栈和场景图

在前面章节中，一直在使用 Canvas2D 中的矩阵堆栈相关功能，并且在 7.1.11 节中自己实现了 Canvas2D 中的矩阵堆栈。事实上，图 10.2 所示的效果，通过矩阵堆栈也能实现，可以按照如下伪代码进行相关绘制操作：

```
平移及旋转矩阵
绘制躯干骨骼
save 当前矩阵
    平移及旋转矩阵
    绘制左臂骨骼
    save 当前矩阵
        平移及旋转矩阵
        绘制左脚骨骼
    restore 矩阵
restore 矩阵
同上类似，分别绘制右臂和右手、左腿和左脚；以及右腿和右脚
```

矩阵堆栈在绘制时的确可以模拟这种层次关系，但是无法进行层次之间的寻址。因为矩阵堆栈属于立即绘制模式，不保留这种层次关系，而精灵系统需要经常获取并操作这种层次关系，所以一劳永逸的解决方案还是自己实现场景图管理模式的精灵系统。

10.1.4　实现场景图精灵系统概述

在第 8 章中已经实现了一个面向接口编程的非场景图类型精灵系统，该系统中可以分为 4 部分，即 IDispatcher 接口、ISpriteContainer 及 ISprite（扩展了 ITransformable 和 IRenderState 接口）和 IShape（扩展了 IDrawable 和 IHittable 接口）。其中 Sprite2DManager 类实现了 IDispatcher 和 ISpriteContainer 接口，而在即将要实现的基于场景图模式的精灵系统中，要全部重新实现 IDispatcher 和 ISpriteContainer 接口相关内容。

对于 ISprte2D 接口实现类 Sprite2D 来说，需要重写 getWorldMatrix 这个方法，让这个方法支持获取从当前精灵到根节点精灵合成的全局矩阵。而对于整个 IShape，以及其所有的形体实现类来说，不需要做任何改动。

至于整个精灵系统的运行及协作流程，在第 8 章中已经全部制定完毕。Canvas2D Application 入口类只和 IDispatcher 接口打交道，只要实现不同的 IDispatcher 接口，就能像积木一样组装或替换，因此不需要做任何修改，这也是面向接口编程的优势所在。

10.1.5　核心的 SpriteNode 类

整个场景图类型的精灵系统的核心类是 SpriteNode 类。顾名思义，该类继承自

TreeNode，而 TreeNode 持有的 data 成员变量（指针）如果指向 ISprite 接口，就能获得两个核心功能，即：
- 通过 TreeNode 提供精灵的层次关系查询和遍历等功能。
- 通过 ISprite 接口获得精灵的几何变换、更改渲染状态、更新、重绘、点选和事件处理等功能。

由于是面向接口编程，在第 8 章中已经定下了 ISpriteContainer 是 ISprite 容器这层关系。但是在第 8 章实现的非场景图精灵系统中，ISpriteContainer 是一种线性容器，其定义的接口方法能兼容树结构儿子列表相关的操作，还缺乏一个获取父亲精灵的接口方法，因此要对 ISpriteContainer 增加这个接口方法，具体代码如下：

```
export interface ISpriteContainer {
    name : string ;                                          // 如有需要，提供一个容器名称
    // 添加一个精灵到容器
    addSprite ( sprite : ISprite ) : ISpriteContainer ;
    // 从容器中删除一个精灵
    removeSprite ( sprite : ISprite ) : boolean ;
    // 清空整个容器
    removeAll ( includeThis : boolean ) : void ;
    // 根据精灵获取索引号，没找到精灵，就返回-1
    getSpriteIndex ( sprite : ISprite ) : number ;
    // 根据索引号，从容器中获取精灵
    getSprite ( idx : number ) : ISprite ;
    // 获取容器中精灵的数量
    getSpriteCount ( ) : number ;
    // 上面这些成员接口方法和属性是在第 8 章非场景图精灵系统中定义的
    // 接着要定义两个支持树结构场景图精灵系统的相关接口成员
    // 获取当前精灵容器的父精灵
    getParentSprite ( ) : ISprite | undefined ;
    readonly sprite : ISprite | undefined ;
}
```

从上面的代码中会看到，增加了一个只读的 sprite 属性，该属性返回当前 ISpriteContainer 所持有的 ISprite 对象，为了更好地理解这个属性，参考如图 10.4 所示的 ISpriteContainer 与 ISprite 关系图。

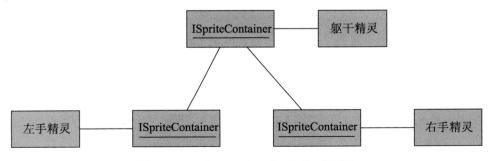

图 10.4　ISpriteContainer 与 ISprite 关系图

通过图 10.4，很清楚地知道了 ISpriteContainer 和 ISprite 之间的关系，ISpriteContainer 是内部包装了 TreeNode < ISprite >相关的操作，获得了树的层次性功能。而 ISpriteContainer 的 sprite 属性可以获得依附在 TreeNode < ISprite >上的 ISprite 接口对象，而 ISprite 接口有一个 owner 属性，可以反过来引用到对应的 ISpriteContainer 接口对象。

10.1.6　实现 SpriteNode 类的接口方法

下面看 SpriteNode 类的接口方法实现，具体代码如下：

```typescript
export class SpriteNode extends TreeNode < ISprite > implements ISpriteContainer {
    // 构造函数
    public constructor ( sprite : ISprite , parent :SpriteNode | undefined = undefined , name : string = "spriteNode" ) {
        // 调用 TreeNode 基类构造函数： 依附哪个精灵、是否将该精灵添加到父亲节点中，以及是否要名字
        super ( sprite , parent , name ) ;
    }

    // 添加儿子精灵
    public addSprite ( sprite : ISprite ) : ISpriteContainer {
        let node : SpriteNode = new SpriteNode ( sprite , this , sprite . name ) ;
        return node ;
    }

    // 移除儿子精灵
    public removeSprite ( sprite : ISprite ) : boolean {
        // sprite 在儿子列表中的索引号
        let idx : number = this . getSpriteIndex ( sprite ) ;
        if ( idx === -1 ) {
            return false ;
        }
        // 存在，则调用基类方法删除
        if ( this . removeChildAt ( idx ) === undefined ) {
            return false ;
        } else {
            return true ;
        }
    }

    // 删除整棵树
    // 参数 includeThis 为 true，将自己也从父亲列表中删除，否则删除所有子孙节点
    public removeAll ( includeThis : boolean ) : void {
        // 删除树节点，要以底向上，深度优先为顺序
        // 使用迭代器方式来操作
        let iter:IEnumerator<TreeNode<ISprite>> = NodeEnumeratorFactory.create_bf_r2l_b2t_iter(this);
        let current: TreeNode<ISprite> | undefined = undefined;
```

```
                while ( iter . moveNext ( ) ) {
                    current = iter .current ;
                    if ( current !== undefined ) {
                        {
                            if (current . data !== undefined ) {
                                if ( current === this ) {
                                    if ( includeThis === true ) {
                                        current . data = undefined ;
                                        current = current . remove ( ) ;
                                    }
                                } else {
                                    current . data = undefined ;
                                    current = current . remove ( ) ;
                                }
                            }
                        }
                    }
                }
            }

            // 获取索引号为 idx 的子精灵
            public getSprite ( idx : number ) : ISprite {
                if ( idx < 0 || idx > this . childCount -1 ) {
                    throw new Error ( "参数 idx 越界!!" ) ;
                }
                let spr : ISprite | undefined = ( this . getChildAt ( idx ) as SpriteNode ) . sprite
                if ( spr === undefined ) {
                    alert ( "sprite 为 undefined, 请检查原因!!!" ) ;
                    throw new Error ( "sprite 为 undefined, 请检查原因!!!" ) ;
                }

                return spr ;
            }

            // 获取父精灵
            public getParentSprite ( ) : ISprite | undefined {
                let parent : SpriteNode | undefined = this . parent as SpriteNode ;
                if ( parent !== undefined ) {
                    return parent . sprite ;
                } else {
                    return undefined ;
                }
            }

            // 获取儿子数量
            public getSpriteCount ( ) : number {
                return this . childCount ;
            }

            // 查询参数精灵在子精灵列表中的索引号
            public getSpriteIndex ( sprite: ISprite ): number {
                for ( let i : number = 0 ; i < this . childCount ; i ++ ) {
                    let child : SpriteNode = this . getChildAt ( i ) as SpriteNode ;
```

```typescript
            if ( child !== undefined ) {
                if ( child . sprite !== undefined ) {
                    if ( child . sprite === sprite ) {
                        return i ;
                    }
                }
            }
        }
        return - 1;
    }

    // 很重要的一个函数, override 基类方法
    public addChildAt ( child : TreeNode < ISprite > , index : number ) : TreeNode < ISprite > | undefined {
        // 调用基类的方法, 这样就添加了 child 子节点
        let ret : TreeNode < ISprite > | undefined = super . addChildAt ( child , index ) ;
        // 如果添加儿子成功
        if ( ret !== undefined ) {
            // 并且儿子有附加的精灵
            if ( ret . data ) {
                // 设置儿子附加的精灵的 owner
                // 这样能从 spriteNode . data 找到精灵对象
                // 而且也能从精灵对象找到其所依附的节点
                ret . data . owner = ret as SpriteNode ;
            }
        }

        return ret;
    }

    public get sprite ( ) : ISprite | undefined {
        return this . data ;
    }
}
```

10.1.7　SpriteNode 的 findSprite 方法实现

接下来要实现一个非常重要的与数学相关的方法，即在层次遍历中进行鼠标命中检测。具体代码如下：

```typescript
// 数学系统支持的方法
// 给定一个点, 查找与该点最先发生碰撞的那个精灵
// 并且如果 localPoint 参数不为 null, 且有选中精灵时, 返回 src 在该精灵坐标系局部表示的点坐标
public findSprite (src : vec2 ,localPoint : vec2 | null = null ):ISprite | undefined {

    //为简单起见, 采取的是从右向左, 深度最前的那个精灵获得点击事件
    let iter : IEnumerator < TreeNode < ISprite > > = NodeEnumeratorFactory . create_bf_r2l_b2t_iter ( this . root ) ;
```

```
        let current : TreeNode<ISprite> | undefined = undefined ;
        let mat : mat2d ;
        let dest : vec2 = vec2 . create ( ) ;
        // 使用迭代器
        while( iter . moveNext ( ) ) {
            current = iter . current ;
            if ( current !== undefined ) {
                if( current . data !== undefined ) {
                    // 获取当前节点对应精灵的世界-局部矩阵
                    mat = current . data . getLocalMatrix ( ) ;
                    {
                        // 将全局表示的点变换到当前精灵所在的坐标系
                        // 这一步是本函数的精华
                        Math2D . transform ( mat , src , dest ) ;
                        // 进行碰撞检测
                        if( current . data . hitTest ( dest ) ) {
                            // 如果碰撞检测成功
                            if ( localPoint !== null ) {
                                // 输出局部表示的点的坐标
                                localPoint . x = dest . x ;
                                localPoint . y = dest . y ;
                            }
                            // 返回碰撞检测发生的精灵
                            return current . data ;
                        }
                    }
                }
            }
        }
        // 到这里说明没找到碰撞的精灵,返回 undefined
        return undefined ;
}
```

10.1.8 递归的更新与绘制操作

下面来看两个算法实现上具有高度相似性的操作。具体代码如下:

```
// 本书以前的递归算法都是函数本身调用函数
// 这里换一种实现方式,采用 update 调用_updateChildren,然后在_updateChildren
中再递归调用 update 的方式
// 由此可见,递归有两种形式:自己调用自己,以及 a->b->a 这种方式
public update ( msec : number , diffSec : number ) : void {
    if ( this . sprite !== undefined ) {
        // 调用精灵的 update 函数,内部触发 PREORDER 类型的 updateEvent
        this . sprite . update ( msec , diffSec , EOrder . PREORDER ) ;
        this . _updateChildren ( msec , diffSec ) ;
        // 调用精灵的 update 函数,内部触发 POSTORDER 类型的 updateEvent
        this . sprite . update ( msec , diffSec , EOrder . POSTORDER ) ;
    }
}
```

```
protected _updateChildren ( msec : number,diffSec : number ) : void {
    for ( let i = 0 ; i < this . childCount ; i++ ) {
        let child : TreeNode<ISprite> | undefined = this . getChildAt ( i ) ;
        if ( child !== undefined ) {
            let spriteNode : SpriteNode = child as SpriteNode ;
            spriteNode . update ( msec , diffSec ) ;
        }
    }
}

public draw ( context: CanvasRenderingContext2D ): void {
    if ( this . sprite !== undefined ) {
        this . sprite . draw ( context ) ;
        this . _drawChildren ( context ) ;
    }
}

protected _drawChildren ( context: CanvasRenderingContext2D ): void {
    // 深度优先，从上到下，从左到右的递归遍历
    for ( let i : number = 0; i < this . childCount ; i++ ) {
        let child : TreeNode < ISprite > | undefined = this . getChildAt ( i ) ;
        if ( child !== undefined ) {
            let spriteNode : SpriteNode = child as SpriteNode ;
            spriteNode . draw ( context ) ;
        }
    }
}
```

至此，SpriteNode 的所有实现源码已介绍完毕，接下来就可以实现 IDispatcher 接口来分发相关命令。

10.1.9　SpriteNodeManager 类

SpriteNodeManager 类实现了 IDispatcher 接口，并且持有一个 SpriteNode 类型的根节点，其 dispatchDraw 和 dispatchUpdate 方法在内部调用根节点相对应的方法，而其他分发方法如下处理：

```
export class SpriteNodeManager implements IDispatcher {
    private _rootNode : SpriteNode ;
    // 所有的鼠标拖动事件都发送到该精灵上
    private _dragSprite : ISprite | undefined = undefined ;

    // 对于_rootNode，让它直接挂接一个 grid 类型的精灵
    public constructor ( width : number , height : number ) {
        let spr : ISprite = SpriteFactory . createISprite ( SpriteFactory .
        createGrid ( width , height ) ) ;
        spr . name = 'root' ;
        spr . strokeStyle = "black" ;
        spr . fillStyle ='white' ;
        spr . renderType = ERenderType . STROKE_FILL ;
        // 无父亲节点，是根节点
```

```typescript
        this._rootNode = new SpriteNode( spr , undefined , spr.name );
        // 设置精灵的 owner
        spr.owner = this._rootNode;
    }

    public get container() : ISpriteContainer {
        return this._rootNode;
    }

    // 鼠标事件分发
    public dispatchMouseEvent( evt : CanvasMouseEvent ) : void {
        if ( evt.type === EInputEventType.MOUSEUP ) {
            this._dragSprite = undefined;
        } else if (evt.type === EInputEventType.MOUSEDRAG) {
            if ( this._dragSprite !== undefined ) {
                if ( this._dragSprite.mouseEvent !== null ) {
                    this._dragSprite.mouseEvent( this._dragSprite ,evt );
                    // 处理后直接退出
                    return ;
                }
            }
        }

        // 调用 rootSprite 的 findSprite 方法，查找到鼠标命中的精灵
        let spr : ISprite | undefined = this._rootNode.findSprite( evt.canvasPosition , evt.localPosition );
        if ( spr !== undefined ) {
            evt.hasLocalPosition = true;
            if ( evt.button === 0 && evt.type === EInputEventType.MOUSEDOWN ) {
                this._dragSprite = spr;
            }

            if ( evt.type === EInputEventType.MOUSEDRAG )
                return ;

            // 触发鼠标事件
            if ( spr.mouseEvent ) {
                spr.mouseEvent( spr , evt );
                return ;
            }
        } else {
            evt.hasLocalPosition = false;
        }
    }

    // 键盘采取最简单处理方式，遍历整个场景图，让每个场景图中具有键盘事件处理函数的
    // 精灵都触发一次键盘事件
    public dispatchKeyEvent( evt: CanvasKeyBoardEvent ) : void {
        // 调用 visit 方法，使用箭头函数作回调函数
        this._rootNode.visit(
            ( node : TreeNode < ISprite > ) : void => {
                if ( node.data !== undefined ) {
                    if ( node.data.keyEvent !== null ) {
                        node.data.keyEvent( node.data , evt );
```

```
                    }
                }
            }
        );
    }

    dispatchUpdate ( msec : number , diffSec : number ) : void {
        this . _rootNode . update ( msec , diffSec ) ;
    }

    dispatchDraw ( context : CanvasRenderingContext2D ) : void {
        this . _rootNode . draw ( context ) ;
    }
}
```

10.1.10　修改 Sprite2D 类的 getWorldMatrix 方法

现在只要修改完 getWorldMatrix 后，场景图精灵系统就完成了。具体的修改代码如下：

```
public getWorldMatrix ( ) : mat2d {
    // 使用 js instanceof 操作符，能判断 this . owner 是不是 SpriteNode 类的对象
    // 如果是，则获取当前精灵到根节点精灵合成的局部-全局变换矩阵
    if ( this . owner instanceof SpriteNode ) {
        let arr: TreeNode < ISprite > [ ] = [ ] ;
        let curr: TreeNode < ISprite > | undefined = this . owner as SpriteNode ;
        while ( curr !== undefined ) {
            //从当前的节点到 root 节点记录在 arr 中
            arr . push ( curr ) ;
            curr = curr . parent ;
        }
        let out : mat2d = mat2d . create ( ) ;
        let currMat : mat2d ;
        // 这时候，arr 中的内容如:[this, parent, ..., root];
        // 但是要进行从局部到全局的矩阵合成操作
        // 因此需要矩阵乘法的顺序为 root * ... * parent * this
        // 所以遍历时，需要以从后向前为顺序
        for ( let i : number = arr . length - 1 ; i >= 0 ; i-- ) {
            curr = arr [ i ] ;
            if( curr . data ) {
                // transform2D 类并没有公开给 ISprite 接口，因此需要用
                // as 关键词进行向下转型操作
                currMat = (curr . data as Sprite2D ). transform . toMatrix ( ) ;
                mat2d . multiply ( out , currMat , out ) ;
            }
        }
        return out ;
    } else {
        // instanceof 返回 false,表示是非场景图精灵，没有层次关系
```

```
        // 与原来的代码一样
        return this.transform.toMatrix();
    }
}
```

10.1.11 让 Sprite2DApplication 类支持场景图精灵系统

为了让原来的 Sprite2DApplication 兼容场景图精灵系统，需要在构造函数中增加一个 isHierarchical 参数，用来指示是否使用层次场景图类型的精灵系统。代码如下：

```
public constructor (canvas : HTMLCanvasElement ,isHierarchical: boolean = true ) {
    // 取消右键上下文菜单，这样就可以使用鼠标右键进行事件处理
    document.oncontextmenu = function () {
        return false;
    }
    super( canvas ); // 调用基类构造函数
    if ( isHierarchical ) {
        this._dispatcher = new SpriteNodeManager ( canvas.width , canvas.height );
    } else {
        this._dispatcher = new Sprite2DManager ( );
    }
}
```

当用如下方式调用第 8 章实现的精灵系统测试 Demo 时，场景图类型的精灵系统可以不做任何修改地运行该 Demo 的代码。具体效果如图 10.5 所示。

```
let canvas : HTMLCanvasElement | null = document.getElementById ('canvas') as HTMLCanvasElement
let app : Sprite2DApplication = new Sprite2DApplication ( canvas , true );
// 切换为 false，使用非场景图精灵系统，true 则为场景图类型精灵系统
new IShape_Event_Hittest_Draw_Test_Demo ( app );
```

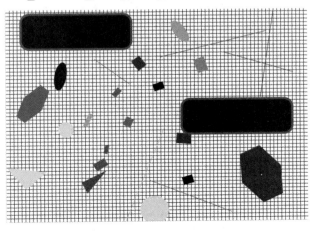

图 10.5 场景图方式运行第 8 章中的精灵系统测试 Demo

下一节将使用场景图类型的精灵系统来实现骨骼层次精灵 Demo。

10.2 骨骼层次精灵 Demo

对场景图精灵系统来说，笔者认为最经典的莫过于太阳系模拟，以及骨骼动画这两个例子。在 5.1.9 节的公转与自转内容中，已经通过 Canvas2D 中的矩阵堆栈模拟了月亮绕太阳公转、太阳及月亮自转的 Demo，我们发现矩阵堆栈就是一个简易版本的无法体现层次关系的场景图实现。那么在本节中，将实现简化版的骨骼动画系统。

真正的骨骼动画（Skeleton Animation）全称为骨骼蒙皮动画，该动画系统由两部分组成：用于控制动作的骨架（Skeleton）部分，以及用于绘制模型的蒙皮（Skin 或 Mesh）部分。骨架与蒙皮分离，在运行时，通过相关算法，动态地将蒙皮部分的顶点坐标映射到对应的骨骼坐标空间中，从而形成骨骼动画的某一帧。在 Demo 中，只实现最简单的层次骨骼（Bone）部分。

10.2.1 实现骨骼形体

如图 10.2 所示，骨架是由一根根具有层次性的骨骼组成的。如图 10.6 所示，每一根骨骼都是由蓝色的圆圈（见标注①，表示骨骼的原点）和一根红色的直线表示（见标注②表示骨骼）。

我们可以专门实现一个形体（IShape）来表示骨骼，具体代码如下：

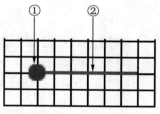

图 10.6　骨骼形体

```
export class Bone extends Line {
    // override 覆写基类 type 只读属性
    public get type ( ) : string { return "Bone" ; }

    // override 覆写基类 endDraw 方法
    public endDraw ( transformable : ITransformable , state : IRenderState , context : CanvasRenderingContext2D ) : void {
        super . endDraw ( transformable , state , context ) ;
        context . save ( ) ;
        let mat : mat2d = transformable . getWorldMatrix ( ) ;
        context . setTransform ( 1 , 0 , 0 , 1 , mat . values [ 4 ] , mat . values [ 5 ] ) ;
        context . beginPath ( ) ;
        context . fillStyle = 'blue' ;
        context . arc ( this . start . x , this . start . y , 5 , 0 , Math . PI * 2 ) ;
        context . fill ( ) ;
        context . restore ( ) ;
    }
}
```

代码很简单，之所以继承自 Line 类，是因为骨骼使用红色的线段来表示，可以重用 Line 类的所有代码。

现在要增加蓝色的圆圈表示骨骼的原点，对于该圆圈，是有一定限制的，需要满足如下两个要求，即：

- 在 Bone 形体所附加的精灵发生平移时，该圆圈要一起平移。
- 在 Bone 形体所附加的精灵发生缩放和（或）旋转现象时，该圆圈不能同时缩放和（或）旋转。

之所以要满足上述第二条要求，原因是：

我们肉眼观察不到圆圈的旋转变化，因此圆圈的旋转是没有任何意义的。

观察图 10.2 和图 10.3 会发现，虽然有 9 个骨骼精灵，但是这九个骨骼精灵具有相同的形状（Bone），只是每个骨骼精灵形状（Bone）朝向和 Bone 中红色线段的长短不同。但是蓝色圆圈却保持同样大小。

要实现蓝色圆圈仅需要平移，不需要旋转和缩放效果的代码，在此再次强调一下这段代码，具体如下：

```
// transformable . getWorldMatrix ( )获得的是精灵的当前局部-全局的矩阵
// 该矩阵的 values [ 0 ]、values [ 1 ]、values [ 2 ]和 values [ 3 ]分量中存储的是旋转和缩放相关的信息
// 将其设置为单位矩阵，就能消除所有旋转和缩放相关的信息
// 同时由于局部-全局矩阵中的 values[4]和 values[5]这两个分量中存储了平移相关信息
// 需要这两个分量信息，让圆圈跟着线段一起移动，保持相对位置的不变性
// 可以使用 context . setTransform ( 1 , 0 , 0 , 1 , mat . values [ 4 ] , mat . values [ 5 ] ) 这句代码实现
let mat : mat2d = transformable . getWorldMatrix ( ) ;
context . setTransform ( 1 , 0 , 0 , 1 , mat . values [ 4 ] , mat . values [ 5 ] ) ;
```

最后要强调的是，Bone 类是一个享元类，在整个应用程序内，只需要一个实列就能被无数个骨骼精灵所共享。这就是一直强调的基于享元设计模式的精灵系统，可以减少内存使用。

10.2.2 SkeletonPersonTest 类

有了 Bone 类，就可以构建如图 10.2 所示的骨架，并且移动或旋转整个骨架，或对该骨架的每根选中的骨骼精灵进行绕原点（蓝色圆圈）旋转操作。

可以将上述需求的实现封装到一个名为 SkeletonPersonTest 的类中。先来看一下其成员变量和构造函数，具体代码如下：

```
class skeletonPersonTest {
    private _app : Sprite2DApplication ;      // 指向应用程序的入口类
    private _skeletonPerson ! : ISprite ;     // 根骨骼精灵（躯干）
    private _bone : IShape ;                  // 被多个精灵所共享的骨骼形体实例
```

```
        private _boneLen : number ;              // 骨骼的基准长度
        private _armScale : number ;             // 左手臂和右手臂精灵x轴方向的缩放系数
        private _hand_foot_Scale : number ;
                                                 // 左手脚和右手脚精灵x轴方向的缩放系数
        private _legScale : number ;             // 左腿和右腿部精灵x轴方向的缩放系数
        // 当前没鼠标点击命中的骨骼精灵
        private _hittedBoneSprite : ISprite | null ;
        public constructor ( app : Sprite2DApplication ) {
            this . _app = app ;
            this . _hittedBoneSprite = null ;
            this . _boneLen = 60 ;
            this . _armScale = 0.8 ;
            this . _hand_foot_Scale = 0.4 ;
            this . _legScale = 1.5 ;
            // 创建初始朝向为x轴，长度为_boneLen个单位的Bone形体实例
            this . _bone = SpriteFactory . createBone ( this . _boneLen , 0 ) ;
            this . createSkeleton ( ) ;            // 创建骨架
            this . _app . start ( ) ;              // 启动动画循环
        }
```

接下来看一下关键的 createSkeleton 方法。该方法创建如图 10.2 所示的左侧图像初始化骨架层次结构，在该方法中调用了一个创建骨骼精灵并形成骨骼层次关系的私有方法 createBoneSprite。先来看一下这个方法的实现，具体代码如下：

```
// scale 参数表示当前创建的精灵x轴方向的缩放系数
private createBoneSprite ( scale : number , rotation : number , parent : ISpriteContainer , name : string = '' ) : ISprite {
    // 使用bone形体实例创建一个骨骼精灵，并设置相关参数
    let spr : ISprite = SpriteFactory . createSprite ( this . _bone ) ;
    spr . lineWidth = 2 ;
    spr. strokeStyle = 'red' ;
    spr . rotation = rotation ;
    spr . scaleX = scale ;
    spr . name = name ;
    // 将当前的精灵作为儿子添加到参数parent中
    parent . addSprite ( spr ) ;
    return spr ;
}
```

有了上面的辅助方法，就来实现创建整个骨架的 createSkeleton 方法，具体代码如下：

```
private createSkeleton ( x : number = 200 , y : number = 200 ) {
    let spr : ISprite ;
    // 根（身躯）
    // 初始化时，骨骼如图10.6所示，需要逆时针旋转-90°，这样垂直朝上
    this . _skeletonPerson = this . createBoneSprite ( 1.0 , -90 , this . _app . rootContainer , 'person' ) ;
    this . _skeletonPerson . x = x ;
    this . _skeletonPerson . y = y ;

    // 头使用一个10个单位做半径的圆圈表示
```

```
        let circle : IShape = SpriteFactory . createCircle ( 10 ) ;
        // 偏移_boneLen 个单位
        spr = SpriteFactory . createISprite ( circle , this . _boneLen , 0 ) ;
        spr . fillStyle = 'blue' ;
        spr . rotation =  0 ;
        this . _skeletonPerson . owner . addSprite ( spr ) ;
                                                    // 将头精灵作为身躯精灵的儿子
        // 左臂精灵在身躯精灵旋转的基础上再逆时针旋转 90°，这样指向左侧（x 轴的负方向）
        spr = this . createBoneSprite ( this . _armScale , -90 , this .
        _skeletonPerson . owner ) ;
           // 左手精灵在左臂精灵的基础上再逆时针旋转 90°
           spr = this . createBoneSprite ( this . _hand_foot_Scale , -90 ,
           spr . owner ) ;
           spr . x = this . _boneLen ;
        // 右臂精灵在身躯精灵旋转的基础上再顺时针旋转 90 度，这样指向右侧（x 轴的正方向）
        spr = this . createBoneSprite ( this . _armScale , 90 , this .
        _skeletonPerson . owner ) ;
           // 右手精灵在右臂精灵的基础上再顺时针旋转 90°
           spr = this . createBoneSprite ( this . _hand_foot_Scale , 90 ,
           spr . owner ) ;
              spr . x = this . _boneLen ;
        // 左腿
        spr = this . createBoneSprite ( this . _legScale , -160 , this .
        _skeletonPerson . owner ) ;
           // 左脚
           spr = this . createBoneSprite ( this . _hand_foot_Scale , 70 ,
           spr . owner ) ;
           spr . x = this . _boneLen ;
        // 右腿
        spr = this . createBoneSprite ( this . _legScale , 160 , this .
        _skeletonPerson . owner ) ;
           // 右脚
           spr = this . createBoneSprite ( this . _hand_foot_Scale , -70 ,
           spr . owner ) ;
           spr . x = this . _boneLen ;
}
```

如果此时调用如上代码就能得到图 10.2 左侧的初始化骨架效果图。

10.2.3 事件处理程序

本节将实现如下一些简单的需求：
- 当鼠标点选中骨骼时，绿色加粗显示选中的骨骼。
- 当选中根（躯干）骨骼并拖动时，可以移动整个骨架到鼠标指针所在位置。
- 当选中某个骨骼时，按 F 键会让选中的骨骼顺时针旋转，按 B 键时，则逆时针旋转。

对于上述前两条需求而言，需要实现一个鼠标事件处理方法，具体代码如下：

```typescript
private mouseEvent ( s : ISprite , evt : CanvasMouseEvent ) : void {
    if ( evt . button === 0 ) {
        // 左键down事件处理
        if ( evt . type === EInputEventType . MOUSEDOWN ) {
            // 如果当前选中的是grid背景精灵（场景图的根节点）
            if ( s === this . _app . rootContainer . sprite ) {
                if ( this . _hittedBoneSprite !== null ) {
                    // 恢复到骨骼的原始状态
                    this . _hittedBoneSprite . strokeStyle = 'red' ;
                    this . _hittedBoneSprite . lineWidth = 2 ;
                }
            } else if (this . _hittedBoneSprite !== s ) {
                        // 理解指针的最好方式,如果这次点中的和上次点中的不是同一个精灵
                // ts最大的好处是强制null检测, 从编译器角度确保程序严谨性
                // 如果上一次点中的精灵不为null,则恢复到原始状态
                if ( this . _hittedBoneSprite !== null ) {
                    this . _hittedBoneSprite . strokeStyle = 'red' ;
                    this . _hittedBoneSprite . lineWidth = 2 ;
                }
                // 由于上次和这次点中的不是同一个精灵,需要重新设置跟踪对象及点中属性
                this . _hittedBoneSprite = s ;
         // 这个因为有设置, 并且s本身一定是不为null,所以下面的代码就不需要null检测
                this . _hittedBoneSprite . strokeStyle = 'green' ;
                                                  // 选中的骨骼以绿色显示
                this . _hittedBoneSprite . lineWidth = 4 ;    // 并加粗
            }
        } // 左键up事件处理
        else if ( evt . type === EInputEventType . MOUSEDRAG )
                                            // 左键drag事件处理
        {
            if ( s === this . _skeletonPerson )
            {
                // 精灵跟随鼠标位置移动
                s . x = evt . canvasPosition . x ;
                s . y = evt . canvasPosition . y ;
            }
        }
    }
}
```

接下来实现选中骨骼的旋转操作,实现keyEvent事件处理方法,具体代码如下:

```typescript
private keyEvent ( spr : ISprite , evt : CanvasKeyBoardEvent ) : void {
    if ( this . _hittedBoneSprite === null ) {
        return ;
    }
    if ( evt . type === EInputEventType . KEYPRESS ) {
        if ( evt . key === 'f'){
            this . _hittedBoneSprite . rotation += 1 ;
```

```
        } else if ( evt . key === 'b') {
            this . _hittedBoneSprite . rotation -= 1 ;
        }
    }
}
```

要让上面的鼠标和键盘事件起作用，需要做一些事件绑定的相关操作。先来看一下鼠标事件如何绑定。

根据 10.1.9 节鼠标事件分发过程的 dispatchMouseEvent 源码知道，每次鼠标点击或移动时，总是先调用 findSprite 方法获得当前鼠标指针下的那个精灵，如果该精灵有 mouseEvent 事件处理函数，就会触发 mouseEvent 事件，一旦触发完成后，立即退出鼠标事件分发流程方法。

这意味着，当前实现的 mouseEvent 方法需要挂接到场景图的每个精灵上，才能正确地响应鼠标事件，因此可以在 createBoneSprite 方法的最后一句代码前增加如下语句：

```
// 精灵挂接事件处理函数，一定要用 bind，出现多次错误就是因为忘记 bind 导致 this 指针报
null 值，切记！！！
spr . mouseEvent = this . mouseEvent . bind ( this ) ;
```

当调用上面代码后会发现，可以正常地点选骨骼，如果点选中的是根（躯干）骨骼精灵，则可以用鼠标拖动到任意位置。但是一旦选中某根骨骼精灵后，除了切换其他骨骼精灵外，无法取消选中的骨骼精灵，这是因为事件机制导致的。为了正确地取消选中骨骼精灵，还需要将 mouseEvent 事件处理函数挂接到根节点的精灵（grid）上，因此要在 createSkeleton 方法体最后增加如下代码：

```
if ( this . _app . rootContainer . sprite !== undefined ) {
    this . _app . rootContainer . sprite . mouseEvent = this . mouseEvent .
bind ( this ) ;
}
```

当单击到非骨骼精灵（也就是点中的是 grid 精灵）时，如果当前有选中的骨骼精灵，则会取消选中它。

接下来看一下键盘事件的运行机制。仍旧参考 10.1.9 节的 dispatchKeyEvent 方法，该方法采取简单直接的方式，遍历整个场景图，让场景图中每个具有键盘事件处理方法的精灵都触发一次键盘事件，只要监听根（grid）节点精灵，就能响应上面实现的 keyEvent 事件。因此 keyEvent 事件也是挂接在 createSkeleton 方法中的根（grid）节点精灵上，具体代码如下：

```
if ( this . _app . rootContainer . sprite !== undefined ) {
    this . _app . rootContainer . sprite . mouseEvent = this . mouseEvent .
bind ( this ) ;
    // 增加监盘处理事件
    this . _app . rootContainer . sprite . keyEvent = this . keyEvent . bind
( this ) ;
}
```

当调用上述代码，就能让精灵系统按照需求正常地运行起来，具体效果如图 10.7 所示。

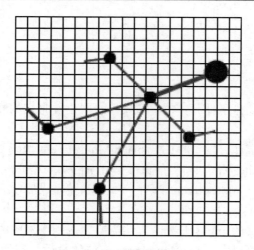

图 10.7　事件系统演示效果

在此需要强调的一点是，在上面的代码中，当多个精灵挂接同一个事件处理方法时，一定要使用函数对象的 bind 方法，否则 this 指针总是报 null 值。

10.2.4　使用 renderEvent 事件

到目前为止，已经了解了 updateEvent（在第 8 章的精灵系统 Demo 中）、mouseEvent 及 keyEvent 这三个事件处理程序，但是一直没涉及精灵系统的 renderEvent 事件处理方式，那么本节就来了解一下在什么时机下，如何使用该事件。

以上面的骨骼层次精灵 Demo 为例子，我们已经实现了 Bone 形体，并且通过基本形体组合方式合成了一个演示用的骨架。例如，根节点原点处的红色圆圈部分，由于左右手臂和左右腿部都在此处连接，会导致多次重绘。

这是因为使用了继承方式实现 Bone 形体类。如果直接使用 Line 类，然后对深度在最前面的右腿精灵挂接 renderEvent 事件处理方法，在 POSTORDER 阶段绘制红色圆圈，只需要绘制一次就能解决问题。

那么来看一下具体如何做呢？首先为了方便起见，在 SkeletonPersonTest 类中增加如下两个成员变量并初始化_line 对象：

```
private _linePerson ! : ISprite ;
private _line : IShape = SpriteFactory . createXLine ( this . _boneLen , 0 ) ;
```

然后参考 createBoneSprite 方法，实现 createLineSprite 方法，该方法代码和 createBoneSprite 的唯一区别在于，创建精灵时使用的是_line 形体，而不是_bone 形体。

同样地，实现 createLineSkeleton 方法，该方法和 createSkeleton 方法一致，只是用的是 Line 类型精灵，而不是 Bone 类型精灵。

不需要修改键盘和鼠标事件，直接运行代码，就能获得如图 10.8 所示的右图效果。

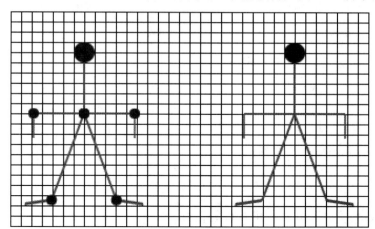

图 10.8　Bone 与 Line 表示的骨骼精灵效果图

接下来就来实现 renderEvent 事件，在 5 条骨骼的交汇处绘制蓝色的圆圈，具体代码如下：

```
private renderEvent ( spr : ISprite , context : CanvasRenderingContext2D , renderOreder : EOrder ) : void {
    // renderEvent 发生在 PREORDER 时，直接退出处理函数
    // 只关心 POSTORDER
    if ( EOrder . PREORDER ) return ;
    // 获取当前精灵局部-世界的矩阵，并调用 origin 属性获取原点坐标
    let orgin : vec2 = spr . getWorldMatrix ( ) . origin ;
    context . save ( ) ;
        // 使用 setTransform 方法，将当前对象的旋转角度设置为 0 度，缩放位置设置为[1,
        1]，而平移设置为[orgin . x , orgin . y]
        context . setTransform ( 1 , 0 , 0 , 1 , orgin . x , orgin . y ) ;
        context . beginPath ( ) ;
        context . fillStyle = 'blue' ;
        // 此时 arc 的原点是[ 0 , 0 ]
        context . arc ( 0 , 0 , 5 , 0 , Math . PI * 2 ) ;
        context . fill ( ) ;
    context . restore ( ) ;
}
```

将上述 renderEvent 事件挂接到右腿骨骼精灵上，就能获得如图 10.9 所示的效果。至于手臂和手掌，以及腿部和脚的连接处，就留给本书读者来尝试实现。

综上所述，使用了两种方式来对绘图功能进行扩展，第一种方式是，基于继承的方式实现了 Bone 形体类，会多次重绘重叠部分的蓝色圆圈。另外一种方式是，基于对象组合，使用 renderEvent 事件对绘图操作进行扩展，在上述的需求情况下，这种方式可能更加有效率（因为可以减少四次重绘绿色圆圈形体）。

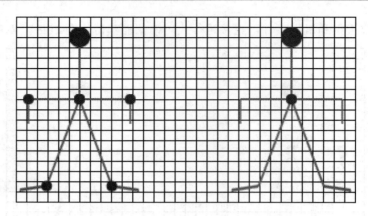

图 10.9　使用 renderEvent 事件后的效果图

但是基于对象组合，并且使用事件方式也有缺点，首先，这是基于对象实列的扩展，而不是类扩展，因此不能全局地影响到其他对象。其次，这种方式具有一定的硬编码性质，因此使用时，要根据实际需求来决定采用事件还是继承方式扩展功能。

10.2.5　本节总结

在本节中，通过一个骨骼层次精灵 Demo 来演示了基于场景图系统的、采取享元设计模式的精灵系统。

整个 Demo 中，仅使用一个 IShape（Bone 或 Line 形体）对象来表示骨骼，所有的精灵通过平移、缩放和旋转，以及设置不同的渲染属性来表示相同的形体，通过精灵与形体相分离的设计，实现共享的目的，这就是享元设计模式的精华所在。通过享元设计模式可以避免大量非常相似的类的开销，减少数据资源（IShape）内存的消耗。

其次演示了本书中精灵系统支持的三个基本事件处理方法：mouseEvent、keyEvent，以及 renderEvent，目的是了解当前事件系统的使用时机点、使用方法，以及优缺点，这样才能进行二次扩展或修正，至于 updateEvent 事件，在第 8 章介绍的 Demo 中有演示。

到目前为止，代码运行后可以显示两个不同的骨架，但是我们会发现，只有使用 Bone 形体（图 10.9 所示的左侧骨架）支持根骨骼跟随鼠标移动，可以在 mouseEvent 事件中修正，具体代码如下：

```
// mouseEvent 事件处理方法中原本拖动代码如下：
else if ( evt . type === EInputEventType . MOUSEDRAG )
                                                    // 左键 drag 事件处理
{
    if ( s === this . _skeletonPerson )
    {
        // 精灵跟随鼠标位置移动
        s . x = evt . canvasPosition . x ;
        s . y = evt . canvasPosition . y ;
```

            }
        }

```
// 可以将 if ( s === this . _skeletonPerson )这句代码修正为：
if( s . owner . getParentSprite ( ) === this . _app . rootContainer . sprite )
// 修正后就能支持两个不同的骨架的拖放功能
```

这是因为骨架 1 和骨架 2 的根（躯干）骨骼精灵都是根节点 grid 精灵的两个儿子，所以上面这句相对寻址代码就能确定是否选中的是根骨骼精灵，只要是根骨骼精灵，不管是属于哪个骨架，都允许鼠标拖动操作。

使用这种基于树结构层次的相对寻址方式，可以将代码写得更加通用，而且可以完全不使用_skeletonPerson 和_linePerson 这两个成员变量。树结构层次寻址在精灵系统的内部及第三方使用时，都会大量地使用，这也是基于场景图（树数据结构）的一种优势所在。

10.3 坦克沿贝塞尔路径运动 Demo

作为本书的最后一个 Demo，将整合前面章节学到的相关知识，实现一个基于基本形体组合而成的坦克，沿着可以自由改变曲率或位置的二次贝塞尔路径对象，朝向正确地移动的动画效果。具体效果图如图 10.10 所示。

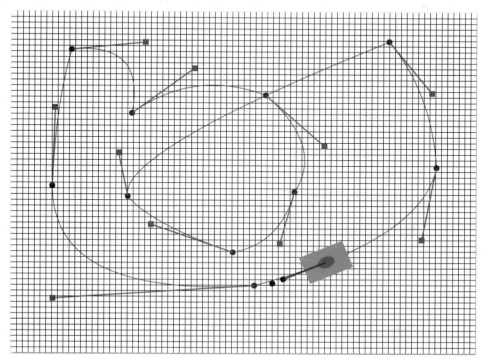

图 10.10 坦克沿二次贝塞尔路径运行动画效果

关于上述 Demo 的需求，在后续章节中详细介绍。在下一节中将介绍如何实现贝塞尔路径形体。

10.3.1　实现 BezierPath 形体类

关于贝塞尔曲线相关知识在第 7 章中已作详细介绍，并且实现了一个实时调整一条二次或三次贝塞尔曲线曲率或位置的 Demo。本节将使用前面学到的贝塞尔曲线的相关知识，实现一个继承自 BaseShpae2D 的 BezierPath 形体类，具体代码如下：

```
export class BezierPath extends BaseShape2D {
    public points : vec2 [ ] ;                    // 指向参数曲线的控制点集数据源
    public isCubic : boolean ;                    // 是二次还是三次贝塞尔曲线路径
    public constructor ( points : vec2 [ ] , isCubic : boolean = false ) {
        super ( ) ;
        this . points = points ;
        this . isCubic = isCubic ;
        this . data = points ;
    }
    // 实现基类抽象方法
    public get type ( ) : string {
        return "BezierPath" ;
    }
    // 不支持贝塞尔曲线的碰撞检测，直接返回 false，这样就会跳过碰撞检测，鼠标点选就
    永远不会点中
    public hitTest ( localPt : vec2 , transform : ITransformable ) : boolean
    {return false ;}
    public draw ( transformable: ITransformable, state : IRenderState ,
    context : CanvasRenderingContext2D ): void {
        context . beginPath ( ) ;
        context . moveTo ( this . points [ 0 ] . x ,this . points [ 0 ] . y ) ;
        if ( this . isCubic ) {
            // 绘制三次贝塞尔路径
            for ( let i = 1 ; i < this . points . length ; i += 3 ) {
                context . bezierCurveTo (this . points [ i ] . x ,
                    this . points [ i ] . y ,
                    this . points [ i + 1 ] . x ,
                    this . points [ i + 1 ] . y ,
                    this . points [ i + 2 ] . x ,
                    this . points [ i + 2 ] . y ) ;
            }
        } else {
            // 绘制二次贝塞尔路径
            // [ a0 , c1 , a2 , c3 , a4 , c5 , a0 ]
            for ( let i : number = 1 ; i < this.points .length ; i += 2) {
                context . quadraticCurveTo ( this . points [ i ] . x ,
                                this . points [ i] . y ,
                                this . points [ i + 1 ] . x ,
                                this . points [ i + 1 ] . y ) ;
            }
        }
```

```
            // 一定要调用基类方法，进行呈现操作
            super . draw ( transformable , state , context ) ;
        }
    }
```

上述代码中要注意以下几点：
（1）使用点的集合来表示二次或三次贝塞尔曲线路径对象。
（2）二次或三次贝塞尔路径对象是一个封闭路径。
（3）以二次贝塞尔路径为例，通过构造函数 points 参数提交给贝塞尔路径对象的顶点数量，以及形式需要满足如下几个条件：

- points 数组中的顶点数量必须要大于 3 个以上。
- points 数组中的最后一个顶点必须和第一个顶点的 x 和 y 分量相同，这样才能首尾相连，形成封闭路径。
- 二次贝塞尔路径顶点数量与形成的线段之间的关系表达式为：(n - 1) / 2，其中 n 为 points 的数量。以两条贝塞尔曲线形成的封闭路径的顶点数组为例：[a0 , c1 , a1 , c2 , a0]，其中 a0-c1-a1 形成一条贝塞尔曲线，a1-c2-a0 形成另外一条贝塞尔曲线，其中数组的开始和结尾都是 a0，从而形成封闭形状，并且满足上述表达式（5 - 1）/ 2 = 2，这个关系式很重要，后续 Demo 中将多次用到此关系式。
- 关于三次贝塞尔曲线顶点与线段数量之间的关系，请读者自己推导了。

（4）由于贝塞尔曲线的点选碰撞检测很耗时，最坏情况下需要进行（n - 1）/ 2 次的点与线段碰撞检测操作，而且 Demo 中用不到点选功能，因此直接在 hitTest 方法中返回 false，表示当前的贝塞尔曲线永远都不会被选中。

10.3.2 需求描述

有了上面实现的 BezierPath 形体类，就能进入 Demo 编写流程了，接下来看一下 Demo 的需求及演示。

（1）当进入运行程序时，仅显示网格背景。
（2）当单击鼠标按钮（左中右任意鼠标键），就会根据二次贝塞尔路径生成规则产生一个蓝色的圆圈（表示贝塞尔曲线的锚点）或一个红色的正方形（表示贝塞尔曲线的控制点），具体效果如图 10.11 所示。
（3）如果当前顶点数量大于 3，并且最后一个顶点是红色正方形表示的控制点，当按 E（表示 end）键时，程序内部会自动封闭整个路径，并创建出一个坦克精灵，然后自动沿着封闭的贝塞尔路径朝向正确地运行。具体效果如图 10.11 所示。
（4）当鼠标左键选中红色的控制点并拖动时，会改变当前贝塞尔路径的曲率，而当鼠标左键选中蓝色的锚点时，则会改变贝塞尔路径的锚点位置。不管改变哪条曲线段上的曲率还是两个锚点位置，坦克都能够朝向正确地沿着改变后的路径继续运行。具体效果如图 10.13 所示，将图 10.12 的贝塞尔路径形状调整为心形，坦克仍旧正确运行。

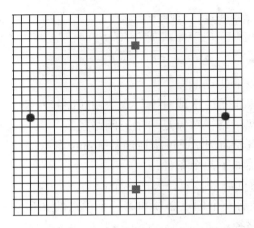

图 10.11　使用鼠标点击事件创建贝塞尔路径对象

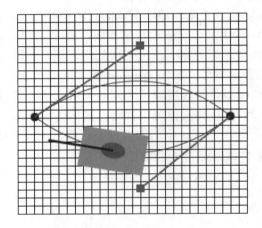

图 10.12　按 e 键自动封闭路径并创建运动的坦克

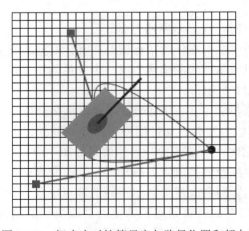

图 10.13　坦克实时计算贝塞尔路径位置和朝向

（5）当按 A 键时，会顺时针转动坦克的炮塔，当按 S 键时，会逆时针转动坦克的炮塔。

（6）当按 R 键时，表示 removeAll，意味着除了根节点（grid）外，所以其他精灵全部被删除掉，然后可以继续按照第（2）步流程周而复始。

10.3.3　Demo 的场景图

在介绍 Demo 的具体实现细节前，先来看一下整个 Demo 的场景图结构，具体如图 10.14 所示。

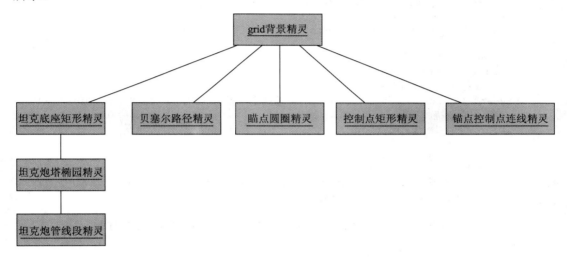

图 10.14　坦克沿路径运动 Demo 的场景图

从图 10.14 中会看到整个 Demo 的层次关系，除了坦克由三层基本形状组合而成外，其他都是处于同一层次的基本形体精灵，其中圆点精灵和正方形精灵为享元对象，被任意多锚点或控制点精灵所共享。

10.3.4　TankFollowBezierPathDemo 类初始化

现在开始介绍上述 Demo 的实现要点。下面来看其成员变量及构造函数，具体代码如下：
```
class TankFollowBezierPathDemo {
    // 指向入口类
    private _app : Sprite2DApplication ;

    // 贝塞尔路径相关成员变量
    private _curvePts : vec2 [ ] ;      //存储所有锚点和控制点的数组
    private _bezierPath !: IShape ;    // 使用上面点集数组创建的二次贝塞尔形体对象
```

```typescript
// 共享的几何 Shape，用来标记贝塞尔曲线的锚点和控制点
private _circle : IShape ;
private _rect : IShape ;

// 标记变量，用来指示曲线封闭路径是否生成了
// 需要使用这个变量来做一些很重要的操作
private _addPointEnd : boolean ;

// 当前坦克沿着路径运行时的速度变量
private _speed : number ;

// 整个 Demo 中，涉及的算法的核心变量，在程序中会详解
private _curveIndex : number ;
private _curveParamT : number ;

// 用于插值计算当前的 tank 位置
private _postion : vec2 ;
// 保留的上一次坦克的位置
// 通过_position 和_lastPosition 这两个位置差，就能调用 vec2 . getOrientation
算出坦克正确的朝向
private _lastPosition : vec2 ;

public constructor ( app : Sprite2DApplication ) {
    this . _app = app ;
    // 初始化为 false，表示目前没有形成封闭的路径
    this . _addPointEnd = false ;

    // 如果生成坦克，初始化时，坦克位于第一条贝塞尔曲线的 t 为 0 的插值点处！！！
    this . _curveIndex = 0 ;
    this . _curveParamT = 0 ;

    // 给两个位置变量分配内存
    this . _postion = vec2 . create ( ) ;
    this . _lastPosition = vec2 .create ( ) ;

    // 初始化运行速度变量为 5，可以自行调整
    this . _speed = 5 ;
    this . _curvePts = [ ] ;

    // 创建两个享元类
    this . _circle = SpriteFactory . createCircle ( 5 ) ;
    this . _rect = SpriteFactory . createRect ( 10 , 10 , 0.5 , 0.5 ) ;

    // 初始化时，在根节点 grid 精灵上挂接鼠标和键盘事件
    if ( this . _app . rootContainer . sprite !== undefined ) {
        // 必须要 bind，否则 this 指向错误
        this . _app . rootContainer . sprite . mouseEvent = this . mouseEvent .bind ( this ) ;
        this . _app . rootContainer . sprite . keyEvent = this . keyEvent .bind ( this ) ;
    }
```

```
        // 启动动画循环
        this._app.start();
    }
}
```

10.3.5　创建锚点、控制点及连线精灵

当单击鼠标时，会生成锚点或控制点精灵，并且会在起始锚点和控制点精灵之间生成一条连接线段精灵，因此需要生成上面精灵的相关方法。具体代码如下：

```
// 生成锚点或控制点精灵代码
private createBezierMarker ( x : number , y : number , isCircle : boolean ) : void {
    let idx : number = this._curvePts.length ;
    // 1. 将当前点坐标添加到_curvePts数组中
    this._curvePts.push ( vec2.create ( x , y ) ) ;
    // 2. 创建精灵对象
    let sprite : ISprite ;
    if ( isCircle ) {
        sprite = SpriteFactory.createSprite ( this._circle) ;
        sprite.fillStyle = 'blue' ;
    } else {
        sprite = SpriteFactory.createSprite ( this._rect ) ;
        sprite.fillStyle = 'red' ;
    }
    // 鼠标点击处的位置坐标
    sprite.x = x ;
    sprite.y = y ;
    sprite.name = "curvePt"+ this._curvePts.length ;

    // 将锚点或控制点加入根节点的儿子列表中
    this._app.rootContainer.addSprite ( sprite ) ;

    // 3. 给精灵对象添加drag事件，使用匿名箭头函数，this指针永远正确！！
    sprite.mouseEvent = (spr: ISprite, evt:CanvasMouseEvent ) :void => {
        if ( evt.type === EInputEventType.MOUSEDRAG ) {
            spr.x = evt.canvasPosition.x ;
            spr.y = evt.canvasPosition.y ;

            // 当拖动时，会实时地更新贝塞尔曲线上对应的锚点或控制点的坐标
            // 这样就能调整当前某条曲线线段的曲率或位置
            this._curvePts[ idx ].x = spr.x ;
            this._curvePts[ idx ].y = spr.y ;
        }
    }
}

// 生成起始锚点和控制点之间连线精灵代码
```

```
// idx 没有特别的意义,只是用于 debug 时显示名字
private createLine ( start : vec2 , end : vec2 , idx : number ) : void {
    let line : ISprite = SpriteFactory . createISprite ( SpriteFactory .
    createLine ( start , end ) , 0 , 0 ) ;
    line . lineWidth = 2 ;
    line . strokeStyle = 'green' ;
    line . name = "line" + idx ;
    this . _app . rootContainer . addSprite ( line ) ;
}
```

10.3.6 创建二次贝塞尔路径及坦克精灵

下面来看创建二次贝塞尔路径精灵的具体代码。

```
private createBezierPath ( ) : void {
    this . _bezierPath = SpriteFactory . createBezierPath ( this .
    _curvePts ) ;
    let sprite : ISprite =  SpriteFactory . createSprite ( this .
    _bezierPath ) ;
    sprite . strokeStyle = 'blue' ;
    sprite . renderType = ERenderType . STROKE ;
    sprite . name = "bezierPath" ;
    this . _app . rootContainer . addSprite ( sprite ) ;

    // 调用上面实现的 createLine 方法创建连线精灵
    for ( let i : number = 1 ; i < this . _curvePts . length ; i += 2 ) {
        this . createLine ( this . _curvePts [ i - 1 ] , this . _curvePts
        [ i ] , i ) ;
    }
}
```

上面的代码还是比较简单的。需要注意的一点是,创建连线精灵时连接的是起始锚点和接下来的控制点。假设现在二次贝塞尔路径点集_curvePts 存储的是[a0 , c1 , a2 , c3 , a0]这些顶点数据,遍历时 i 从 1 开始,那么_curvePts [i - 1]是起始锚点,而_curvePts [i]是控制点,下一个起始锚点和控制点的步进间隔是 2。

继续来看一下通过基本形体精灵合成坦克精灵的代码,具体如下:

```
private createTank ( x : number , y : number , width : number , height :
number , gunLength : number ) : void {
    // 从上到下构造 tank
    // tank 底盘
    let shape : IShape = SpriteFactory.createRect(width ,height , 0.5 , 0.5);
    let tank : ISprite = SpriteFactory.createISprite (shape , x , y ,0 ,1 ,1);
    tank . fillStyle = 'grey' ;
    tank . name = "tank";
    this . _app . rootContainer . addSprite ( tank ) ;

    // 坦克炮塔,作为底盘的儿子
    shape = SpriteFactory . createEllipse ( 15 , 10 ) ;
    let turret : ISprite = SpriteFactory . createISprite ( shape ) ;
```

```
        turret . fillStyle = 'red' ;
        turret . name = "turret" ;
        turret . keyEvent = this . keyEvent . bind ( this ) ;
        tank . owner . addSprite ( turret ) ;

        // 坦克炮管，作为炮塔的儿子
        shape = SpriteFactory . createLine ( vec2 . create ( 0 , 0 ) , vec2 .
        create ( gunLength , 0 ) ) ;
        let gun : ISprite = SpriteFactory . createISprite ( shape ) ;
        gun . strokeStyle = 'blue' ;
        gun . lineWidth = 3 ;
        gun . name = 'gun' ;
        turret . owner . addSprite ( gun ) ;
    }
```

10.3.7 键盘事件处理方法

到目前为止，完成了本 Demo 需要用到的所有精灵的创建方法，现在是时候实现需求描述中相关的内容了。我们发现，需求描述中的第 3、第 5 和第 6 条相关内容都是和键盘事件相关联的，下面就来实现这些需求，具体代码如下：

```
    private keyEvent ( spr: ISprite , evt: CanvasKeyBoardEvent ) : void {
        // 发生在 keyUp 时
        if ( evt . type === EInputEventType . KEYUP ) {
            // 下面是实现需求 3 的代码
            // e 表示 end，用来封闭贝塞尔曲线段路径
            if ( evt . key === 'e' ) {
                if ( this . _addPointEnd === true ) {
                    return ;
                }//要求最少四个点才能封闭一个二次贝塞尔曲线段路径
                if ( this . _curvePts . length > 3 ) {
                    // 算法[ 锚点，控制点，锚点，控制点，锚点，... , 锚点]
                    // 只有最后一个是锚点，才能自动封闭 n 条曲线为形体
                    if ( ( this . _curvePts . length - 1 ) % 2 > 0 ) {
                        // 最后一个点如何处理呢？答案是将第一个顶点 push 到_curvePts 的尾部
                        // 这样首尾相连，然后设置_addPointEnd 为 true，说明控制点添加完毕
                        this . _curvePts . push ( this . _curvePts [ 0 ] ) ;
                        this . _addPointEnd = true ;
                        //调用 createBezierPath 生成贝塞尔精灵，以及与控制点及锚点的连线精灵
                        this . createBezierPath ( ) ;
                        // 在初始化路径后，将坦克当前的位置位于第 1 条曲线段 t=0 处
                        this . _postion . x = this . _curvePts [ 0 ] . x ;
                        this . _postion . y = this . _curvePts [ 0 ] . y ;
                        // 创建沿着贝塞尔曲线运行的坦克精灵
                        this . createTank ( this . _postion . x , this . _postion .
                        y , 80 , 50 , 80 ) ;
                    }
                } // 下面是实现需求 6 的代码
            } else if ( evt . key === 'r' )
                            // 按键 r，表示 remove，将坦克及曲线路径全部清除 {
```

```
                if ( this . _addPointEnd === true ) {
                    this . _addPointEnd = false ;        // 标记_addPointEnd 为 false
                    this . _curvePts = [ ] ;             // 路径顶点数组清空
                    this . _app . rootContainer . removeAll ( false ) ;
                                    // 遍历场景图中的所有节点, 全部 remove 掉
                }
            }
        }
        else if ( evt . type === EInputEventType . KEYPRESS )
                                                        // 发生 keypress 时
        {   // 下面是实现需求 5 的代码
            // 炮塔每次转动顺时针 / 逆时针 5 度
            if ( evt . key === 'a' ) {
                if ( this . _addPointEnd === true ) {
                    if ( spr . name === 'turret') {
                        spr . rotation += 5 ;
                    }
                }
            } else if ( evt . key === 's' ) {
                if ( this . _addPointEnd === true ) {
                    if ( spr . name === 'turret') {
                        spr . rotation -= 5 ;
                    }
                }
            }
        }
    }
```

当完成了键盘事件处理方法后,需要知道将键盘事件挂接到哪些精灵上。从第 3 条和第 6 条需求描述可以确定该键盘事件由根节点 grid 精灵来处理,而从第 5 条需求描述可知键盘事件应该由炮塔精灵来处理。

10.3.8 鼠标事件处理方法

虽然实现了键盘事件方法,但是程序要运行,需要先用鼠标点击事件来创建锚点和控制点精灵,然后使用按键 E 来生成曲线路径及坦克精灵,之后才可以通过拖动锚点或控制点精灵来调整贝塞尔路径中某条曲线段的曲率或位置。这也是需求描述中第 2 条和第 4 条的内容,那么来看一下第 2 条需求实现的鼠标事件处理方法,具体代码如下:

```
private mouseEvent ( spr: ISprite, evt: CanvasMouseEvent ) : void {
    if ( evt . type === EInputEventType . MOUSEDOWN ) {
        if (spr === this . _app . rootContainer . sprite  /* spr . name ===
'root'*/ ) {
            // 根节点点击事件只有在贝塞尔曲线路径建立前才有用
            if ( this . _addPointEnd === true ) {
                return ;
            }
            // 每次点击,根据偶奇性来确定创建锚点还是控制点
            if ( this . _curvePts . length % 2 === 0 ) {
                this . createBezierMarker ( evt . canvasPosition . x , evt .
```

```
            canvasPosition.y, true );
        } else {
            this.createBezierMarker( evt.canvasPosition.x, evt.
            canvasPosition.y, false );
        }
    } s
  }
}
```

根据第 2 条需求描述，可知上面的 mouseEvent 事件应该由根节点 grid 精灵来响应创建锚点或控制点精灵，因此需要将 mouseEvent 绑定到根节点 grid 精灵上。

关于第 4 条需求，是通过鼠标来操作锚点或控制点精灵的。因此这个事件处理程序应该绑定到锚点或控制点精灵上，使用箭头匿名函数，将鼠标事件处理程序写在 createBezierMarker 方法中，在前面的 createBezierMarker 方法中已经实现了这个箭头函数，大家可以参考 10.3.5 节的内容。

到目前为止，如果运行上述代码，我们会发现能够满足第 1、第 2、第 3、第 5 和第 6 条的全部需求，以及第 4 条的部分需求，剩下没实现的需求是：坦克安静的待在第 1 条曲线段 t 为 0 处的位置上。

10.3.9　坦克沿路径运动的核心算法

要让坦克动起来，就要实现一个 updateEvent 事件处理方法，然后将该事件处理方法绑定到坦克上，这很容易理解。

问题是不单单让坦克动起来，而且要沿着多条曲线段围成的贝塞尔路径，朝向正确地运动，这需要解决几个关键的问题：

（1）如何将坦克当前的位置映射到正确的曲线段上？
（2）如何计算坦克当前的位置？
（3）如何计算坦克当前的朝向？

幸运地是，关于第（2）个和第（3）个问题，可以使用第 6 章向量及第 7 章贝塞尔曲线学到的相关知识来解决。现在来解决第（1）个问题，首先需要一个辅助方法，从当前的锚点和控制点的数量计算出曲线段的数量，该算法在 10.3.1 节中讲述过，现在来实现一下，具体代码如下：

```
// n 个顶点具有 ( n - 1 ) / 2 条曲线
private getCurveCount() : number {
    let n : number = this._curvePts.length;
    if ( n <= 3 ) {
        throw new Error( "顶点数必须要大于 3 个！！！" );
    }
    return ( n - 1 ) / 2;
}
```

实现一个名为 updateCurveIndex 的方法，该方法能够获得坦克当前位置所对应的曲线

段的索引号,是整个 Demo 的核心算法,具体代码如下:
```
private updateCurveIndex ( diffSec : number ) : void {
    // 根据前后帧时间差获取 t
    this . _curveParamT += this . _speed * diffSec ;
    // 如果 t >= 1.0 说明要进入下一条曲线段了
    if ( this . _curveParamT >= 1.0 ) {
        this . _curveIndex ++ ;              // 进位,意味着进入下一根曲线段
        this . _curveParamT = this . _curveParamT % 1.0 ;
                                             // 取模操作,又变成[0 , 1]取值范围
    }
    // 如果当前的曲线索引超过了曲线数量,那就再从头开始,周而复始
    if ( this . _curveIndex >= this . getCurveCount ( ) ) {
        this . _curveIndex = 0 ;
    }
}
```

10.3.10　让坦克动起来

接下来就让坦克朝向正确地运动起来,具体代码如下:
```
// 只有 tank 加入显示列表后才会调用 update 事件
private updateEvent ( spr : ISprite , mesc : number , diffSec : number ,
travelOrder : EOrder ) : void {
    // 在前序遍历时进行更新操作
    if ( travelOrder === EOrder . PREORDER ) {

        // 关键算法封装在 updateCurveIndex 方法中
        // 解决上面第一个问题
        this . updateCurveIndex ( diffSec * 0.1 ) ;

        // 二次贝塞尔曲线线段索引和顶点之间关系
        let a0 : vec2 = this . _curvePts [ this . _curveIndex * 2 ] ;
                                             // 第一个锚点
        let a1 : vec2 = this . _curvePts [ this . _curveIndex * 2 + 1 ] ;
                                             //控制点
        let a2 : vec2 = this . _curvePts [ this . _curveIndex * 2 + 2 ] ;;
                                             // 第二个锚点

        // 使用贝塞尔曲线知识解决上面第二个问题
        // 将当前的_position 记录到_lastPosition 中
        vec2 . copy ( this . _postion , this . _lastPosition ) ;
        // 然后更新当前的_position
        Math2D . getQuadraticBezierVector ( a0 , a1 , a2 , this . _curveParamT ,
        this . _postion ) ;
        // 更新坦克精灵的位置
        spr . x = this . _postion . x ;
        spr . y = this . _postion . y ;

        // 使用向量知识解决上面第三个问题
        // 通过坦克当前位置和上一次的位置计算出旋转角度
```

```
        // 然后更新坦克的 rotation，这样坦克就能朝向正确地沿着贝塞尔路径运行
        spr . rotation = vec2 . getOrientation ( this . _lastPosition , this .
_postion , false ) ;
    }
}
```

在 createTank 方法中，将 updateEvent 事件绑定给刚创建出来的坦克精灵，然后运行整个程序，会发现坦克按照需求描述，正确地运行。

最后，为了测试 renderEvent 的 PREORDER 和 POSTORDER 相关内容，来实现一个需求，即在坦克的炮管顶部绘制一个正方形和圆形，具体代码如下：

```
// 纯粹为了演示 renderEvent 事件
private renderEvent ( spr : ISprite , context : CanvasRenderingContext2D ,
renderOrder : EOrder ) : void {
    if ( renderOrder === EOrder . POSTORDER ) {
        context . save ( ) ;
            context . translate ( 100 , 0 ) ;
            context . beginPath ( ) ;
            context . arc ( 0 , 0 , 5 , 0 , Math . PI * 2 ) ;
            context . fill ( ) ;
        context . restore ( ) ;
    } else {
        context . save ( ) ;
            context . translate ( 80 , 0 ) ;
            context . fillRect ( -5 , -5 , 10 , 10 ) ;
        context . restore ( ) ;
    }
}
```

运行代码，效果具体如图 10.15 所示。

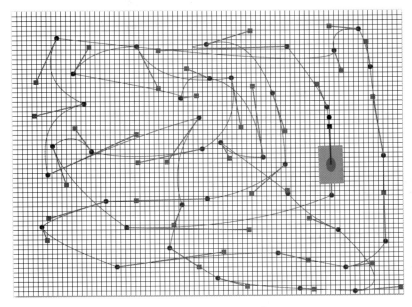

图 10.15　最终效果图

10.4　让精灵系统支持裁剪操作

精灵系统在利用 Canvas2D 作为底层绘图 API 时，内置了裁剪功能。看一下 interface.ts 文件中的 ERenderType 枚举结构，内部有 CLIP 枚举值，并且在 shapes.ts 文件的 BaseShape2D 抽象基类的 draw 和 endDraw 方法中实现了针对 CLIP 枚举值处理的代码。

本节在坦克沿贝塞尔路径运行 Demo 的基础上如何正确地使用裁剪相关内容呢？下面在 createTank 方法中，增加如下两句代码：

```
private createTank ( x : number , y : number , width : number , height : number , gunLength : number ) : void {
    ..............................................................................
    tank . renderType = ERenderType . CLIP ;
    ..............................................................................
    tank. owner . addSprite ( SpriteFactory . createClipSprite ( ) ) ;
    ..............................................................................
}
```

来看一下 SpriteFactory 类的静态方法 createClipSprite，以及相关的代码，具体如下：

```
// 静态变量
public static endCLipShape : IShape = new EndClipShape ( ) ;
public static createClipSprite ( ) : ISprite {
    let spr : ISprite = new Sprite2D ( SpriteFactory . endCLipShape , name ) ;
    spr . renderType = ERenderType . CLIP ;
    return spr ;
}
```

接下来在 shapes.ts 文件中输入如下代码：

```
export class EndClipShape implements IShape {
    public data : any ;
    public hitTest (localPt : vec2 ,transform : ITransformable ):boolean {
        return false ;
    }
    public beginDraw ( transformable : ITransformable , state : IRenderState , context : CanvasRenderingContext2D ): void {

    }
    public draw ( transformable : ITransformable , state : IRenderState , context : CanvasRenderingContext2D ) : void {

    }
    // 恢复渲染状态堆栈
    public endDraw ( transformable: ITransformable, state : IRenderState , context: CanvasRenderingContext2D ): void {
        context . restore ( ) ;
    }
    //子类必须override，返回当前子类的实际类型
```

```
    public get type (): string {
        return "EndCLipShape";
    }
}
```

可以看到，EndClipShape 形体最关键的一句代码是在 endDraw 方法中调用恢复渲染堆栈的操作，由此可见，通过 EndClipShape 目的是在裁剪操作中平衡渲染堆栈。

现在调用坦克 Demo 的应用，会得到如图 10.16 所示的效果。

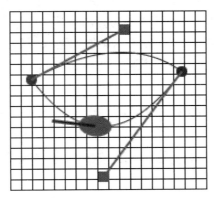

图 10.16　坦克底座使用 CLIP 枚举值的效果

图 10.16 显示的是正确的效果。因为坦克底座矩形部分设置为 CLIP 模式，则不会被绘制，它的所有子孙节点在底座矩形之外的部分会被裁剪。而定义子孙的裁剪影响范围依靠的是 ClipSprite，为了更好地理解，提供如图 10.17 所示的坦克裁剪时的场景图来说明原理。

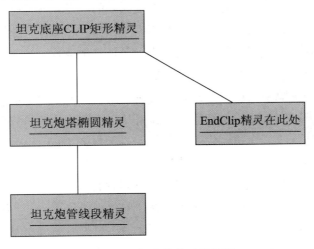

图 10.17　坦克裁剪时场景图

从图 10.17 中可以看到，之所以需要增加 EndClip 精灵，是因为 Canvas2D 中的裁剪区域也是受渲染堆栈管理的。

BaseShape2D 的 beginDraw 中存储了当前的渲染状态，如果当前精灵设置 CLIP 枚举值，在 endDraw 中并没有渲染状态的 restore 操作，这是因为希望将 CLIP 操作用于当前设定裁剪的精灵的所有子孙节点。如果直接在 endDraw 中调用 restore，则只能裁剪该精灵本身形体，根本无法裁剪其子孙精灵，所以为了能够层次裁剪，特提供 EndClip 精灵让渲染堆栈 save 和 restore 保持配对调用。

其次，关于 EndClip 放在哪里的问题，根据渲染和裁剪时都是按照深度优先、从上到下（先根／前序）、从左到右的顺序树遍历算法，那么答案显而易见：

EndClip 精灵最佳的位置是 Clip 精灵（坦克底座）的最右儿子节点，因为它总是在坦克底座影响范围内最后遍历到的节点，让它来关门是最合适的，这也体现了树结构的强大之处。

由此可见，Canvas2D 中的渲染堆栈可以影响渲染属性、变换矩阵，以及裁剪区域。

10.5　本 章 总 结

本章是本书的最后一章，主要对前面章节所介绍的各种技术进行了综合应用。回顾本书的整个流程会发现，其主要目的是以面向接口的编程方式实现一个具有必要功能（更新、重绘、裁剪及事件分发和响应）的、使用场景图类型的、支持精确点选的、基于非立即渲染模式（保留模式）的、采取享元设计模式的，并兼容第 8 章实现的非场景图类型的精灵系统。

首先在 10.1 节中讲解了非场景图类型的精灵系统的缺点，并将树结构与非场景图系统中实现的精灵子系统结合起来，在修改最少代码的基础上完成基于树结构的场景图精灵系统。新实现的精灵系统能很好地兼容原来的代码，正常地运行。

然后，通过两个与变换相关的例子来测试场景图系统。

先介绍了经典的骨骼层次精灵 Demo 的实现。在该 Demo 中，能够通过鼠标点选中骨骼，通过键盘事件来控制骨骼的旋转运动，如果是根骨骼的话，可以响应鼠标拖动事件，从而让整个骨架移动到我们想要的位置。另外，为了测试 renderEvent 事件，还扩展了相关例子。可以说，本 Demo 是享元设计模式优点的最佳体现。

接着介绍了第二个 Demo 的实现。该 Demo 通过鼠标按键可以生成封闭的贝塞尔路径，并且能够通过点选控制点或锚点动态地更改路径的曲率或位置，同时在其上运行的坦克能实时地响应路径变化，以正确的方向一直运行。当然也可以清屏重复前面的步骤。

10.4 节在坦克跟随路径 Demo 的基础上演示和实现了一个基于 Canvas2D API 的裁剪系统，并详细地解释了实现的原理。

最后给出精灵的系统架构图，如图 10.18 所示。

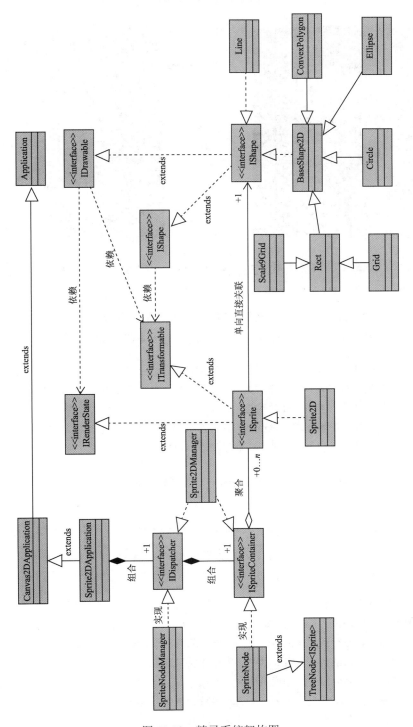

图 10.18 精灵系统架构图

推荐阅读

深度学习与计算机视觉：算法原理、框架应用与代码实现

作者：叶韵　书号：978-7-111-57367-8　定价：79.00元

全面、深入剖析深度学习和计算机视觉算法，西门子高级研究员田疆博士作序力荐！
Google软件工程师吕佳楠、英伟达高级工程师华远志、理光软件研究院研究员钟诚博士力荐！

　　本书全面介绍了深度学习及计算机视觉中的基础知识，并结合常见的应用场景和大量实例带领读者进入丰富多彩的计算机视觉领域。作为一本"原理+实践"教程，本书在讲解原理的基础上，通过有趣的实例带领读者一步步亲自动手，不断提高动手能力，而不是枯燥和深奥原理的堆砌。

　　本书适合对人工智能、机器学习、深度学习和计算机视觉感兴趣的读者阅读。阅读本书要求读者具备一定的数学基础和基本的编程能力，并需要读者了解Linux的基本使用。

深度学习之TensorFlow：入门、原理与进阶实战

作者：李金洪　书号：978-7-111-59005-7　定价：99.00元

磁云科技创始人/京东终身荣誉技术顾问李大学、创客总部/创客共赢基金合伙人李建军共同推荐
一线研发工程师以14年开发经验的视角全面解析TensorFlow应用
涵盖数值、语音、语义、图像等多个领域的96个深度学习应用实战案例！

　　本书采用"理论+实践"的形式编写，通过大量的实例（共96个），全面而深入地讲解了深度学习神经网络原理和TensorFlow使用方法两方面的内容。书中的实例具有很强的实用性，如对图片分类、制作一个简单的聊天机器人、进行图像识别等。书中每章都配有一段教学视频，视频和图书的重点内容对应，能帮助读者快速地掌握该章的重点内容。本书还免费提供了所有实例的源代码及数据样本，这不仅方便了读者学习，而且也能为读者以后的工作提供便利。

　　本书特别适合TensorFlow深度学习的初学者和进阶读者作为自学教程阅读。另外，本书也适合作为相关培训学校的教材，以及各大院校相关专业的教学参考书。

推荐阅读

数字图像融合算法分析与应用

作者：刘帅奇 郑伟 赵杰 胡绍海　书号：978-7-111-59302-7　定价：59.00元

详细介绍了数字图像融合领域的一些基本概念和常见算法
涵盖多聚焦图像融合、红外与可见光图像融合、医学图像融合和遥感图像融合

　　本书全面介绍了数字图像融合的基本概念和一些常见算法，便于读者了解和学习数字图像融合领域的一些前沿知识，以适应现代信息技术的发展。书中对不同传感器获得的数字图像进行了分类，并对不同类型的数字图像分别介绍了不同的图像融合算法，可以给读者提供有效的帮助和指导。

　　本书分为8章，主要内容包括图像融合简介、基于小波和轮廓波的多聚焦图像融合、基于剪切波和Smoothlet的多聚焦图像融合、红外与可见光图像融合、医学图像融合、基于仿生算法的医学图像融合、遥感图像融合等，最后简要介绍了数字图像融合的发展趋势。

　　本书适合计算机视觉、卫星遥感和医学图像等相关领域的研究人员、工程技术人员和算法爱好者阅读，同时也适合各大院校电子信息专业的本科生、研究生和教师作为教材或教学参考书使用。

推荐阅读

Unity与C++网络游戏开发实战：基于VR、AI与分布式架构

作者：王静逸 刘岵　书号：978-7-111-61761-7　定价：139.00元

游戏开发资深专家呕心沥血之作，分享10年实战经验
摩拜联合创始人、中手游创始人等7位重量级大咖力荐

权威：资深技术专家倾情奉献，7位重量级大咖力荐
全面：涵盖图形学、仿真系统、网络架构和人工智能等众多领域
系统：全流程讲解大型网络游戏及网络仿真系统的前后端开发
实用：详解一个完整的仿真模拟系统开发，提供完整的工业级源代码

　　本书以Unity图形开发和C++网络开发为主线，系统地介绍了网络仿真系统和网络游戏开发的相关知识。本书从客户端开发和服务器端开发两个方面着手，讲解了一个完整的仿真模拟系统的开发，既有详细的基础知识，也有常见的流行技术，更有完整的项目实战案例，而且还介绍了AR、人工智能和分布式架构等前沿知识在开发中的应用。

　　本书共21章，分为4篇。第1、2篇为客户端开发，主要介绍了Unity基础知识与实战开发；第3、4篇为服务器端开发，主要介绍了C++网络开发基础知识与C++网络开发实战。

　　本书内容全面，讲解通俗易懂，适合网络游戏开发、军事虚拟仿真系统开发和智能网络仿真系统开发等领域的开发人员和技术爱好者阅读，也适合系统架构人员阅读。另外，本书还适合作为相关院校和培训机构的培训教材使用。